animals & men
THE JOURNAL OF THE CENTRE FOR FORTEAN ZOOLOGY

THE CALL OF THE WILD
Issues 11 – 15 of *Animals & Men* magazine
April 1996 – July 1997

Edited by Jonathan Downes

Scanned back issues of Animals & Men by Oliver Lewis
Typeset by Jonathan Downes,
Cover and Layout by Mark North for CFZ Communications
Using Microsoft Word 2000, Microsoft , Publisher 2000, Adobe Photoshop CS.

Photographs © 2007 CFZ except where noted

First published in Great Britain by CFZ Press

**CFZ Press
Myrtle Cottage
Woolsery
Bideford
North Devon
EX39 5QR**

© CFZ MMVII

All rights reserved. Without limiting the rights under copyright reserved above, no part of this publication may be reproduced, stored in or introduced into a retrieval system, or transmitted, in any form of by any means (electronic, mechanical, photocopying, recording or otherwise), without the prior written permission of both the copyright owners and the publishers of this book.

ISBN: 978-1-905723-07-2

PREFACE

In 1996 I was aware that Jon was running an outfit called the CFZ, but I had little to do with it. I was more familiar with his music-related activities: performing around the pubs with his band, the *Amphibians from Outer Space*, and selling a music magazine called ISMO; a magazine which he and his wife Alison produced in A5 format jigsaw-fashion, adhering panels of computer printout to sheets of paper. The resulting "master" would be photocopied commercially. And this is how *Animals & Men* initially was produced, too.

One day in 1996, a sea-change in the affairs of the CFZ occurred when Alison unexpectedly left him. I was one of several people to drop in that day... and spent much of the afternoon wondering how usefully to commiserate with someone whose wife's suddenly walked out. There are times when a man needs a drink – and I accepted the offer of a bottle of wine with alacrity. Coincidentally, Jon needed a drink or two as well, and that evening was spent "knocking them back" and musing in a rambling fashion about the events of the day.

After a couple of days, it became clear that a return to matrimonial normality wasn't swiftly going to occur, and Jon's attention returned – albeit fragmentally – to the needs of the CFZ. Whether there even *would* be a CFZ in the future was unclear... but there was a magazine in mid-production – *Animals & Men* issue 10 – and a book – *The Owlman and Others* – and it seemed a good idea to press on with their production. Actually, with Jon's financial future suddenly cast into severe doubt, the sooner the better.

Since I was unemployed at the time, and have always been interested in science, and was always willing to partake of free booze, I assisted with these matters. This was back in the days when daytime drinking seemed a fun thing to do; and, although the tasks are administratively fairly simple, it does help to have a clear head. However, clear heads were at a premium at the CFZ around this time.

A mainstay of CFZ finances at this time was book sales: a mail-order service for second-hand crypto and fortean books trawled for at bookfairs and shops. Regrettably, hangovers and general preoccupation with the impending divorce rather got in the way of administrative efficiency.

There were points of interest. I saw an amphiuma for the first time – which I rather uncharitably likened to the proverbial turd in a swimming pool; and I got to play with the most powerful computer I'd ever used: Jon's Amiga 1200, with a dazzling 14MHz CPU speed and *two* whole megabytes (2MB) of onboard memory.

Weeks passed, and it became clear Alison wasn't going to return. I assisted with much of the admin relating to the CFZ and to the looming divorce. My employment background was in the area of admin, so it wasn't particularly difficult. Except when all-night drinking sessions left their mark the next day, that is.

With 1997 looming, we started work on putting *Animals & Men* issue 11 together, and I became deputy editor. However, the only piece I personally wrote for that issue was about a meteorite from Mars that allegedly bore signs of extraterrestrial life. I knew very little about cryptozoology back then (and, to be honest, I'm not an expert even now) so I mainly stuck to proofreading, and recasting contributions where needed.

Individual pages were assembled on the still-powerful Amiga computer and then printed out in A4 form, then adhered to A3 master sheets in the correct order for eventual photocopying. Jon wrote an editorial about the upheavals of 1996, and enthused about the year to come, and there was little outward sign that the future of the CFZ was still deeply uncertain. It still wasn't clear whether he'd lose his house or even be driven into bankruptcy – nasty things can happen during divorce proceedings.

In the course of 1997 it emerged that the doomsday scenario of losing the house wouldn't occur. However, CFZ finances were still looking distinctly dodgy at times.

Things began to look up, in the summer.

There was the emerging prospect of Jon – and possibly me – being taken by a film crew to Mexico and Puerto Rico to look for a chupacabra. "What's a chupacabra," I asked Jon. He dug around for a book and showed me the relevant portion – we weren't on the Internet in those days, and, indeed, I only had the haziest idea of what the Internet was.

My first experience of the Internet was later in 1997, when a friend initiated a dial-up connection and then announced to Jon and me that his computer was "on the Internet." I looked at the unchanged screen display and asked why I couldn't see anything, and was told, "Oh, you need a browser. I haven't got one."

I did *not* feel, at that point, that I'd had a glimpse of the future. If someone had told me then that, ten years down the line, the bulk of the CFZ's activities would involve the Internet in one way or another, I'd have reacted with derision.

If the CFZ didn't have a presence on the Internet – that was to come in mid 1998 – we were often on the radio. Indeed, Jon and I started doing a fortnightly slot on BBC Radio Devon, called *Weird about the West*, in which we and the afternoon presenter discussed such things as UFOs, local folklore, crop circles and the like.

Talk of a Mexico trip continued, and I went to meet the film crew for them to check me out. It remains the only time ever that I've gone to a "job interview" while wearing a t-shirt saying "too drunk to f---." However, they gave me the all-clear to go along, and Jon bought a video camera which I enjoyed experimenting with, each night that I walked home.

1997 drew to a close with preparations for the Mexico trip hotting up.

Previous CFZ expeditions had involved one or two people driving 50 or 100 miles to places like Cornwall or Dartmoor for a day or so. Three weeks in Mexico threatened to put a severe strain on the CFZ, as helpers were rather thin on the ground in those days. Although we had a network of representatives and article writers around the UK and the world (issue 12 of *Animals & Men* lists 15 reps in the UK) we didn't, unlike in 2006, have a pool of volunteers and field operatives from which to draw cover for operations back at base. However, we – as always – managed... and the foray across the Atlantic in search of *el Chupacabra* is now part of CFZ history.

Graham Inglis
(Deputy Director, CFZ)
Woolfardisworthy
Bideford
North Devon
December 2007

INTRODUCTION

Over Christmas 2002 I read the authorised biography of my hero Gerald Durrell. It was both heartening, and slightly distressing to find out quite how much we had in common. I do not wish, for one moment to suggest that I am equal to him. Durrell was undoubtedly a giant amongst men, and was undoubtedly one of the greatest naturalists of the past 300 years. His achievements rank with those of Linneaus, Fabre, Attenborough and Scott, and in comparison my humble efforts to enrich the sum total of man's knowledge about the natural world pale so far into insignificance that there is no real comparison. However, like Durrell, I hope that the sum total of my work will have made some difference to the world in which we live. On a personal level, however, we are far more similar. We have similar vices and are haunted by similar demons.

I was saddened when I read the account of his divorce from his first wife. The circumstances were very similar to that of my own divorce in 1996. When I read how he tried - for quite understandable reasons - to rewrite the history of his great organisation and attempted to expunge the very significant role, which his first wife had held it in its formation, I was determined not to do the same thing. In writing this, the introduction to the second volume of reprints of *Animals & Men* I have to cover, not only the years when the Centre for Fortean Zoology started becoming a real force to be reckoned with, rather than a back bedroom publication written for and by hobbyists, but the years when my own marriage fell apart.

Like Durrell, it is hard for me not to draw the inference that my single-minded obsessiveness with making the Centre for Fortean Zoology a success, contributed in great part to the dissolution of my marriage. I formed the Centre for Fortean Zoology in 1992, and launched the first issues of its journal in 1994. Jan has gone on to do other things, and again, I'd like to pay tribute to the immense volume of work that she put into the CFZ during the first few years of its existence.

It was issue six of the journal, which really took us on to a new level. The first five issues (which are collected together in *Volume One: In the Beginning*) were essentially a magazine finding its feet, finding its identity, and working harder to create a foothold within the fortean zoological community. Issue 6 was the first magazine to contain any groundbreaking fieldwork, and in many ways set the seal on how the CFZ would operate over the next decade. It included a number of topics that would - in many people's eyes - become synonymous with the Centre for Fortean Zoology over the years to come. It included the first accounts of "Gavin's" sighting of the Owlman of Mawnan, and also the first contributions from the notorious Tony Shiels.

It also included the first expedition report from the man who was eventually to become our honorary life President: Colonel John Blashford-Snell. It set the tone for the future of the magazine - a potent mix of hard science, meticulous fieldwork, and surrealchemy. The twin icons of the distinguished Colonel and the notorious Wizard were to spur the CFZ on to greater and more impressive feats over the years.

It was with this issue that we also began making inroads into the zoological establishment.

animals&men

THE JOURNAL OF THE CENTRE FOR FORTEAN ZOOLOGY

Chris Moiser - a man who over the years has become a close friend and trusted ally - came on board with this issue as did Clinton Keeling (another contributor who is no longer part of the CFZ family.) It is probably Tony Shiels, however, who had the greatest effect on the CFZ during the years between 1995 and 1997. He taught me to see the world through different eyes, and in many ways it is true to say that he had as great an effect on me as I approached middle-age, as Gerald Durrell had done when I first read his books as a small child. The story of Tony's and my alcohol-fuelled adventures can be read in my book *The Owlman and Others*, and I am happy to say that we are still friends today.

However, I have no wish to rewrite history. If it hadn't been for the support of my first wife, Alison, the Centre for Fortean Zoology would never have gone off the ground, and all of us involved with the project today - over 10 years after its first inception - owe her a debt of gratitude for her encouragement and hard work during the early days. We are no longer in contact with each other, and it is probable we never shall be again, but it would be wrong of me not to pay tribute to her in this introduction. Another person, no longer involved with the CFZ is Jan Williams, who - again - was pivotal in launching the magazine. Without those two women, there would be no CFZ. .

Three other people, but for whom there would be no CFZ, appeared on the scene towards the end of this volume: Mark North, an artist from Weymouth started to contribute cartoons, an impoverished young student from Leeds called Richard Freeman appeared in the letters pages, and a certain Graham Inglis - at that time the roadie for my notorious art-rock band - began receiving the credit for his carrying out similar duties for the CFZ. It is difficult, so many years after issue 10 was published, to imagine the Centre for Fortean Zoology without either of these two extraordinary, eccentric, and very gifted men.

On July 28th, 1996 my marriage effectively ended. What I did not realise at the time, was that my unexpected and completely unwanted change in circumstances would, in time, lead to the Centre for Fortean Zoology actually becoming what I had always claimed it had been - the most important, the biggest, and the best Cryptozoological research organisation and the world.

Slainte Mhor

Jon Downes
(Director of The CFZ)
Exeter
Devon
22nd July 2001

ISSUE 11
OCTOBER 1996

animals&men

THE JOURNAL OF THE CENTRE FOR FORTEAN ZOOLOGY

This was the first issue that Graham and I put together in the wake of Alison's and my separation. This is not a time in my life upon which I look back with any pleasure at all, but it is heartening to remember how everybody pulled together as a team, and rallied around to help.

This was the first of our `theme` issues - this issue's theme was walruses. It is particularly notable for a wonderfully bizarre cartoon by Mark North of a walrus Elvis impersonator, and the late Clin Keeling's memorable admission that he found large aquatic mammals abhorrent.

However, I am particularly proud about the article on out of place walruses in the British Isles; a subject which remains obscure, but may well be the truth behind some of the most notorious sea-monster sightings.

Animals & Men

The Journal of the Centre for Fortean Zoology

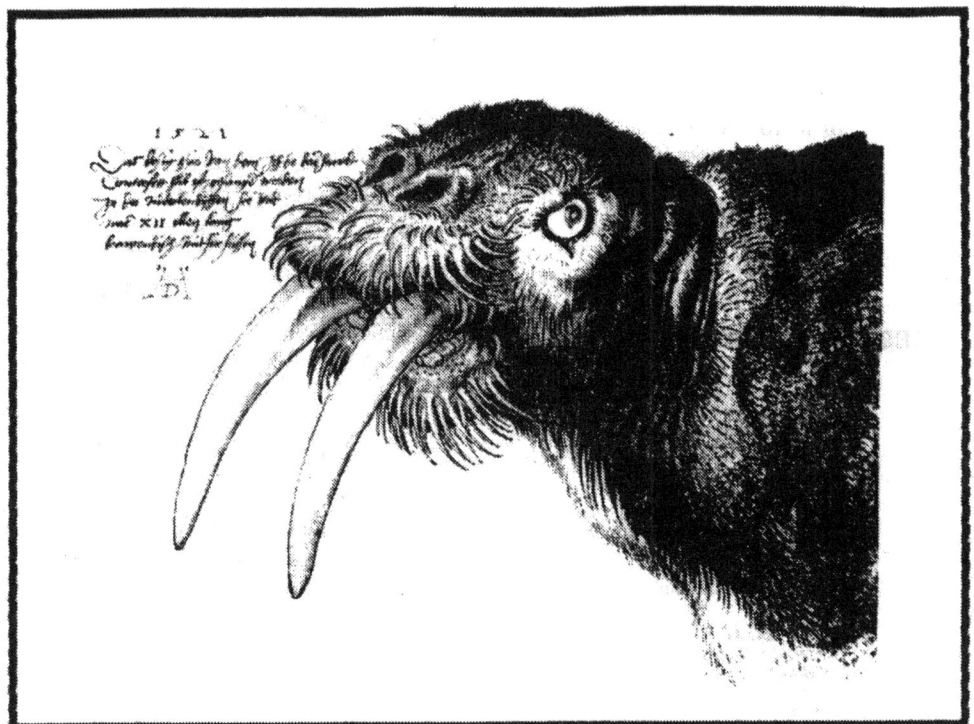

Wondrous Walruses; Initial Bipedalism; Mystery Whales; The Loch Lochy Monster; Ground Sloth Survival in North America.

Issue Eleven   £2.00

animals & men

THE JOURNAL OF THE CENTRE FOR FORTEAN ZOOLOGY

Animals & Men — Issue Eleven

The ever changing crew of the 'Animals & Men' mothership presently consists of:

Jonathan Downes: Editor
Graham Inglis: Deputy Editor
Jan Williams: Newsfile Editor
Mark North: Artist
Richard Muirhead: Newsagent from Nowhere

CONSULTANTS

Dr Bernard Heuvelmans
(Honorary Consulting Editor)
Dr Karl P.N.Shuker
(Cryptozoological Consultant)
C.H.Keeling
(Zoological Consultant)
Tony 'Doc' Shiels
(Surrealchemist in Residence)

REGIONAL REPRESENTATIVES

U.K.
Scotland: Tom Anderson
Surrey: Nick Smith
Yorkshire: Richard Freeman
Somerset: Dave McNally
West Midlands: Dr Karl Shuker
Kent: Neil Arnold
Sussex: Sally Parsons
Hampshire: Darren Naish
Lancashire: Stuart Leadbetter
Norfolk: Justin Boote
Leicestershire: Alaistair Curzon
Cumbria: Brian Goodwin
S.Wales/Salop: Jon Matthias

EUROPE
Switzerland: Sunila Sen-Gupta
Spain: Alberto Lopez Acha
Germany: Wolfgang Schmidt & Hermann Reichenbach
France: Françcois de Sarre
Denmark: Lars Thomas and Eric Sorenson
Eire: The Wizard of the western world.

OUTSIDE EUROPE
Mexico: Dr R.A Lara Palmeros
Canada: Ben Roesch
New Zealand: Steve Matthewman

DISCLAIMER
The Views published in articles and letters in this magazine are not necessarily those of the publisher or editorial team, who although they have taken all lengths not to print anything defamatory or which infringes anyone's copyright take no responsibility for any such statement which is inadvertantly included.

CONTENTS
P 3. Editorial
P 4. Newsfile
p. 12 New Projects
p.12 I am the walrus
p 19. Ground Sloth Survival
p 23. Initial Bipedalism
p 27. Is there life on Mars?
p. 28. Whales in Limbo
p 34. Loch Lochy Monster
p.35 Now that's what I call Crypto
p.35 North of the Border
p 36. Letters
p 38 Omens of Misfortune
p 39 Book Reviews.
p 41 Crypto Shop
p 47. Periodical Reviews
p.48 Cartoon.

Contributors to this issue:
Darren Naish, Mark North, François de Sarre, Justin Boote, Clinton Keeling, Sunila Sen Gupta, Steve Johnson, Ben Roesch, Erik Sorensen, Richard Carter, Neil Nixon, Tom Anderson, Neil Arnold

SUBSCRIPTIONS
For a Four Issue Subscription
£8.00 UK
£9.00 EEC
£12.00 US, CANADA, OZ, NZ
(Surface Mail)
£14.00 US, CANADA, OZ, NZ *(Air Mail)*
£15.00 Rest of World
(Air Mail)

'Animals & Men'

THE CENTRE FOR FORTEAN ZOOLOGY,
15 HOLNE COURT,
EXWICK, EXETER.
EX4 2NA

Tel 01392 424811

THE GREAT DAYS OF ZOOLOGY ARE NOT DONE

My Dear Friends,

Welcome to issue eleven. As the last issue of 1996 it seems mildly appropriate that the unwritten theme of this issue seems to be very much 'Ring out the old, ring in the new'. As most of you know Alison and I are in the process of getting divorced. This was not something that I ever wanted - indeed I am finding it very difficult to come to terms with, but much against my will it is happening and there is nothing that I can do about it. As many of you know, Alison and I always worked together as a team, and now because of the current situation that team is no longer operable. This means that inevitably changes have had to take place in my life, and as the greater part of my life is involved with forteana and fortean zoology in particular, this has meant great changes at The Centre for Fortean Zoology.

The most important change is that my long time friend and colleague Graham Inglis, who will be familiar to anyone who was at this year's Zoologica or either of the last two UnConventions, has now become Assistant Director of the CFZ, and together we have launched several new projects. These are detailed elsewhere in the issue, but in the words of Tony Shiels "Fun and Games are afoot!"

Many thanks to everyone in the A&M readership and team who has supported me with visits, telephone calls and letters during the past two months, which, as you are all aware has not been easy for me. I won't single any of you out for special thanks because to do so would take up half the magazine, but you know who you are, and you have my heartfelt gratitude. It is a hoary old platitude, but at times like this - (and I won't pretend that much of the last three months has been anything but hell), you find out who your friends are.

I found that out very quickly and most of them are readers and/or subscribers to this magazine. We are sorry that this issue is again a little late, but we are getting back on track. I am sure that you will forgive us. I am not too sure, however, that I can forgive Mr Naish (an artist and cryptocetologist of this parish) for his utterly libellous picture of me being carried away by a giant eagle which appears in the bottom right hand corner of this page, but I am sure that I can learn to live with it.

May we be the first to wish you all a successful and peaceful 1997.

Best wishes,

Jonathan Downes
(Editor)

Animals & Men — Issue Eleven

This is a particularly peculiar newsfile because although there is quite a lot of news, it doesn't fall into the categories that we usually use, so several of the charming logos prepared for us by Mark North have to remain unused this time....

ATTACKS

RELEASE THE BATS (2)

In a scenario sounding somewhat similar to the recent Chupacabras incidents forty villagers in El Pozon, a small village in eastern San Salvador have been attacked by vampire bats. Local experts claim that this unusual state of affairs is due to a lack of more normal food sources for the tiny animals. *Boston Globe 3.4.96*

MAN EATING WOLVES

Rumours of werewolves are sweeping through the Indian state of Uttar Pradesh. Thirty three children have been killed and a further twenty seriously mauled since April. Investigations state that the killings, which centre on the Ganges basin are the work of wolves.

Modern research suggests that most accounts of attacks on human beings by wolves are folkloric, and that with the exception of rabid animals, wolves are relatively harmless to man. This view contrasts strongly with the native oral tradition in Europe, Asia and North America.

The naturalist Ernest Thompson Seton, writing in the 1920s, believed that the wolf's avoidance of man was a recent development due to its fear of guns, and that Native-American legends of man eating wolves were true.

In 19th Century India the wolf had a deadly reputation of being 'deadly to children', but there have been no reports of wolf attacks of anything like this scale for more than a hundred years.

The wolf has been exterminated from vast tracts of its former territory, and the population in Northern India is greatly reduced. Only ten wolves have been tracked down in the region, despite hunts by thousands of villagers and police officers.

In comparison, hunters searching for the Beast of Gevaudan in 18th Century France shot around 2,000 wolves. Some villagers are claiming that werewolves are responsible for the killings, others that the children are being murdered by infiltrators from Pakistan wearing wolf skins. *Cincinnati Enquirer 1.9.96*.

NOW PLEASE WASH YOUR HANDS...

Denver Zoo: 50 people contracted salmonella in the space of 5 days from Indonesian Komodo Dragons - some from touching the enclosure and some from direct handling.

No-one died. The US Centre for Disease Control subsequently recommended that zoos erect double barriers to keep animals and humans apart, and provide *"facilities for people to wash their hands"*...*New Bedford Standard-Times 2.3.96 via COUDi*

ESCAPEES

FOLLOWING IN BERT'S FOOTSTEPS

In a belated tribute to the most (only?) famous Capybara of all time, two of the world's largest rodents made a break for freedom last summer.

Reg, a male of the species went AWOL from Malvern Zoo at the end of August. His rather laid back escape attempt ended after only five days, when he was tracked down with the aid of a BBC Radio 5 Hotline, and found wallowing in a pond only half a mile from the zoo.

A female capybara from a diabetes research farm run by Oxford University has proved a more successful survivor. The exact date of her escape seems uncertain, but reports of a four foot long rodent frightening unsuspecting anglers around the village of Stanton Harcourt, (between Oxford and Witney), have been surfacing since early August. Research staff have searched the area, and enlisted the help of local villagers, but have failed to track her down. Experts have said that she is unlikely to survive the local winter. Let's wait and see shall we? *BBC Radio 5, 3.9.96; Sunday Times 11.8.95; Daily Mail 13.9.96.*

Editorial Query: Does anyone in the wide and almost impossibly eclectic A&M readership have the slightest inkling of an idea why a capybara would be any use for diabetes research?

THE EAGLE HAS(N'T) LANDED

A rare eagle has escaped from the UK Falconry Centre in Thirsk, North Yorkshire. Olga, a Russian Steppe Eagle with a six-foot wingspan, took off during a flying exhibition in August. *Daily Mail 21.8.96*

GREAT SNAKES!

A Royal Python was confiscated by customs officers at Heathrow Airport after causing chaos on a British Airways flight. The python belonged to a Polish man, who had bought it in a pet shop in Seattle. He showed it to airport security staff at Seattle airport and was allowed to board the plane with the sleepy snake in his hand baggage. The man fell asleep during the flight but the python woke up and wrapped itself around the ankles of a woman passenger. After some disruption, cabin staff caught the snake, and confined it securely in an in-flight metal container.

A staff member at Seattle airport was sacked. *Mail on Sunday 29.9.96.*

RUDOLPH STRIKES BACK

Marmite, a reindeer from Pennywell farm animal centre, found that freedom has some unusual hazards. Straying into a garden in Buckfastleigh, Devon, it got its antlers entangled in a rotary washing line and had to be rescued by police. *The Times 22.8.96.*

AVIAN ANOMALIES

BLOWING IN THE WIND

South westerly gales at the beginning of October blew a rich crop of rare birds to the westcountry.

Guesthouses on the Scilly Isles reported a boom in business from twitchers anxious to see birds including a semi-palmated sandpiper, a buff bellied pipit, a northern oriole, a red eyed vireo and two pectoral sandpipers. *ITV Teletext 2.10.96.*

COME IN NUMBER 57 YOUR TIME IS...

A female fulmar, known only as Number 57 has been named as the oldest known bird to survive in the wild since records began.

The seabird is more than fifty years old, and still breeds every year on the uninhabited island of Eynhallow in the Orkneys. *Daily Mail 19.8.96*

DEAD PELICANS SOCIETY

The Salton Sea National Wildlife Refuge in California announced that thousands of pelicans and other rare sea birds were being killed at their refuge by indiscriminate use of pesticides in the locality and they called for an urgent review of the practice. *Boston Globe 28.7.96 Via COUDi.*

OUT OF PLACE

FERRETING OUT THE TRUTH

Confirming what our beloved editor wrote in his most recent book the Vincent Wildlife Trust have claimed that: "Britain's Polecat population is expanding rapidly and has spread to most parts of the country". They estimate that there are about 15,000 animals living outside their known stronghold in Wales. At the CFZ we can only say: WE TOLD YOU SO! *Daily Mail 30.9.96.*

IN THE SOUP.

Valerie Taylor an Australian shark expert warned that whale sharks - the world's largest known species of fish, were facing extinction due to exploitative fishing by Chinese fishermen in search of their fins for soup. "The Chinese will kill a whole creature so they can use its fins for a bowl of soup. It sickens me to think of this, and I just hope that enough attention can be drawn to this to put a stop to it". *Daily Mail 16.9.96.*

MISSING LYNX

A 20lb Bobcat strolled into a carpet warehouse in Dalton, Georgia where it caused a certain amount of havoc before being humanely captured by an employee armed with a noose on the end of a long stick. It became stuck between giant rolls of carpeting, and apparently the store was in turmoil for about thirty minutes. History doesn't relate what happened to the Bobcat. *Atlanta Journal 24.2.96 via COUDi*

MYSTERY CATS

Buckinghamshire

A cat-like animal crossed the road fifty yards from John Taylor of Akeley as he drove from Akeley to Maid's Norton near Buckingham on the 28th July.

"It was about three feet long", he said, "I couldn't believe it. It was like a big cat, light grey in colour, and with a big, long, bushy tail. It walked just like a leopard or a tiger". *Buckingham and Winslow Advertiser 2.8.96.*

Sally Jones saw a big cat in Pounden, near Bicester, on the fourth of August. She walked to within fifty yards of the three foot long animal before it ran of.

She said: "It had a fairly large head and a face just like a cat's, with short pointed ears". *Buckingham and Winslow Advertiser 9.8.96.*

Northamptonshire.

A large cat came within three feet of witness Peter Jenkins in Brackley Pocket Park on 4th August. He described it as three and a half to four feet long, with green eyes and pointed ears.

It bounded off into the undergrowth. *Buckingham and Winslow Advertiser 9.8.96.*

Yorkshire

A brown cat-like animal with green eyes was seen by David Rose on the outskirts of Walkington. He described it as being similar to a puma.

This is the latest in a series of sightings in that part of Yorkshire. Pawprints have also been found. *Hull Daily Mail 26.8.96 Via COUDi*

Dorset.

Giles Eastwood aged 33 was walking his dog about half a mile from Burton Bradstock when he looked over his shoulder and saw a huge, puma-like creature slinking across a freshly-cut cornfield.

"It had a very dark, silky body, which certainly looked like that of a puma" he said., "I feel privileged to have seen it!" *Dorset Evening Echo 8.9.96*

A week or so later a similar animal was seen just outside Bridport *"slinking along just yards from the mutilated carcass of a young deer"*. The carcass was fresh and the head was completely missing. *Dorset Evening Echo 12.9.96.*

Mark North, our Dorset Correspondent and in-house cartoonist has compiled a map of other recent ABC sightings in the county:

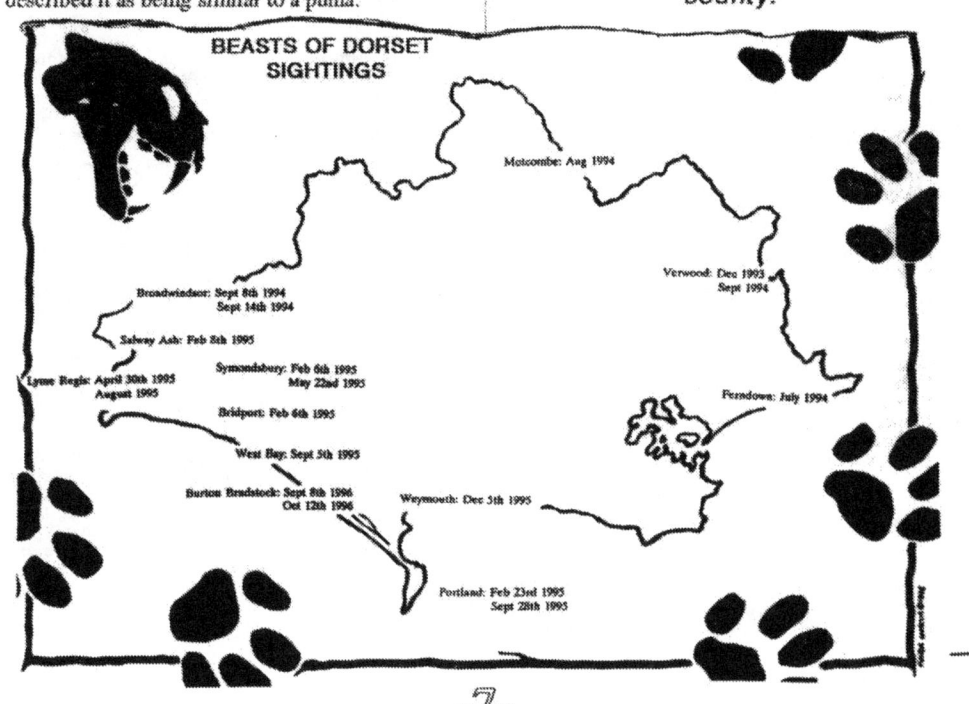

RESCUES

MONKEY MAMA

Binti Jua, a female gorilla at the zoo in Brookfield, Illinois, rescued a three year old boy who had fallen eighteen feet onto rocks in what the newspaper described as a 'gorilla pit' at the zoo. The child was released from hospital three days later, and anthropologists are divided as to whether this seeming act of kindness was carried out as a result of her maternal feelings for the child of a fellow primate, or whether the incident would not have happened if she had not been hand reared by human foster parents.

Those who, like the editor, were raised on a diet of Edgar Rice-Burroughs will not be at all surprised by these revelations. We would like to know, however, whether this is a case of nature imitating art? Or perhaps it is the other way around?

The Tarzan connection did not escape american fans of the books. They presented Binti Jua with 'the first ever Kala award' (named after Tarzan's fictional foster mother), with the following citation:

"...for her rescue; care and return of a male human child ... and for displaying extraordinary alertness, compassion and bravery..."

This display of literary inspired anthropomorphism was put into some sort of proportion by Thomas Insel, Director of the Yerkes Primate Centre in Atlanta:

We shouldn't expect gorillas to be these awesomely fierce, aggressive creatures. I really don't find this so surprising, that a lactating female would pick up an injured infant from a related species"...

From the point of view of the fortean zoologist, however, this is an interesting episode if only for the corroborative evidence that it gives to stories of lost children being 'brought up' by anthropoid apes in the wild. *USA Today 21.8.96. Via COUDi. Boston Globe 28.8.96 Via COUDi, St Louis Post Dispatch 26.8.96 via COUDi*

GOING TO THE DOGS.

In a similar story, two stray dogs were credited with saving the lives of a ten year old boy with Down's syndrome who was lost in the woods. According to the County Sheriff's office the child would have certainly died from exposure if the dogs had not been there to 'lend a helping paw'. In a slightly fortean scenario somewhat reminiscent of one of the more soppier episodes of a TV show like 'Lassie', the family of the boy wanted to adopt the dogs, but they have not been seen since the boy's recovery. *St Louis Post Dispatch 10.3.96. Via COUDi*

THE RETURN OF FLIPPER

Jumping figuratively from one cultural reference involving crassly anthropomorphic childrens television from the 1960's to another one we must now cast our minds back to a dreadful TV series called 'Flipper' in which a dolphin of that name repeatedly rescued two annoying red blooded american kids from smugglers, helicopter crashes and shark attacks.

Such rescues of humans by dolphins are not unknown. Indeed the exploits of one specific dolphin nicknamed 'Beaky' by the Westcountry Press in 1976/7 appear in the editor's new book: 'The Owlman and Others'. A similar event happened this summer when a British tourist was attacked by a shark in the Red Sea (afficianados if classic childrens literature will note the Tintin reference here), and was in imminent danger of being eaten alive when a group of three Bottle Nosed Dolphins came. 'Flipper' like to the rescue. They had obviously been watching the recent sunday morning re-runs on Channel 4 and hoped to secure lucrative film careers of their own. *New Bedford Standard-Times 26.7.96 via COUDi*

ANOTHER WONDER DOG

Lyric, an Irish Setter saved the life of her mistress Judi Bayley when her oxygen mask became unplugged. Although the headlines claimed that she had telephoned '911', she had actually been trained to 'hit' a panic putton with her paw if she heard an alarm bell go off.

Whilst one can only congratulate all involved for the succesful outcome of this episode this story is hardly fortean. Th implications, however are very important to the fortean researcher.

The headline read "Dog Telephones 911 to save owner" and is accompanied by a photograph which apparently shows the dog doing exactly what the headline claims. As we have seen, the truth is far more prosaic, but it is exactly this type of sensationalist and misleading reporting which causes so many spurious stories to enter the canon of fortean belief. Caveat lector! *USA Today 13.3.96 via COUDi*

OTHER STORIES (WHICH DON'T FIT INTO ANY ESTABLISHED CATEGORY)

FISHMEN OF THE APOCOLYPSE?

Dr. Antonio Yapha, chairman of the Seby Provisional Health Committee in the Philippines was, in July, planning to visit a family who claim that they can breathe underwater. Segundina Jimena claims that her three children can stay under water for up to six minutes 'breathing' through gill slits in their necks.

She also claims that her late husband had the same ability. Various doctors were claiming that although these 'gills' might indeed be vestigial remains of branchial clefts - primitive gills possessed by foetuses, they would not be functional, and indeed could cause grave medical problems if they became infected!

We do have to ask, however, if these 'gill slits' are nothing more than holes in the neck, how can these people 'breathe' underwater? Is it all a silly hoax in a country not known for its reserved and rational outlook on life, or is there perhaps some truth behind this seemingly silly story.

Interested readers are referred to 'The Magic Zoo' by Peter Costello in which similar cases, some of them in Europe, are discussed. *Atlanta Journal 28.7.96 Via COUDi.*

GET AHEAD IN GIBRALTAR

A Neanderthal skull found in Gibraltar has been dated by British scientists to less than 30,000 years ago. A team lead by Chris Stringer of the British Museum (Natural History), in London, is dating skulls found in a cluster of limestone caves. They have concluded that Neanderthals were forced to migrate southwards across Europe by Cro-Magnon Man, and that the two species did not interbreed. Stringer says:

"If you look at northern and southern Spain you have a picture of Neanderthals surviving later in the south than in the north which suggests that southern Spain, including Gibraltar, could be amongst the last footholds of Neanderthals". *Sunday Times 29.9.96*

I'D RATHER BE A LEMMING...

For the third year in succession suicidal walruses have been lining up to hurl themselves from a two hundred foot high cliff in Alaska. Sixty male walruses have plunged to their death this summer from the cliff at Togiak National Wildlife Refuge. *Daily Mail 14.9.96*

ART FOR ART'S SAKE

In the wake of Damien Hirst the animal mutilation ethic within art seems to becoming more and more prevalent.

In one particularly revoling episode artist Vincent Gothard dipped forty live baby mice in resin and then cut them into cubes as part of an undefined 'art project'. He was charged with animal cruelty and faces a large fine and a prison sentance if convicted. *Boston Globe 6.4.96.*

TALK TO THE ANIMALS

One of the editor's best, (but most juvenile and retarded) friends pointed out a cartoon in a recent issue of Viz Comic. In a spoof of a well known musical and series of children's books it tells the story of Doctor Poolittle - who talks to the animals about constipation.

This has nothing to do with the main content of this newsfile item, but your editor has just won a £5 bet by including it..

The real story in this section is that researchers at Toledo Zoo are recording the sounds made by their two hippos in order to try and establish whether or not these enormous creatures have a language. *News Advertiser (Toledo) 3.3.96 via COUDi*

NEWSFILE EXTRA: BIRDS OF A FEATHER

Feathers are unique to birds. Or are they? A new fossil of a small, meat eating dinosaur from Liaoning Province, China, suggests that feathers actually appeared long before birds themselves.

There is no longer any serious doubt that birds are truly flying dinosaurs - a vast body of data now supports such a statement. Not only does this mean that dinosaurs are alive and well, it may also mean that features we associate with birds were also found in extinct dinosaurs. Feathers are the ultimate example, for if small dinosaurs were 'warm blooded', they would probably require insulation, and would therefore probably have been feathered. Skin impressions of large dinosaurs show that they had scaly skin as do crocodiles.

But amongst mammals, large forms are generally naked skinned - what if mammals were extinct, and we only had skin impressions of elephants and rhinos? Excepting the famous Archaeopteryx - usually considered to be a bird because of its feathers - the evidence for feathery dinosaurs has thus far been ambiguous.

Despite this, many dinosaur experts have long considered feathery dinosaurs to be a likelihood: the idea was first proposed in the last century, and in the 1970s a number of artists began to draw small dinosaurs clothed in feathers. Some have even suggested that the young of all dinosaurs were down-covered.

News of the latest discovery first appeared in the Japanese press on the 9th October. It was announced to the community (via the internet's dinosaur discussion group), by Dr. Paul Davis, an expert in the fossilisation of feathers at the National Science Museum, Tokyo.

Dr. Phil Currie, a noted authority on predatory dinosaurs, or theropods, had been in China at the time of the discovery.

He verified that a small (about 1m long), well preserved theropod with feathers had been found.

The fossil is from the Lower Cretaceous (about 135 million years old). Photographs, posted on the www, show the animal in classic theropod 'death pose' with the neck bent back and the hind limbs partly folded. The eye is represented by a dark circle of organic matter, and there are also traces of soft tissue in the lower abdomen.

The animal has a long, heavy tail. In photographs of the specimen the feathers do not show up well. They appear as a dark streak along the length of the backbone, and thus it is impossible to verify whether they are feathers or not. However, Currie examined the actual specimen and seems in no doubt that they are feathers. He reports that they are relatively short (about 1cm long), and are preserved over most of the body.

The timing of the discovery was excellent, as New York's annual meeting of the Society for Vertebrate Palaentology was held the following week. The majority of theropod experts were therefore able to view high-quality photographs of the specimen, and they were all very impressed.

Unfortunately, the skull is crushed and the forelimbs are disarticulated. However, a number of characteristics suggest that the new dinosaur - given the awful name *Sinosauropteryx prima* by Chinese scientists - is one of the compsognathids. These are small, long legged, running dinosaurs that seem to have frequented shorelines and lakesides. *Sinosauropteryx* is from lake deposits that are already famous for the number of early bird fossils they have yielded. Many of these birds also have feathers preserved.

While it would now seem that a truly feathered non-avian dinosaur has at last been found, microscopic analysis is needed to demonstrate beyond doubt that these structures really are feathers. If they are, then we now know that feathers appeared before birds did. (*Sinosauropteryx* belongs to a group of dinosaurs that diverged from other theropods before the immediate ancestors of birds).

Many questions remain unanswered.

Did feathers first evolve for a purpose other than flight? (e.g. as insulation?) Does the presence of integument necessarily prove warm bloodedness? Were feathers widespread amongst dinosaurs, or just restricted to bird-like theropods and birds? The great dinosaur debate goes on..

Darren Naish.

Animals & Men — Issue Eleven

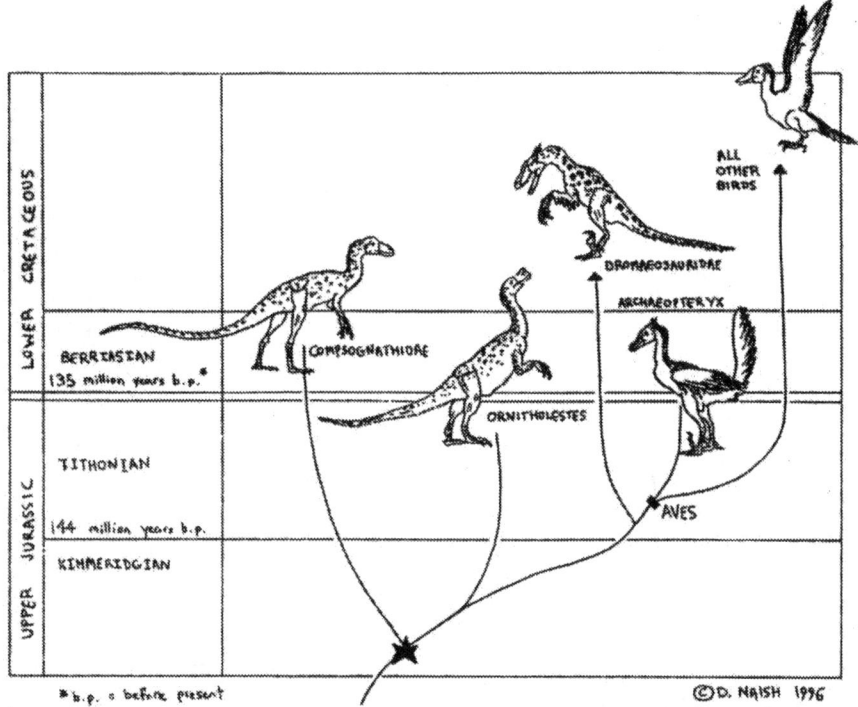

The discovery of feathers in a compsognathid would suggest that feathers were present in the common ancestor of compsognathids and living birds (that common ancestor would be where the star is), and all the descendants of that common ancestor.

Newsfile Correspondents

Mark North, Chris Moiser, Tom Anderson, Darren Naish, John Allegri, Sally Parsons, Ginny Ware, COUDi

Editorial note: Frank Gibbons, for so many years the mainstay of the South West Herpetological Society died as we were putting this magazine together. I would like to dedicate this issue to his memory, and also to the memory of Tony Shiels' mother, and Jessica Mitford - who despite being one of my favourite journalists was one of the only members of her family whom I have never written a song about... The world will be poorer for your passing.

NEW PROJECTS FROM THE CENTRE FOR FORTEAN ZOOLOGY.

'Research Kits'.

Our Fortean Zoological archives have been expanding at an alarming rate over the past two years, and although we have always intended to make this material available to the general public, until now we have not been able to do so.

We have now started the task of cataloguing the entire archive, in a comprehensive database. We are now in a position to be able to provide photocopies of our B.H.M. and Lake Monsters material at a nominal cost; other categories will be "on-line" hopefully within a month.

If you are interested please send a stamped addressed envelope and the details of the material that you are interested in (Lake Monsters, Big Cats etc). This can be as specific or as wide as you like. We will then send you a printout detailing the material currently available on that subject. (If the printout runs to more than five sheets, we shall have to charge 50p for each extra five sheets - this will be invoiced to you.)

If, upon receiving the printout, there is material in our possession that you want, then we can provide photocopies at a rate of 20p per sheet. (You may wish to send the printout back with the items in which you are interested indicated.)

There is no minimum order but there is an administrative charge of £1.00 per order to cover postage, archive retrieval time, etc.

If you are in the fortunate (or possibly unfortunate) position of owning an Amiga computer we can provide the database on disc for you at the cost of £3.00 including p&p.

This is a new service, but one which we hope will expand and augment our web site on the internet which will be set up early in 1997.

New Magazine.

We have also launched our long-awaited sister magazine 'The Goblin Universe' which deals with non-zoological fortean phenomena in a similar style to this magazine and featuring many of the same writers. There have been issues of 'The Goblin Universe' before but they have been freebie promotional mags used merely to publicise the activities of a certain seven piece rock band from Exeter. The subscription rates are the same as for this magazine, but we have a special introductory offer for A&M subscribers.

If you are a subscriber to A&M you can subscribe to our new magazine for £6.00 (UK/EC) £9.00 (Rest of world).

If you wish to renew your subscription to Animals and Men (whether or not it is due), you can subscribe to BOTH magazines for £12.00. This offer ends on the 31st December 1996.

The first issue includes pieces on Cattle Mutilation, secret Government projects, Cornish witches, the weirdest village in the south-west and considerably more.

The 1997 Yearbook is nearing completion and will be available in mid November. It is, we think at least, bigger and better than before. Details of the contents are found elsewhere in the magazine, as are details of my newest book 'The Owlman and Others' which is also available now!

We have also become involved in the Exeter Strange Phenomena (ESP) Research Group and produce a newsletter with the stomach churning title 'The EXE Files'. For details send us a SAE.

The only mildly vexing fly in the fortean zoological ointment, is that because of increased printing and postage costs we have had to increase the cost of Animals & Men by 25p a copy. We are sorry about this but it is the way that the proverbial cookie crumbles.

I AM THE WALRUS

Well, I'm not actually, and neither to the best of my knowledge are any of the other people on the editorial team, (except possibly Aberdeen's Mr Entertainment), but it's a great song and the following miscellany of articles is probably the best excuse we shall ever have to use this headline....

in mitigation.....

Something that all good forteans will readily accept is that phenomena, or even items of mild interest never seem to occur singly. In August this year Richard Muirhead was staying with me and over breakfast (a polite euphemism for the cold pizza and beer we were consuming just before the time when most people have their lunch. he pointed out an interesting paragraph in the *Daily Telegraph*. Alleging that what appeared to be a walrus had been sighted off the coast of Scotland. the news item rang some figurative bells in my mental database, and after a few telephone calls we had what could probably be described as *'Everything you wanted to know about crypto-walruses but were afraid to ask'*...

THE NORFOLK WALRUS.
By Justin Boote.

There are two sub-species of walrus - the Atlantic species *(Odobenus r. rosmarus)*. and the Pacific walrus *(O.r.divergens)*. The majority of the population is situated in the Arctic circle. The Atlantic species extends to Hudson bay and the northwest coast of the Baffin Isles, but lives mostly along the northern coasts of Greenland and Baffin Bay. The Pacific species lives along the north-east coast of Siberia across the Bering strait and around the north-west coasts of Alaska. Fossil remains have been found in the south of England and Belgium on the east of the Atlantic. and South Carolina on the west.

They migrate north in the summer and south in the winter, travelling vast distances on great Ice Floes. You would not, therefore, expect to see one on the north coast of Norfolk. Yet one did turn up alive and well in 1981.

It was first spotted in the morning off the Lincolnshire coast on the 13th September. and was not seen again for the rest of the day. Then. on the 14th, it followed a Conservancy Board barge in from The Wash in north Norfolk and then entered the River Ouse at Kings Lynn. It continued up the river for 15 miles and finally settled at Salter's Lode, south of Downham Market.

It was seen again on the 15th, by which time the word had got around and a large crowd of astonished locals had amassed to witness the creature. More fun was to come when a helicopter flew overhead, carrying a team of specialists and a few volunteer helpers who attempted several times unsuccessfully to capture

PICTURE: THE NORFOLK WALRUS © LYNN NEWS. USED WITH KIND PERMISSION.

it. Eventually the walrus grew tired of the attention it was receiving and slipped back into the water and made its own way back to sea. It was next seen at Skegness where a more successful mission resulted in its capture.

It was taken to the local marine zoo where it was identified as the Atlantic species and in perfect health for its troubles. It spent another day at the zoo recuperating until it was finally crated and flown from London airport back to its native Greenland.

EDITOR'S NOTE: There are interesting discrepancies between this animal, both as described by Justin, and in the original source material from the King's Lynn Newspaper, and the description of 'Wally' the walrus, also from 1981 given in Darren Naish's article printed in this issue. Perhaps there were two walruses in East Anglia that year? If so, surely this is a quasi fortean event of some significance?

THE WHALE-HORSE
by Clinton Keeling

This, to me, nightmarish-looking Seal takes its popular name from two Norse words meaning "Whale Horse".

Probably England's first living Walrus as a - what for want of a better term I'm going to call "zoological specimen" - was exhibited during the reign of Alfred the Great. And now you know precisely as much as I do, as nothing else - ie where and/or how it was kept, how it was fed, or how long it lived - has been recorded about it. Alfred's reign, by the way, was from 871 until 899.

As far as can be ascertained (as I am literally twenty-five per cent of the country's serious zoological historians much of my work is by

PICTURE: ANOTHER VIEW OF THE NORFOLK WALRUS © LYNN NEWS. USED WITH KIND PERMISSION.

no means easy), the next Walrus to put in an involuntary appearance here was one in the days of James I - whose reign, as far as England's concerned, was 1603-1625. Here we have rather more information, to the extent that it was shown in London.

Eventful centuries roll by; it's 12th October 1853 - and a young female Walrus, caught *"in the Spitzbergen Seas"* by Captain Henry of Peterhead, arrives in a moribund condition at the twenty-five year old London Zoological Garden; she dies the following day.

Another, a young male from Davis's Strait, was purchased by the same institution on 1st November 1867 - for £200; a vast sum in those days. Apparently he had been caught on 28th August by the crew of the steam-whaler Arctic (Captain: Richard Wells), and for some days had been tied to a ring-bolt on deck. Not surprisingly, he refused food for some time, but gradually was induced to take strips of boiled pork.

When the Shetlands were reached a supply of fresh Mussels (a far more suitable diet) was obtained and he took these avidly until Dundee was reached; whereupon he was transferred to the S.S.Anglia and conveyed to London under the personal care of the Zoological Society's almost legendary superintendent, A D Bartlett. He died on 18th December of that year.

Walruses have never thrived for long in British zoological gardens, but several places such as the New York Aquarium and Hamburg Zoological Garden have had considerable success with the species.

A friend of mine - who is a frequent correspondent to this publication - had several in his care when he worked at Hamburg, a job I wouldn't have undertaken for all the proverbial tea in China (although as a non tea-drinker I wouldn't give you a thank-you for it). I'm unfortunate enough to be a claustrophobe, who regards water as an extremely alien, indeed hostile, element: daft, I know, but I cannot help it.

Consequently I must be the only cryptozoologist brave enough, or honest

enough, to admit that I find large marine mammals repellent and repulsive - in fact they have very much the same effect on me that snakes have on many people - again, I'm the first to admit it's idiotic but there's nowt I can do about it. Anyway, I'm deviating somewhat. My friend assures that the Walrus is hyper-intelligent, a terrific personality and has a wonderful sense of humour.

But, of course, he would; wouldn't he...

EDITOR'S NOTE: We had a tremendous response to our 'phone in quiz last time. The answers, by the way were Nick Cave and the Birthday Party (*Release the Bats*), and the immortal Pop Will East Itself (*Defcon One*). This issue's 'phone in quiz is much simpler. Apart from the obvious one, name two rock songs with the word 'walrus' within the lyrics. The first person to telephone me with the answer wins a free subscription to this wonderful magazine.

WALRUSES IN TIME AND SPACE
by Darren Naish

As with so many other groups of extant animals, the sole living specimens of walrus, *Odobenus rosmarus*, represents but a fraction of the diversity that exists in its fossil relatives. The living walrus is of near circumpolar distribution, being essentially an animal of the Arctic pack-ice, and with populations in the north Pacific (Bering Sea and adjacent Arctic Ocean), north Atlantic (eastern Canadian Arctic, Greenland and Novaya Zemlya), and the Laptev Sea, north of Siberia. These three populations are often recognised as subspecies, though there is debate as to the validity of the Laptev Sea population (*O.r. laptevi*). The Pacific (*O.r. divergens*) and Atlantic (*O.r. rosmarus*) walruses are morphologically distinct - Pacific walruses being longer, heavier, and with longer tusks and wider skulls than Atlantic ones - and their rank and subspecies is not merely for taxonomic convenience, as it so often is. Incidently, the Pacific form was given the subspecific name divergens (by Illiger in 1815) as its tusks were originally stated to have been more curved than those of the Atlantic form. If this observation is valid, then Illiger's type specimen (since lost or destroyed!) was different from known Pacific walruses (whose tusks are not more curved than those of the Atlantic form). Laptev Sea walruses seem to be most like the Pacific form.

Interestingly, the living walrus is truly native to the north Atlantic, and seem to have moved, via Arctic waters, into the north Pacific within recent geological times (within the last 30,000 years). This is especially significant as the earliest walruses were Pacific animals that moved (via the Central American seaway about 9 million years ago) into the Atlantic: Pacific walruses thereafter became extinct, and the Pacific was devoid of walruses until the recent invasion of Odobenus. Early walruses called dusignathines, which flourished in east Pacific waters from mid Miocene to early Pliocene times (15-3 m.y.a.), were more like sealions than the living walruses and were not specialised mollusc-eaters (=durophagores), as is the living walrus. Their upper and lower canines were of equal size, and they probably ate fish. Imagotaria, from the mid-late Miocene of California, was a typical dusignathine (fig. 1). Odobenine walruses, which first appeared in the Pacific about 6.5 m.y.a. began to enlarge their canines shortly before their migration into Atlantic waters, where the rest of their history occurred. By Pleistocene times, 1.6 m.y.a., all walruses bar Odobenus had died out.

There are Pleistocene fossils of the living walrus on both sides of the Atlantic: they occur as

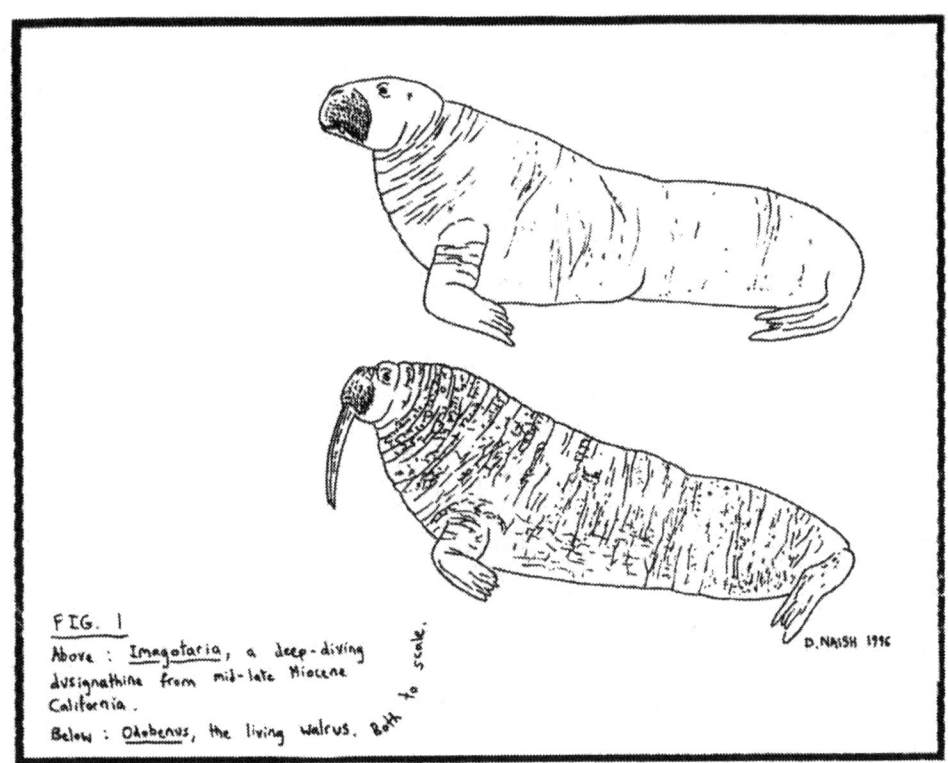

FIG. 1
Above: Imagotaria, a deep-diving dusignathine from mid-late Miocene California.
Below: Odobenus, the living walrus. Both to scale.

far south as the Carolinas in the west, and Paris in the east. In modern times, however, we expect walruses to be restricted to the Arctic. In the European region they are regularly seen around Spitzbergen and occur as vagrants around Iceland (where they have been seriously depleted by hunting) and Norway (31 records between 1900 and 1967). What about further south? A walrus captured in Dutch waters was sent, in 1520, to Pope Leo X in Rome. A year later, what appears to be the same specimen was illustrated by Albrecht Durer. We know that the specimen sent to the Pope was a walrus head, rather than a whole walrus, and this surely explains why Durer's illustration stops at the neck. In 1940 a walrus was reported from Lubeck Bay, Germany, and in 1977 P.J.H. van Bree reported an individual that made stops on the coasts of the Netherlands and Belgium.

As for the British Isles, stragglers have visited our shores for as long as records have been kept. In the seal 'bible', Judith King's Seals of the World (Oxford Uni. Press, 1983), King lists around thirty British walrus reports made between 1456 and 1954 (fig. 2). These include records from the Thames (1456), Orkneys (1527), Severn (1839) and Shannon, Ireland (1897), but with the majority (25 records between 1815 and 1954) from the Scottish coasts. The most recent of the latter are a dead animal washed up at Gairloch, Ross-shire,

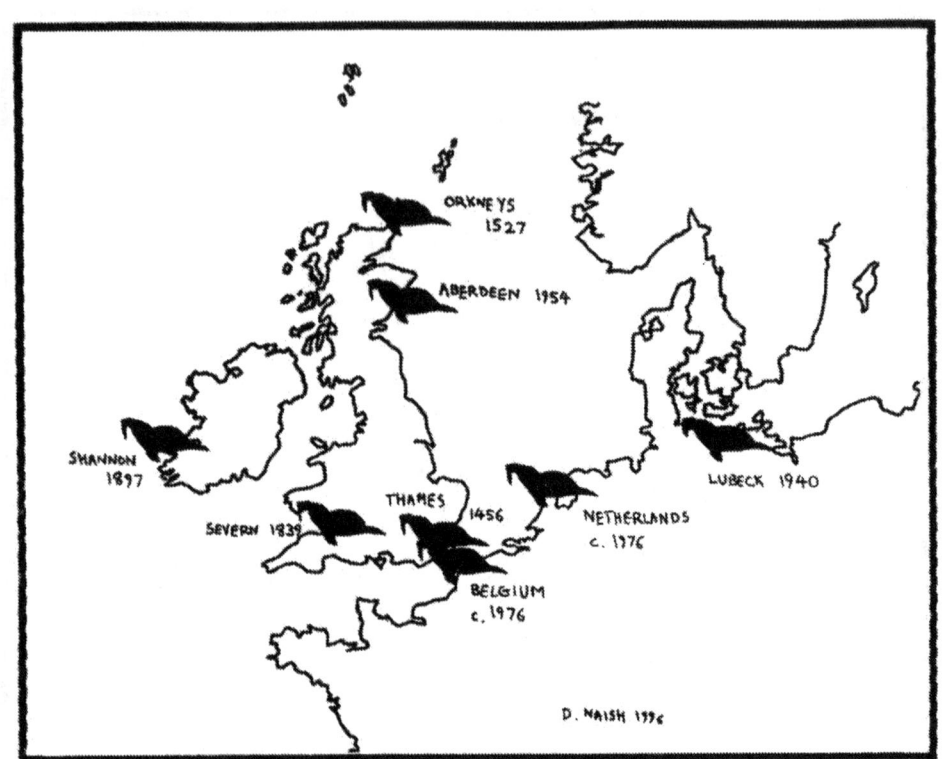

in 1928, and a sighting on the shore at Collieston, Aberdeen, in 1954. Since King's book was published, two walruses have been seen on Shetland and one on Orkney (all three in 1986). The best known of all such vagrants, however, is 'Wally', apparently a young male walrus with a broken left tusk who, in 1981, made several stops on the English east coast. 'His' last visit, on the River Ouse north of Cambridge, was surprisingly far inland: here 'he' was captured and flown to Iceland. However, shortly afterward, a young walrus, also apparently with a broken left tusk, was seen off the Norwegian coast. 'Wally's' return?

As a variety of Arctic marine mammals - Harp seals *(Phoca groenlandica)*, Hooded seals *(Cystophora cristata)* and Narwhals *(Monodon monoceros)* among them - have turned up around British coasts, even as far south as the English Channel in cases, the frequent visitations of walruses should not be so surprising.

Britain is one of those areas bridging biogeographical zones, and we have a rich and diverse fauna to prove it. Walruses included.

Ground Sloth Survival in North America

by Ben S. Roesch

The ground sloths were large, hairy, ground dwelling Edentate (Xanarthra) mammals. Edentates possess very simple or no teeth, and the order is composed of three sub-orders: Pilosa (including the ground sloths); Sloths; and Anteaters.

The name "Giant Ground Sloth" (or just "ground sloth"), is the common name for three supposedly extinct families of Pilosa: Megatheridae, Mylodontidae and Megalonachydae. The members of these families originated and mainly evolved in South America but spread outwards to North America during the late Cenozoic, with some species even reaching the Northwest Territories in the Pleistocene. Many species were large and robust, with some megatheriids reaching the size, and sometimes even surpassing that of today's elephants.

The ground sloths supposedly died out 10,000 years ago in the Late Pleistocene extinctions, which also claimed many of the other 'ice age' mammals, notably the mammoth and the sabre toothed tiger [1]. However, there has been an ongoing controversy over the possibility that these beasts may not have completely died out, and that some remnant populations of the giant mammal might have been able to survive to this day, or at least into historical times.

Reports of creatures that seem to represent ground sloths in modern times are well documented from South America [2]. These reports are still occurring, as the recent expeditions to the Amazon, led by American ornithologist, David Oren, in search of a "hairy and smelly" creature called the mapinguari, which Oren thinks is a ground sloth, have demonstrated [3]. With all the reports from South America of ground sloth type creatures, we might also expect some from North America. After all, ground sloths are a characteristic find from North American late Pliocene and Pleistocene faunas and they first arrived in North America when South America became connected in the late Pliocene (4), giving much time for possible traditions and reports among the Native Americans and white settlers to crop up.

However, there is a remarkable absence of such reports. To my knowledge no-one has ever written on the subject, with the exception of Michael Bradley (which we shall discuss below), and even then only very little. Including Bradley's account, I have only been able to find two incidents, both from the Native American tradition, that could be interpreted 'off the bat' as surviving ground sloths. They are, however, rather interesting accounts.

We will first discuss Michael Bradley's report which comes from the traditions of the Micmac Indians of Nova Scotia, Canada. Bradley writes about this in 'More than a Myth', a book which is actually about his search for an elusive monster in Muskrat Lake (close to Ottowa, Ontario) [5], and states:

"[The] Micmacs remember a time when they were plagued by 'giant squirrels'. Micmacs live in bark teepees and these pesky giant squirrels sometimes descended on Indian villages to devour these teepees or wigwams. The giant squirrels didn't harm the people, but they ate up their houses. Eventually these giant squirrels disappeared and the Micmacs were left in peace".

Bradley goes on to describe how this legend seems absurd at first, but which, if we look at the habits and anatomy of ground sloths, seems to make a lot of sense. After all, from what we know, ground sloths had long, furry tails, ate bark and other vegetation, (to a ground sloth an Indian teepee would be perfect fodder), and in fact looked very much like "giant squirrels". Certainly, if the above story is true, then it would seem that a group of very strange animals were living in Nova Scotia, and a ground sloth seems to be a good match. However, this clashes with currently accepted scientific paradigms of when the ground sloths died out.

Micmac legends occur no further back in history than about 500 A.D [6], and therefore if the 'giant squirrel invasion' really occurred, it would provide excellent evidence that ground sloths were still surviving, perhaps precariously, at about the same time that the Vikings were plundering and pillaging - despite the fact that ground sloths were supposed to have become extinct 10,000 years ago.

In any case, as I mentioned above, there are other traditions, again from Native Americans of potential surviving ground sloths.

These traditions come from the Northern Tutchone nation, which originate from the Yukon, Canada, and have a very rich oral tradition spanning back many hundreds of years. Those traditions of interest to us were first revealed in newsletter #14 of the British Columbia Scientific Cryptozoology Club [7], in which the editor, Paul LeBlond, a cryptozoologist of high standard, presented an interview he had conducted with Ms. Dawn Charlie, who lived in the Yukon and knew of several Indian stories of strange creatures.

Ms. Charlie first told how she had moved to the Yukon from Toronto, Ontario in the early 1970s and soon met up with her future husband, Wilfred Charlie, a Northern Tutchone. Soon they were living in Carmacks, Yukon, and Ms. Charlie began to hear stories of strange animals reported in Tutchone tradition, as recounted to her by her husband and his relatives. One of these animals was the 'satoychin' [7], or 'Beaver Eater'.

According to Ms. Charlie the 'Beaver Eater':

"...is a large animal that eats beavers. It reaches under a beaver house, flips it up and eats the animals inside. When natives of the area were shown a book on extinct animals put out by the Smithsonian Institute, they identified the ground sloth as the Beaver eater".

There is no doubt that eating beavers seems strange behaviour for a ground sloth! As far as we know all species are 100% vegetarian. However, there is an explanation for this strange behaviour if, indeed, the 'saytochin' is a ground sloth. As we saw in the Micmac tradition previously examined, several alleged ground sloths descended on a small village to eat the bark teepees as this was easy-to-get food. Is it possible that when the Beaver-Eater was tearing apart a beaver lodge, it was not really after the beavers within, but rather the stock-pile of branches, bark and other vegetation that the beavers had industriously collected to make their lodge? Again, this would essentially be a 'salad' just waiting for a lucky ground sloth to chance upon, and if a Native American spied upon the Beaver-Eater in the act of tearing open a beaver lodge, they doubtless would have assumed a likely carnivorous nature for the strange beast. The feat of tearing apart a beaver lodge would not have been a very hard task for a ground sloth; they had very large claws on their fore-feet which could easily have been employed for such a task.

In any case, the story of the saytochin has much more to it, as Ms. Charlie continues in the interview:

"Saytochin stories are .... very old but there are some recent reports. The latest report was from Violet Johnny, my husband's sister, who was fishing with her husband and her mother at the head of Tatchun Lake four or five years ago. An animal came out of the woods, eight or nine feet high, bigger than a grizzly bear. It was a 'saytochin' and it was coming towards them. They panicked, fired a few shots over its head and finally managed to get the motor going and took off. There are other reports. There is also a report that a white man shot one in a small lake in the area. Beaver-Eaters are supposed to live in the mountainous areas east of Frenchman's Lake."

This passage is very interesting indeed. In it Ms. Charlie cites an encounter with the 'saytochin' that took place in 1988 or 1989! The report isn't that specific but if they say that it was a 'saytochin' then it could only be a ground sloth! What else could it be? It is obviously not a known, currently recognised species such as a moose or a grizzly-bear, as the Tutchones would not make such a mistake. They have been familiar with the animals of the area since times immemorial. The report sounds like one of a sasquatch, but this doesn't seem possible as the Tutchones would be likely to have separate legends and stories about giant man-like creatures, as do many other tribes in North America [9]. It really does seem that the best solution is a ground sloth!

However, this brings us back to the problem of the accepted extinction of the ground sloth about 10,000 years ago.

The Micmac legend of 'giant squirrels' is reason enough to speculate that ground sloths may have survived as recently as 500 AD, at least in the Nova-Scotia region of Canada. However, since no other similar reports have occurred since the arrival of white men (despite the fact that the region is now well populated with whites), and because the Micmacs mentioned that the giant sloths disappeared after a short while, it is easy to say that the giant sloth is now extinct in that area.

However, the Tutchone legends from the Yukon provide a completely different story altogether. It seems as if ground sloths could very well still be resident in the area today, as Johnny's report from only eight years ago shows. Certainly the great wilderness of the Yukon would provide ample space for a small population of such animals to live in harmony, and without many encounters with man.

This report may also explain the lack of reports of ground sloth like animals in North America. If they exist, they exist far from the reaches of modern man. Even so, it still seems strange that the two reports just discussed are the only two that I could find. However, it is possible that we have more reports - but under the name of a different cryptid.

I will probably start a debate about this, but I must ask:

Is it possible that some bigfoot/sasquatch reports from North America are really cases of a misidentified ground sloth? Let us briefly list the possible evidence for this:

1. The ground sloth could stand upright on its hind legs at least when feeding. It would then have looked like an upright animal resembling a bear. When coupled with its hairy body, it would certainly inspire thoughts of sasquatch in some frightened human who accidentally stumbled across it.

2. The ground sloth has been mistaken for a hairy humanoid before. The mapinguari of the Brazilian Amazon was thought of both by Ivan T Sanderson (10), and Bernard Heuvelmans [11] to be a sasquatch-like creature, yet as David Oren's recent quests and research have pointed out, the mapinguari is most probably a ground sloth.

3. About 140 (5.6%) of North American sasquatch reports tell of a nauseating or overwhelming stench [13]. The mapinguari, now thought to be a ground sloth also possesses such a smell [14].

Certainly, I am not proposing that all sasquatch reports have really been of ground sloths, but I am presenting some evidence to suggest that a small percentage of sasquatch/bigfoot reports could possibly be surviving ground sloths. Also, there is a good chance that the small fraction of 'smelly' sasquatch reports are actually ground sloths. I cannot, of course prove this last supposition besides the comparison with the equally smelly mapinguari. A strong smell would also be a more likely trait in a surviving ground sloth than it would be in a large anthropoid. This is mainly to do with the way that these animals were likely to have lived. They were large, slow and ponderous, and despite having large claws with which to protect themselves, they were probably not fast or agile enough to use them effectively. (It is thought that they were usually used for feeding). Having a strong smell, therefore, which they could 'turn on and off' at will would certainly be a very possible evolutionary step [15].

However, in 1967, when Roger Patterson was frantically filming a surprised sasquatch, featured in the now famous 'Patterson Film', he noted a smell akin to a 'dog rolling in wet manure'. [16]. This could mean, if we accept that the 'Patterson Film' is not a hoax, that the alleged sasquatch is capable of producing a powerful smell similar to that of the mapinguari. Here my theory loses some credibility because there is no way to explain the animal in the 'Patterson Film' (if it is real) as a ground sloth.

In any case for now the 'Ground Sloth as Sasquatch' idea is simply an open-minded theory, and therefore it does have problems. Besides the one discussed in the previous paragraph, one might ask how a ground sloth, probably a slow and lumbering beast, living in geographical areas constantly being searched for sasquatch has managed to avoid detection. One possibility is that they not only live in remote areas but have a very small population. Another is that they are more adept at hiding than we would suppose - especially the smaller ones. If there are any surviving ground sloths in North America they are likely not to be much bigger than a black bear.

Despite sizes and measurements, the evidence of continuing survival of ground sloths in North America, particularly in the Yukon is rather intriguing. The Tutchone stories and legends are hard to dismiss as they also have legends of very real occurrences such as the time "ash fell from the sky" [17], in this case referring to the eruption of a nearby volcano in 700 AD. And if not ground sloths it is certainly something inexplicable.

Finally, as I finished writing this article, an interesting report came to my attention. In a recent edition of the American sasquatch newsletter Track Record, a letter written by Don Peterson was published as follows:

"Then the day the Space Shuttle blew up (Challenger, Jan 28th 1986), I was on a five state radio chat-about programme out of Bemidgi (Paul Bunyan County) in northern Minnesota, and one call-in discussion from the year previous told of something that sounded like a South American ground-sloth walking upright, that they drove out of a very woodsy area". [18]

However exciting this report sounds it doesn't add up to much hard evidence. I am not sure as to the reliability of Don Peterson, and it should be noted that he mentioned the National Inquirer in the above quoted letter. As well, it would seem strange that Mr. Peterson would think that the creature "sounded like a South American ground-sloth walking upright", instead of just a bigfoot or another ape like creature. Just for the record Don Peterson is a sasquatch hunter from Washington State (with no scientific background), who actually worked with Ray Wallace, a rather questionable figure in the sasquatch world indeed (it is believed that he hoaxed several alleged sasquatch photos, among other things) [19]. In any case, I don't see why Mr Peterson would take the trouble to hoax a story about a giant sloth in Minnesota when his main quarry is sasquatch. For now, at least, the above report will remain merely food for thought until further and more reliable information regarding it is turned up.

Certainly nothing can be said for sure until someone launches a serious expedition to either Minnesota or the Yukon to check out these rumours of living ground sloths, and to even try and bring home the evidence that will seal the fate of the story of ground sloth survival in North America - bones, skin or a body - dead or alive!

Acknowledgments

Thanks to Glen Kuban, Peter Massaro, John Moore and Ron Schaffner for valuable comments and material.

References and Notes

1. COLBERT, E.H. 1955. *Evolution of the Vertebrates.* New York: John Wiley and Sons. pp. 264-7. Also [12].
2. HEUVELMANS, B. 1958. *On The Track of Unknown Animals.* New York. Hill and Wang. pp. 253-83.
3. PEARSON, S. 1995. 'Load the Stun Gun, pass the Old Spice'. Outside 20. (November): p.34.
4. COLBERT, E.H. Op.Cit.
5. BRADLEY, M. 1989. *More than a myth - the search for the monster of Muskrat Lake.* Willowdale, Ontario, Hounslow press. pp.21-23.
6. ibid
7. LEBLOND, P. 1993. 'Yukon Cryptids'. BC Scientific Cryptozoology Club Newsletter 14 (April), pp.5-6.
8. Another variation on the spelling is 'saytoechin'.
9. Interestingly, the Tutchones have other legends of potential cryptids. One such is the 'urchow' which is allegedly "*somewhat camel like and has large bottom teeth with which it scrapes the bark off trees*". (Quoting LeBlond, 1993). Could this also be a ground sloth?
10. SANDERSON, I.T. 1961. *Abominable Snowmen - Legend come to Life.* Philadelphia: Chilton Book Co. pp. 357-8.
11. HEUVELMANS, B. Op. Cit. pp. 324-7.
12. OREN, D.C. 1993. 'Did ground sloths survive to recent times in the Amazon region'. Goeldiana Zoologia no. 19. (20 August).
13. GREEN, J. *Sasquatch: The Apes Among Us.* Vancouver, Hancock House. P 444.
14. PEARSON, S. Op. Cit.
15. KUBAN, G. 1996. Pers. Comm. April 5th.
16. PETERSON D. 1996. Letter: Track Record 55 (March). p.15. (Letter was written on March 13).
17. SCHAFFNER. R.1996. Pers. comm. July 16.

NEW FROM CFZ PRESS

"THE OWLMAN & OTHERS"

BY JONATHAN DOWNES

"the first comprehensive look at The Owlman of Mawnan"...

£10.00

Initial Bipedalism: A Theory of Human Evolution

by

Francois de Sarre

Our species is commonly believed to be four million years old, or thereabouts, and is commonly thought to have evolved from some kind of an ape.

How wrong, however, is this Natural History of mankind as defined with such solemnity by 'classical' anthropology? I would answer that it is completely untrue and that the whole story of our alleged Simian ancestry has been invented, and has, since the days of Darwin, been imposed on us through scholarly precepts relayed to us through the newspapers and television.

An australopithecine (a very ancient ape), nicknamed 'Lucy' is presently considered by most investigators to be 'mother of humanity', but she merely represents a figurative 'missing link' between an early terrestrial biped and the quadrupedal chimpanzees of today.

For the last century at least, extraordinary anomalies have become widespread 'knowledge' and have thus become integrated into contemporary creeds of scientific belief

BIPEDAL GAIT AS THE EARLIEST FORM OF MAMMALIAN LOCOMOTION.

Such an assertion seems to be a direct contradiction of what we perceive to be 'proven' facts from our daily experience. It is currently believed by most observers that four legged animals remain in a more archaic state than ourselves. This is because until now Palaentology has dictated that we allegedly evolved from some early quadrupeds.

As a zoologist I should surely take notice of what this science teaches; a science whose purpose is to explore the archives of our past from excavated remains, and interpreting these artifacts as the relictual supplies of ancient fauna.

I do not believe that life on earth is more than a few million years old, but this is not the purpose of this paper.

The concept that our current biodiversity of forms and structures has resulted purely from accumulated random changes caused by minute mutations, natural selection and adjusted adaptation over enormous periods of time is in itself scientifically flawed.

An important piece of my evidence comes from Embryology and comparative anatomy.

Every day, through experimentation, we can observe directly where we have come from, with the help of the real biological material that we possess. Not from stony, altered and evanescant substrats but from our own bodies.

We humans are fundamentally the same creatures which once lived and left the oceans by entering the land.

In previous articles I have, at great length, developed the theory that Man's large brain is not an indication of Simian ancestry. Indeed, we and the quadrupedal mammals all stem from a very primitive stock of bipedal mammals. Therefore, we all carry the same basic genetic properties as our ancestors possessed.

We steadily inherited the character traits of typical pioneers.

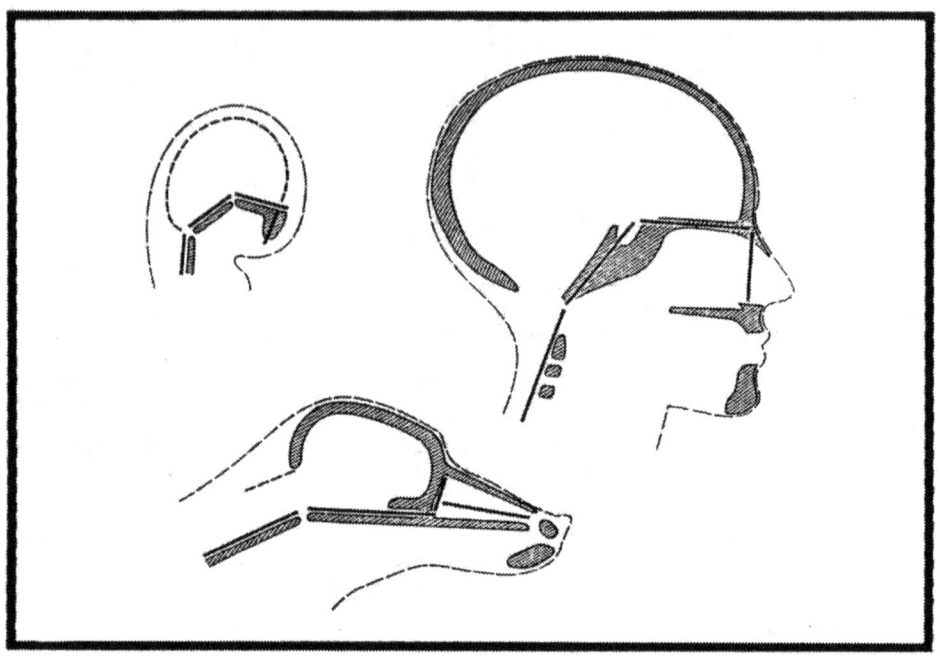

Figure One

Bending of the skull basis (after BOLK 1926).

below: dog's horizontal deck.
above: (left) original bending in mammalian embryos.
(right) bending of the skull basis in adult men.

At early stages of their embryonic development, all quadrupedal animals have the same 'big head' as people. Figure one, after the Dutch Biologist L.Bolk (1926), shows that the typical human bending of the basis of the skull is a prior phase in Mammalian octogenesis.

As far as quadrupedal mammals like dogs are concerned the skull basis forms a horizontal deck in relation to the axis of the body and the horizontal plane of the ground on which the animal is moving. In consequence, the position of the insert of the vertebral column at the back of the skull has become raised.

It would seem that, at a certain point in the embryogenesis, animal foetuses carry on developing beyond the point where the human embryo has become complete.

Furthermore, in bipedal Man, the volume and the weight of a primordial big brain has locked the bent skull base into the original embryonic position.

This is why, amongst all other vertebrates, Man has kept an upright position whilst standing or walking!

APES ORIGINATED FROM HUMAN ANCESTRAL STOCK.

The key error of today's evolutionists is to assume that man has descended from quadrupedal animals that resembled monkeys. It seems very easy to scan the fossil record for 'ancestors that fit' and then to design some ape-like creatures for our hypothetical line of descent.

Inversely, another explanation would be that apes such as Proconsul or the recently unearthed Australopithecus ramidus retained relict human features by specialising to an arboreal existence, where they made use of their prehensile arms, instead of, as commonly believed, having evolved into a man-like form.

Indeed, it appears that the apes should be considered as vestiges of the human lineage rather than as our antecedents.

Figure two is very convincing in the way that it shows how a monkey foetus resembles a human foetus rather than the other way around. Logically it appears that human evolution has never passed through a stage like that which characterizes the tree dwelling apes. In fact, Simians appear to have originated from a form similar to our own body shape.

We must therefore proceed with a complete re-evaluation of the common point of view as regards this topic. Let us therefore contemplate figure three - a scheme conceived by Adolf Schultz, a well-known American primatologist who made some very important discoveries in this field concerning the different relative proportions of the head, trunk and members during the stages of the growth of both Man and several of the apes from the foetal stage until maturity.

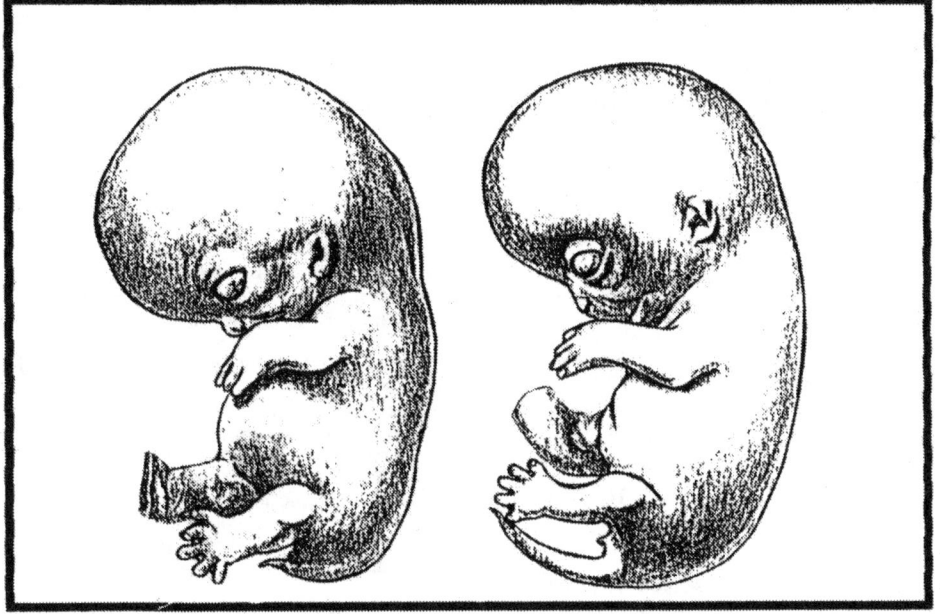

Figure Two. A Comparison of the foetuses of Man (left) and of Macaque (right) at comparative stages of their development (Man:49 days, Macaque:44 days).

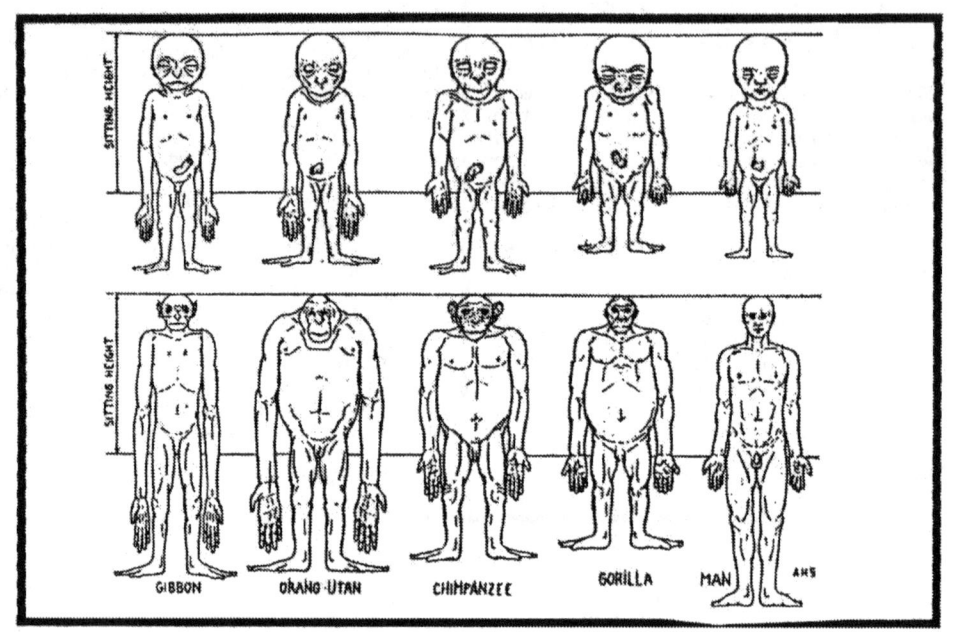

Figure Three.

Bodily proportions of the anthropomorphous apes and man in the foetal and adult stages. (After SCHULTZ, 1926). The foetuses (above) are arranged in the same order as the adults (below), all being given identical vertico-perineal sitting heights.

Professor Max Westenhøfer, a German anatomist, who in the 1920s was a main instigator of the theory of Initial Bipedalism, claimed that, if we do not hold any sectarian preformed opinions, this diagram will in itself be sufficient proof to make us abandon any preconceptions of a supposed Simian ancestry to Man.

THE FABULATION ABOUT OUR FOSSIL 'ANCESTORS'.

Critically-minded readers would certainly allege that today's scientists are in position of 'intangible proofs' of our evolution from monkeys. Look out, however! Lucy, the world famous East African australopithecine is NOT the mother of humanity, or even our figurative great aunt. She is no more than an ape!

Dehominised forms like the australopithecines appear as a result of over-specialization to arboreal life. They evolved from our direct ancestors, but they might as well have split up today into the contemporary chimpanzee species. The question that I would like to ask to contemporary professional palaeontologists like the ones who discovered Australopithecus ramidus is, when will they stop getting so excited about discovering the remains of apes?

I am sure that the famous idea of Man's Simian ancestry will soon join the other forgotten archives of natural history, where are stored the 'eminently scientific' orations and discourses of

16th Century naturalists like Conrad Gessner in Switzerland, and the Italian professor Aldrovani, about the lives and therapeutic qualities of their supposedly contemporary dragons!

BIBLIOGRAPHY

FRECHKOP, Serge (1949). *Le crane de l'Homme en tant que crane de Mammifere.* Bull. Mus. R. Hist. Nat. Belg., 25: 1-12 Bruxelles.

GEE, Henry (1995): Uprooting the human family tree. Nature 373: 15. 5th January.

GRIBBIN, J. and CHERFAS, J. (1981). *Descent of Man - or ascent of the ape?* New Scientist, 91: 592-595.

HEUVELMANS Bernard. (1966): *Le Chimpanze descend-il de L'Homme?* Planete, 31: 87-97. Paris.

KURTEN, Bjorn (1972). *Not from the apes.* (Victor Gollancz, London).

SARRE, Francois de (1994): The theory of initial bipedalism on the question of human origins. Biology Forum, 87 (2/3): 237-258.

SARRE, Francois de (1995): *Die 'Affen-Abstammung' des Menschen: Eine illusion geht zu Ende.* Efodon Synesis, 10: 24-27.

SERMONTI, Guiseppe (1988). *Dopo l'Uomo la Scimmia.* Abstracta, 3: 74-81, Roma.

SNOO, Klass de (1942): *Das Problems der Menschwerdung im Lichte der vergleichenden Gebertschilfe.* Gustav Fischer Verl., Frankfurt.

WESTENHOFER, Max (1948): *Die Grundlagen meiner Theorie vom Eigenweg des Menschen.* C. Winter Verl., Heidelberg.

* * * * *

OUT OF THIS WORLD

Xenobiology by Graham Inglis

LIFE ON MARS?

The first indications that life has existed on a planet other than our own were outlined in August after a two year NASA study of a pear-sized Antarctic meteorite, called ALH 84001, believed to have originated from Mars.

It is now suggested that bacteria-like life developed on Mars three billion years ago (at about the same time life is believed to have begun on Earth) - but then fizzled out.

Most meteorites that hit our planet appear to be primordial remnants 'left over' from the formation of the solar system. It is believed that, around 16 million years ago, an object such as an asteroid hit Mars, scattering ejecta - some of which was flung into space. Current best estimates are that the item of ejecta now known as ALH 84001 encountered the Earth some 13,000 years ago, falling in Antarctica.

Three possible indicators of ancient life were found within this meteorite:

» Microscopic elongated tube-like structures: which possibly are fossils of bacteria-like entities.

» Deposits of a carbonate - a mineral product of some Earth bacteria and some geological processes.

» Aromatic hydrocarbons - organic molecules produced by some bacteria and by the processes of stellar evolution.

Individually, these three indicators are hardly conclusive: it is the presence of all three in one small rock that collectively constitutes the best evidence yet of life having existed elsewhere in the solar system.

The astronomers Hoyle and Wickramasinge have suggested that an earlier such rock-borne 'visitation' could have seeded Earth with its life. Effectively, we all could be Martians - and perhaps ALH 84001 bears some of our long-lost brethren.

MULTITUDINOUS ENIGMATIC CETACEANS, OR *"WHALES IN LIMBO"*

Part 4 in a 7 part series.

by Darren Naish

Unavoidably, cryptozoology is a narrow-minded science, for it is concerned almost solely with the mere identity of animals that people have seen. Its icons are typically creatures that, if real, are totally unique in the modern fauna. This contrasts with, say, a sighting of an unusual cetacean, an animal belonging to a group already consisting of almost 80 recognised living species. In this article I have attempted to review some of the more notable (and therefore better known) cryptocetaceans of which I am at present aware, and make conclusions on their possible identification and/or validity.

As will be explored in a future article, field identification and specific identity amongst known cetaceans is already a messy business, and many of the sightings alluded to here highlight the difficulties inherent in poor sighting records. However, do not despair. A number of the cetaceans reported here are so odd that they cannot be confused with anything else. If records of these animals truly represent a reality, they also reveal new diversity amongst living cetaceans.

A diversity of fins

The existence of, let us say, a sperm whale with a tall dorsal fin would indeed prove this point. Such an animal has indeed been seen. Like several other apparently new species discussed in this article, this animal is already well known in the cryptozoological community. Seen off the Shetland Islands in the 17th century, apparently several times, it was described by cetological pioneer Sir Robert Sibbald in 1692 [1]. It was of great size - apparently 15 m long (the biggest authenticated sperm whale bulls reach 18 or 19 m).

Sibbald commented upon the fact that teeth were present solely in the lower jaw, and though sperm whales are unmistakable animals in any case, this observation further shows that the animal was a sperm whale of some kind. Exactly how Sibbald came to be sure of this dentition I am not sure, as sperm whales' teeth are only visible at close range, and obviously only if the animal is lifting its head and opening its jaw. Carl von Linne, the great taxonomist, was obviously impressed enough with Sibbald's sperm whale to include it in his catalogue of organisms, the *Systema Naturae* (first published in 1758) [2]. He treated it as a species and gave it the name *Physeter tursio* - 'sperm whale dolphin': the 'dolphin' alluding to the similarity of fin structure to that of dolphins. Fig. 1 is a highly conjectural restoration of this animal.

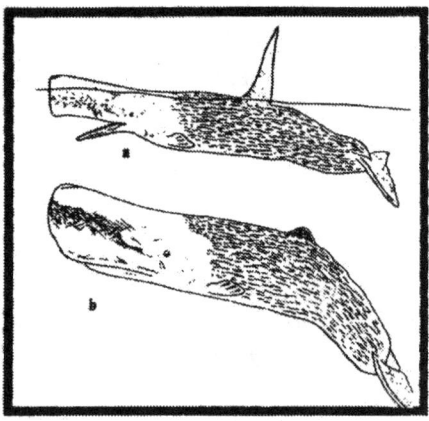

Fig. 1a. The conjectural restoration of *Physeter tursio*. 15m long in life.

Fig 1b. Adult male *P. catadon* for comparison. 18m - 19m long in life.

A possibility is that *Physeter tursio* is not a species, but was perhaps an abnormal individual. A dolphin with two dorsal fins was recorded from the coast of Cornwall in 1857 [3], but that it was amongst a school of otherwise normal animals strongly suggests a deformity (more about two-finned cetaceans in a moment!). However, it is hard to explain exactly why a sperm whale would accidentally end up with a tall dorsal fin as an abnormality. Other such external features, including tails in humans and external hind limbs in sperm whales can always be explained as expression of an ancestral character still 'hidden' in the genome. Nobody knows if *Physeter* has the genetic 'memory' to generate such a dorsal fin, but if it does we may have an alternative theory.

More remarkable still are, as mentioned above, those cetaceans that would seem to have two dorsal fins! Michel Raynal, working in one case with Jean-Pierre Sylvestre [4], has already [5] reviewed reports of such cetaceans. Consequently, I will not analyse the area in depth.

Raynal and Sylvestre began by eliminating the confusion caused by a 'monstrous fish' (bearing two dorsal fins), reported by Mongitore to have stranded on the Sicilian coast in September 1741. Mongitore illustrated this animal - quite certainly inaccurately - and it bears scant resemblance to any cetacean. Treating it as such, however, in a publication from 1814, the French-American naturalist Constantine Samuel Rafinesque-Schmalz thought it to be a two-finned cetacean and named it *Oxypterus mongitori*. Raynal and Sylvestre regard the account as too vague to be useful, and I agree with them that the Sicilian animal was probably a large shark. In his 1989 book, *There are Giants in the Sea* [6], Michael Bright was obviously unaware of the original account of '*Oxypterus*', and erroneously refers to it as a 'twin-dorsal-finned black-and-white spotted dolphin', which it clearly was not.

In fact, Bright had confused '*Oxypterus*' with other animals, this time ones that do indeed appear to be two-finned cetaceans [7]. Perhaps the best-known of all cryptocetaceans - one even appears on the front cover of the magazine Exotic Zoology - these were apparently seen between Hawaii and Australia by the French naturalists Jean-Rene Constant Quoy and Joseph Paul Gaimard in 1819 [10]. They described and illustrated them (fig. 2) as being about double the size of a porpoise [11], marked with black and white spots, and with two curved dorsal fins, one set just behind the head. It would seem that they were viewed at close range, for some length of time, and furthermore by many of the crew members [10]. Quoy and Gaimard named their new dolphins *Delphinus rhinoceros*.

Having myself seen video footage where several dolphins gel together and create what appears to be a large multi-finned animal [12], I think it is tempting to speculate that such an explanation lies behind, at the very least, aboriginal myths of multi-finned cetaceans [13], as Raynal himself has also noted [5].

Quoy and Gaimard's sighting, however, is too precise to allow for such a possibility. Taking their accounts at face value, they clearly depict an unknown cetacean: one that must also represent a new genus of dolphin. In accordance with this, Raynal and Sylvestre have proposed the name *Celodipterus* [4].

What would appear to be another two-finned cetacean, but this time a mysticete, has also been thoroughly discussed by Raynal and Sylvestre [5]. Following his sighting of this animal between Peru and Chile in 1867, the Italian naturalist Enrico Hillyer Giglioli named it *Amphiptera pacifica* [14]. Giglioli's extremely competent illustration (on which fig. 2b is based) eliminates the need for description. *Amphiptera* is clearly remarkable. Such an animal would deserve its own genus, and Giglioli's own suggestion that the balaenopterid subfamily Amphipterinae be created for it seems entirely reasonable to me. Sightings of what would seem to be other specimens of *Amphiptera* have been reported [15], one even from 1983 [16].

Fig 2a. A moderately speculative rendition of Quoy and Gainard's two finned *Delphinus rhinoceros*. In life length would be 4-5m.

Fig 2b. Giglioli's *Amphiptera pacifica*, an 18m long baleen whale; unlike rorquals, it lacks throat grooves and has two dorsal fins.

As will become evident in the rest of this article, the majority of cryptocetaceans are, by cetacean standards, small. Joining *Physeter tursio* and

Amphiptera, however, are two large cryptic forms of right whale (family Balaenidae), both from the Arctic. For reasons of space, however, they will be discussed elsewhere, as will the cryptic whales of the Southern Hemisphere. These include yet another tall-finned toothed whale, a variety of monodonts [17], and a grey and black dolphin probably related to Risso's dolphin (*Grampus griseus*). A preliminary analysis of what may be another cryptocetacean, that seen by Sir Peter Scott in the Magellan Straits, has already been presented [18,19].

Morzer Bruyns and his cryptocetaceans

Ironically the best known of all mystery cetaceans, more doubts are entertained as to the validity of the following four species than for any others. These animals are described in a popular book written by W.F.J Morzer Bruyns in 1971 [20], and as such are generally more accessible then obscure papers and research notes. They are well known in the cryptozoological community as all four are included in Heuvelmans' 1986 checklist of cryptids [8].

Morzer Bruyns made important observations on whales the world over, many of which are of interest to cryptozoologists, and was bold enough to present his four 'unknowns' as potential new species. He even gave them species names, albeit rather makeshift ones.

1) The Alula whale. This is a large orca-like dolphin, sepia brown, but marked with star-shaped scars, apparently seen by Morzer Bruyns in the eastern Gulf of Aden.

He describes it as 6-7 m long and perhaps 1800 kg in weight. It has a rounded head 'similar but not quite as round as in *Globicephala* [the pilot whales]', and a prominent orca-like dorsal fin (fig. 3).

Importantly, Morzer Bruyns reports that several other officers besides himself watched one of these whales from their ship; a number of them made sketches, and later a painting was prepared. It would also seem that Alula whales were seen by Morzer Bruyns and his crew on more than one occasion, and also that more than one whale was seen at a time. On one occasion it is reported that the whales were seen to pursue a school of smaller dolphins, and furthermore, 'devouring of live whales by predators in this area was witnessed'. The predators doing the devouring could not be identified, but the inference is that they were Alula whales.

Fig 3a. Alula Whale based on rendition in (20). In life apparently 6-7m in length.

Fig 3b. Adult male Killer Whale (*O.orca*) for comparison. Between 8-10m in length.

If we are to assume that the Alula whale is real, and I really don't know whether or not we should, it is clearly a new species. Perhaps it is enough like the Killer whale (*Orcinus orca*) to belong in the same genus [21]; it is clearly very much like that species in its morphology and behaviour. It is interesting that colour variants of the Killer whale could, conceivably, be regarded as 'new' whales when seen: to my knowledge, however, all such variants have always been identified correctly. Watson reports the existence of all-black killer whales [22] and all-white individuals have been seen several times [23] (in 1970, one was captured for British Columbia's 'Sealand of the Pacific'. Called 'Chimo', it lived for 2 years [24,25]. Other variants of *Orcinus* will be discussed in the next article in this series. However, the existence of aberrant orcas is irrelevant to the identity of the Alula whale: it is distinct from them.

2) The Greek dolphin. Of all living cetaceans, the small dolphins of the genus *Stenella* have been the biggest headache for taxonomists. Five living species are currently recognised [27] but a bewildering variety of morphs and hybrids amongst these species ensure that some confusion remains.

Furthermore, if Morzer Bruyns is right, there are

another two species of this genus awaiting description! The 'Greek dolphin' is one of these: Morzer Bruyns claims to have seen it on several occasions, and says that it is very much like the Striped dolphin (*S. coeruleoalba*) but without the distinctive black stripes [50]. Though there is quite some variation in the pigmentation and patterning of Striped dolphins, nowhere else have they been said, to my knowledge, to ever lack the eye-to-anus stripe.

3) The Senegal dolphin. This is Morzer Bruyns' second as-yet-unknown Stenella species. He described it as differing slightly from the pantropical spotted dolphin (*S. attenuata*), the cosmopolitan one of the two known spotted Stenella species. As anyone who has seen the things in the field will know, differentiating the two species of spotted dolphins in the wild can be very difficult, and identifying a third form where two already occur (as they do, as it happens, off the Atlantic coast of Africa) is, in my opinion, beyond the data ordinarily available. While the idea that new species of spotted dolphin may exist out there amongst the presently recognised 'species' of spotted dolphin is a fair possibility, given the very broad variation already known to be present in these species, I do not believe it reasonable to recognise a distinct form, viewed only in the field, as representative of a new species.

This does not necessarily mean that Morzer Bruyns is incorrect - it does, however, show that his distinctions may be on shaky ground.

4) The Illigan dolphin. This is perhaps the most remarkable of the cetaceans Morzer Bruyns claims to have seen, this time in the Mindanao Sea, amongst the Philippines. It has a brown back, yellow flanks, a pink belly and, most interestingly, appears to be very similar in size and shape to the Melon-headed whale (*Peponocephala electra*).

As Heuvelmans has observed [8], it is particularly noteworthy that Morzer Bruyns seems to have been aware of a Peponocephala-like whale before Peponocephala itself was deemed to be truly distinct from other dolphins. As Morzer Bruyns did not publish these observations until after Peponocephala was described as a new genus [28] however, the best piece of supporting evidence is void!

I also find it hard to believe in a very brightly coloured dolphin when all of its closest relatives (either the killer whale - pilot whale group [29], or the Lagenorhynchus species) are always marked in blacks, whites and greys.

However, there is no real reason why an Illigan dolphin cannot exist. But what are we to make of Morzer Bruyns and his claimed sightings? Besides having supposedly encountered the four cetacean species discussed above - all as yet unseen by other cetologists - Morzer Bruyns also writes of several sightings of unidentified ziphiids - beaked whales, including a rust-brown species seen in the Gulf of Aden. Then there is a 'South China Sea dolphin' and a 'Malacca dolphin': animals that, Morzer Bruyns suggests, may be similar to Fraser's dolphin (*Lagenodelphis hosei*) - yet again, an animal unknown from complete specimens at the time of Morzer Bruyns' purported observations [26,30].

Short of accusing Morzer Bruyns of fabricating evidence, I am totally unsure as to how seriously we should regard his accounts. Stephen Leatherwood, the world authority on cetacean field identification, has referred to Morzer Bruyns as one who has 'muddied the waters' of field cetology [31].

The most elusive of ziphiids

Ziphiids are by far the most elusive and poorly known of cetaceans, probably by virtue of their deep-sea lifestyle. Within this century, seven new species have been named, the most recent of these in 1991 [32]. Currently hardly reported in the zoological press is another new species, again from the eastern Pacific [33]. As I write, the formal description of this species awaits publication [34]. Yet again, here is proof that new cetacean species await discovery. So might there be other unknown ziphiids out there? [35]

Again as of writing, Longman's beaked whale (*Mesoplodon [=Indopacetus] pacificus*) has been described solely on the basis of skeletal remains [36]. We just don't know what the complete animal, dead or alive, looks like. This has led to some interesting speculations, as there are several ziphiid sighting records that do not match known species: might these have been live specimens of Longman's beaked whale?

For example, Mark Carwardine reports that two large mesoplodonts were seen, by 'experienced observers', in waters around the Seychelles in 1980 [37]. These animals agreed in size and morphology with what is known of Longman's beaked whale (namely, two skulls) and were light grey. Their appearance was reconstructed in 1995 [38] and appears as fig. 4 here. Another suggestion is that the rust-brown ziphiids seen by Morzer Bruyns in the Gulf of Aden may themselves have been Longman's beaked whales as may have been a group of large ziphiids photographed near Christmas Island[22].

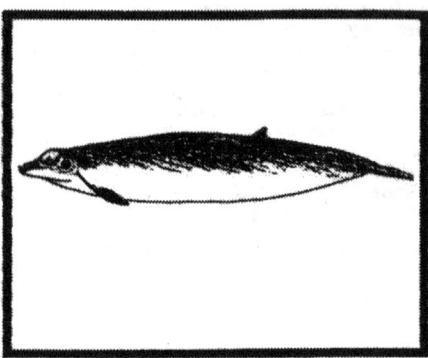

Fig 4. Restored appearance of ziphiid seen off the Seychelles in 1980. Based on [38].

Fig 5. 'Male' specimen of an unidentified ziphiid seen and photographed in the eastern Pacific. In life 5-5.5m in length. Based on illustrations and photographs in [39] and illustration on pp.112-3 of [37].

Presently the most widely reported of all cryptocetaceans is that reported by Pitman, Aguayo and Urban in 1987[39].

This animal agrees in morphology with known mesoplodonts and is notable for occurring in two distinct colour morphs that, on occasion, have been seen together. One is larger, has a broad whitish band sweeping diagonally from behind the head to the tailstock (fig. 5) and displays numerous scratches and scars.

These are typical of male mesoplodonts which, unlike the females and juveniles, have protruding teeth and seem to use them in fights with other males. It can therefore be deduced that this form is the male. It has several times been seen associated with a second, less distinctive, uniform grey-brown animal.

Probably female, this second form can only really be identified as the same species as the 'male' when it is seen together with it. Indeed Pitman and colleagues write: 'The brown animals are relatively nondescript and separating them from other mesoplodonts in the field may not be possible' [39].

As regards the absolute identity of this animal, three possibilities exist. Firstly, it may be a brand new species and, if so, as it has been photographed and seen on more occasions than certain mesoplodonts which are known from specimens, it is now certainly the best documented. It is even included in both recently published popular [37] and technical [41] field guides. Secondly, it may be a race of an already known mesoplodont, in which case we are dealing with a new subspecies. I consider this unlikely, but it remains a possibility (two other poorly known but morphologically distinct mesoplodonts - Hubbs' (*M. carlhubbsi*) and Andrews' beaked whales (*M. bowdoini*) - are considered by some experts to be subspecies of the same species.

Thirdly, this may be an already known species, but one not yet positively identified alive, or known from material other than a skull. Here we return to Longman's beaked whale. That this species has only been recorded with certainty from Queensland and Somalia does not rule out the possibility of it occurring in the east Pacific: ziphiid species are notorious for stranding miles away from their expected ranges. The best example of this is probably Gray's Beaked Whale (*M.grayi*). This seems to be a southern form, with a number of strandings restricted to the latitudes of New Zealand, Tierra Del Fuego and South Africa.

In 1927 however, a specimen stranded at Kijkduin in the Netherlands! It is not therefore inconceivable that Longman's Beaked Whales might be frequently observed off the west coast of Central America. However a true disclaiming characteristic might be size: the Pacific species is small (5 - 5.5 m) whereas Longman's is large (7 - 7.5 m, this length extrapolated from the skulls).

Further complications arise in view of suggestions that (1) the Seychelles whales, or (2) the Gulf of Aden whales, or (3) the Christmas Island whales, were themselves Longman's beaked whales, for if any of them are, the Pacific form must be something else again. Conversely, if the Pacific form is Longman's beaked whale, then the Seychelles, Gulf of Aden and Christmas Island beaked whales will still be unidentified, and all we will know is that they are not specimens of *Mesoplodon pacificus*. (Add to all this the fact that other as-yet-unidentified ziphiids have been referred to in the literature.)

Finally, it is even possible that the new Chilean beaked whale may be something to do with Pitman et al's form. Publication of the description will reveal whether such a speculation is valid [12].

Less obviously, but still most likely, ziphiids were the 9 m long whales watched by famous nineteenth century naturalist Philip Gosse. He described animals with elongate bodies - black dorsally, white ventrally - with white pectoral fins and long pink-tipped snouts [10]. Taken at face value, his account most certainly describes an unknown cetacean and, reconstructed with a mesoplodont body as template (fig. 6), this is rather obvious. With no other recorded sightings, let alone specimens or photos, this animal's purported existence remains, like others we have seen, in limbo and, at this stage, nothing more can be said.

Fig 6. A tentative restoration of the 9m long ziphiid apparently seen by Gosse in the North Atlantic.

The cryptocetaceans only touched on in this article, as well as others, will be presented in another article.

Acknowledgements

Special thanks to Michel Raynal for his help and interest in my work, and to Alexandre Zerbini for literature on new ziphiids and monodonts. Thanks also to Richard Ellis for discussion of poorly known ziphiids, to Richard Muirhead for his uncovery of obscure documents, to Jon Downes for access to CFZ archives, and to Stephen Leatherwood for comments. As always, thanks also to my friends Ben Roesch and John Moore for interminable discussions, helpful comments and a constant stream of literature. This article is dedicated to the memory of Bobs Fett the budgie.

References and notes

1. SIBBALD, R. 1692. *Phalainologia Nova, Sive Observationes de Rarioribus Quibusdam Balaenis In Scotiae Littus Nuper Ejectis*, etc. Typis Joannis Redi (Edinburgh): 13-19
2. LINNAEUS, C. 1758. *Systema Naturae per Regna Tria Naturae. Impensis Laurentii Salvii (Holmiae)*, vol 1 p 77
3. COUCH, J. 1857. Remarks on the species of whales which have been observed on the coasts of Cornwall. Trans. Royal Polytechnic Soc, Cornwall 24th report: 27-46
4. RAYNAL, M and SYLVESTRE, J-P. 1991. Cetaceans with two dorsal fins. Aquatic Mammals 17: 31-36
5. RAYNAL, M. 1994. Do Two-finned Cetaceans Really Exist? The INFO Journal 70: 7-13
6. BRIGHT, M. 1989. *There are Giants in the Sea*. Robson Books (London)
7. Erroneous association of 'Oxypterus' with Quoy and Gaimard's rhinocerous dolphin also occurs on p. 5 of ref 8 and p. 36 of ref 9.
8. HEUVELMANS, B. 1986. Annotated checklist of apparently unknown animals with which cryptozoology is concerned. Cryptozoology 5: 1-26
9. HEUVELMANS, B. 1968. *In the Wake of the Sea-Serpents*, Rupert Hart-Davis (London)
10. QUOY, J.R.C and GAIMARD, J.P. 1824. *Voyage Autour du Monde Execute sur les Corvettes de S.M. l'Uranie et la Physicienne Pendant les Annees 1817, 1818, 1819 et 1820* Publie par M. Louis de Freycinet: Zoologie. Pillet Aine (Paris) vol 1: 86
11. Unfortunately, we do not know if they were referring to true porpoises (*Phocoena phocoena*), or to dolphins, for whom the name porpoise has often been used. Either way, it makes little difference to the approximate size of the two-finned species.
12. NAISH, D W. In press. Analysing video footage purporting to show the 'migo' - a lake monster from Lake Dakataua, New Britain. The Cryptozoology Review 2.
13. STEWART, H. 1979 *Looking at Indian Art of the Northwest Coast*. Douglas and MacIntyre (Vancouver and Toronto). Remarked on also in ref 5.
14. GIGLIO I. H H 1870. *Note Intorno la Distribuzione della Fauna Vertebrata nell'Oceano Prese Durante un Viaggio Intorno al Globo 1865-68*. Giuseppe Civelli (Florence)
15. p 364 of ref 9.
16. MAIGRET, J. 1986. Les cetaces sur les cotes ouest-africaines: encore quelques enigmes! Notes Africaines 189: 20-24. An Amphiptera-like beast is described in this article. See also comments in ref 5.
17. 'Monodonts' are belugas, narwhals and their extinct relatives. An article devoted to cryptic monodonts will be published elsewhere.
18. NAISH, D W. 1996 Cryptocetology: the page 254 story. A&M 8: 23-29

19. I presently think it most likely that Scott's dolphins were specimens of *Cephalorhynchus eutropia* after all.
20. MORZER BRUYNS, W.F.J. 1971. *Field Guide of Whales and Dolphins*, Mees (Amsterdam)
21. Sylvestre has gone further than this, and both includes the Alula whale in his entry on Orcinus orca, and refers to it as a 'sort of killer whale' (pp 125-126) In SYLVESTRE, J-P. 1993. *Dolphins and Porpoises: A Worldwide Guide*. Sterling (New York)
22. WATSON, L. 1988. *Whales of the World*. Hutchinson (London)
23. A remarkable 74 records of all-white or mostly white killer whales, seen between 1923 and 1959, are listed in CARL, G C. 1960. Albinistic killer whales in British Columbia. Prov. Mus. Nat. Hist. Anthrop. Rep. 1959: 1-8
24. HOYT, E. 1990 *Orca: The Whale Called Killer*. Robert Hale (London). 'Chimo's' capture, medical problems and death are discussed on pp 115-122. Worthy of note is that all of the orcas in 'Chimo's' pod were malformed in some way; Hoyt mentions speculations that they may have been a team of rejects that had teamed up together.
25. COUSTEAU, J-Y and DIOLE, P. 1972. *The Whale: Mighty Monarch of the Sea*. Cassell (London). Photos of an albino killer whale of the Juan de Fuca Strait (near Seattle)' are included on pp 232-233. Ellis [p 169 of ref 26] notes that 'the circumstances under which the animal was photographed are not clearly explained'. The whale is probably the same as 'Chimo': Hoyt [p. 117 of ref 24] mentions Cousteau's interest in 'Chimo' and Seattle is in close proximity to Victoria, the location of 'Sealand'.
26. ELLIS, R. 1983 *Dolphins and Porpoises*. Robert Hayle (London)
27. JEFFERSON, T.A, LEATHERWOOD, S and WEBBER, M.A. 1993. *Marine mammals of the world*. FAO (Rome)
28. NISHIWAKI, M and NORRIS, K. 1966. A new genus *Peponocephala* for the odontocete cetacean species *Electra electra*. Sci. Rep. Whales Res. Ist. 20: 95-100
29. In 1866 Grey recognised this group as a family distinct from other dolphins and named it Globicephalidae. Some cetologists today recognise this group while others treat it as a subfamily of the Delphinidae. Still others reject it, believing the different globicephalids to be independently derived. I am unaware of any good characters that prove its monophyly.
30. *Lagenodelphis hosei* was described from skeletal material in 1956: FRASER, F.C. 1956. A new Sarawak dolphin. Sarawak Mus. Jour. 7: 478-503. Specimens in their entirety began turning up in 1971, revealing for the first time the external appearance of this species. Two papers were published in 1973: PERRIN et. al. Rediscovery of Fraser's dolphin (*Lagenodelphis hosei*). Nature 241: 345-350. TOBAYAMA et. al. *Records of the Fraser's Sarawak dolphin in the western North Pacific*. Sci. Rep. Whales Res. Inst. 25: 251-363
31. LEATHERWOOD, S. Pers. comm. 1995
32. REYES, J.C, MEAD, J.C and Van WAEREBEEK, K. 1991. A new species of beaked whale *Mesoplodon peruvianus sp. n. (Cetacea: Ziphiidae)* from Peru. Marine Mammal Science 7: 1-24
33. Van WAEREBEEK, K. 1996 *New beaked whale off Chile*. Marine Mammal Society Newsletter 4 (2): 3
34. According to ref 33, publication will be in REYES, J, Van WAEREBEEK, K, CARDENAS, J-C and YANEZ, J. Boletin del Museo! Nacional de Historia Natural de Chile. It will be written in English.
35. An unabridged discussion of cryptic ziphiids will be published in The Cryptozoology Review.
36. LONGMAN, H.A. 1926. New records of Cetacea. Mem. Queensland Mus. 8: 266-278
37. CARWARDINE, M. 1995 *Whales, Dolphins and Porpoises*. Dorling Kindersely (London)
38. Illustration by Martin Camm on pp 134-135 of ref 37
39. PITMAN, R.L, AGUAYO, A and URBAN, J 1987. *Observations of an unidentified beaked whale (Mesoplodon sp) in the eastern tropical Pacific*. Marine Mammal Science. 3:345-352.
40. GOSSE, P.H. 1851. *A Naturalist sojourn in Jamaica*. Longman Brown, Green and Longman's. London.

Letter from the Lochside

by Richard Carter

'LIZZIE':

THE LOCH LOCHY MONSTER

'Lizzie' of Loch Lochy was seen on 11 September by staff and guests of the Corriecour Lodge Hotel situated on the shore of Loch Lochy. Sandra Turner who was working in the hotel dining room saw three large dark humps about one mile away.

Joined by Kate Allan they, together with 16 guests, watched the object move in what they both described as an unusual manner. The tail end whipped from side to side very fast; the front end was large and rounded and it moved as fast as a dinghy. Both gave the size as about the same as a dolphin but it became obvious it could not be, by its side to side movement.

Dolphins were reported in Loch Linnhe, the sea loch, but this is separated by the canal locks. Sandra Fraser made inquiries around the Loch and one or two local fishermen had reported unusual wakes in the last couple of days.

The sighting ended when the three humps moved out of sight. No-one else from around the Loch reported seeing anything strange.

Another selection from the legendarily non existent compilation album:

NOW THAT'S WHAT I CALL CRYPTO
by Neil Nixon

America - Monster

After their first single - 'Horse With No Name' had peaked at number 3 in the UK, the folk/rock band America meant very little in this country. In the USA they were eclipsed only by The Eagles in terms of record sales during the seventies heyday of country rock. The band's strengths were direct melodies and harmonies wrapping apparently simple lyrics.

In 1977 the band cut 'Monster' as part of their final album as a trio - 'Harbour'. The track has the deceptive simplicity of 'White Album' period Beatles. In the opening lines we discover that a 'monster' has crawled from underground. The creature was 'burning his face as he ran through the town'.

Each line seems simple enough but an overall scan of the two-minute song throws up a series of riddles. 'Hard court and harmony just ain't the same'... 'Red and blue want to like a hole in the head.' etc.

Built around an improbably simple chord progression, this little gem meanders to an open end. Superficially, it's nonsense; ultimately, it's what you make it; probably, it's a riddle.

Using a monster motif and title as a clue, it throws in a solution halfway through with the line 'got all these pictures, just need the frames.' In other words, the thing that will make complete sense of the whole puzzle is missing for the singer too. Lacking a literal truth the song is a clever exercise in using a monster motif as a metaphor for all the riddles and choices we face in life.

America were frequently slagged as lame copyists and purveyors of high school poetry. This song wouldn't have been out of place on The Beatles' 'White' album or 'It's a shame about Ray' by The Lemonheads.

NORTH OF THE BORDER

Caledonian Curiosities from the artist formerly known as Tom Anderson

Edinburgh Zoo authorities were worried about their cheetah's sedentary lifestyle and its inherent detrimental affect upon their livers. This is due to that organ's need for energy bursts and not, as you may have supposed, to their having gained access to the lounge bar of the adjacent hotel.

Researchers have developed a training machine constructed of, among other things, pulleys, cable and a dead rabbit....

A seven foot, four inch leatherback turtle has been washed ashore at Europic Ness, Isle of Lewis. Emanating from the mid Atlantic or Caribbean, it was badly decomposed and, according to the coastguard who found it, had ridden the current of the Gulf Stream for some time. It was the first stranding of this species locally for ten years. At the time of writing (3rd October) it remains on the beach, it's removal dependant on the tides.

There seem to be strange things afoot in the waters around the Orkneys. Half the islands are whale watching now following the upsurge in activity lately - fifteen Minkes stranded on an Orcadian beach, Humpbacks, Orcas, a Beluga, Basking Sharks are common (possibly local hyperbole, I know they're not rare in these parts).

The marine situation up here is in a state of flux at the moment, things appearing where they haven't before and migrations around the east coast of bottle nosed dolphins, previously unknown in Aberdeen Bay.

There is obviously a long term pattern emerging here. As yet we don't know whether its climatic, nutrient or hydrologically influenced, or down to something as yet unknown. Watch this space...

(Did I write that? Stupid damn phrase! Blame the Special Reserve 8.5%. I always do...)

Editorial Note: We usually blame the 'White Lightning' or Grahams unaccountable taste for 'Hawkwind' circa 1972.

Animals & Men — Issue Eleven

As most of my recent postbag has been prefaced with the words "Without Prejudice" it is a joy to be able to present a rather bizarre selection of:

LETTERS TO THE EDITOR

EDITORIAL DISCLAIMER

The Editor would like to stress that opinions expressed by individual letter writers, and indeed by contributors to the magazine under any capacity, do not necessarily correspond with those held by the editorial team. Whilst every attempt is always taken not to infringe anyone's copyright or to print anything libellous any such infringement's are the responsibility of the individual author!

THE WONDER HOUSE

If one will excuse the seeming paradox, I absolutely, albeit reluctantly, endorse the Editor's wry comments about the current surly unhelpfulness of so many of the staff at the British Museum (Natural History) - or as they prefer to dub it, the Natural History Museum.

There was a time, up to the 1960s, when it was indeed a pleasure to research at this place, and to seek the enthusiastic assistance of its knowledgeable and cheerful staff - as indeed I have commented upon in more than one of my books.

Now it's all changed - even to the extent of a standard question barked at you as soon as you request information from any department: "Why do you want to know this? Is it for commercial purposes?" If you think this sounds so preposterous that I must obviously be making it up as part of my own *"Let's hate the B.M.(Nat Hist) Week"*, just try a simple experiment for yourself: give them a ring and ask something.

As has become fashionable in so many places these days, many of the staff are over-confident, over-educated, bossy and mouthy young-to-youngish women, and these seem to take particular and peculiar delight in being unhelpful.

Some time ago I had an unfortunate conversation with such a person at the Kent County Record Office, and was so annoyed I made an official complaint to the head of department - who turned out to be another woman. To my accusations of truculence and aggressiveness I got *"Well, can you blame her - a woman trying to make her way in a man's world?"*

So now you know.

CLINTON KEELING

By popular demand we are proud to bring you:
THE RETURN OF ERIK SORENSON

1. Fortean Zoologists,
Cryptic creatures
and friends of the dragons.
Chasing wild geese,
heresy, hearsay.

2. Braving ridicule
from the negative orgasm
of blind hatred.
In rigor mortis minds
with bigger monsters.

3. Reward? The chase the search
for observation, the son of anecdote.
The fight, so hard
like herding cats.
The wins, the losses.

4. Why do it?
In noble few hundred
in brotherhood of men

with desire
to know.

5. In new Middle Ages,
and future degenerations.
Pattern in darkness
torchbearers in network,
waiting for dawn.

SWISS MISS

Dear Jon,

.... a few comments on issue #10.

To the article by Neil Arnold on Wild Boars. In Switzerland the boar is rather well represented and it's also hunted and eaten (though I never tasted it). If you want to know some more just tell me.

The letter from Tom Anderson reminded me of a similar experience that I had at school, only in reverse. My teacher had bought a greenfinch and told us it was a "Green Tit". I was only six but I knew she was wrong although I never admitted it! I wonder what a greentit looks like?

Also, to the letter of Mr Keeling, I wanted to say that I absolutely love the "Now that's what I call crypto" column and I find it utterly funny. Of course if someone like Mr Keeling doesn't have the slightest interest in rock music - he (or she) won't understand it! Please keep it where it is....

Best wishes and good luck,

Sunila sen Gupta.
Switzerland.

Editorial Note: Sunila also asked us to point out that her cartoon which was included in issue 10 originally appeared in French in a wonderfully bizarre little book called "The Dahu", which was published by:

Editions de la Giraffe
Musee d'histoire naturelle
CH-2300, La Chaux-de-Fonds
Switzerland.

Her memories of her six year old 'greentit' experience triggered memories of a certain teacher in Hong Kong who told your editor (aged seven) that a pony was a baby horse...

Were we all taught by idiots?

Whilst in 'rock music mode' Pink Floyd afficianados will be sad to hear that in the last issue there was only one wild boar article ready because I have been wanting to use the headline "Pigs (Three different Ones)" for ages!

A RAINBOW WARRIOR

Dear Mr Downes

I was interested in your editorial remarks on the "Moth Menace" on p10 of the last issue, where you wondered if climate change was responsible for the infestations.

I have increasingly been thinking about the area of common interest that green activists (like me) and cryptozoologists share. The hunt for cryptids often involves previously-thought-extinct animals, whilst Greenpeace and other environmental organisations seek to prevent further extinctions. Also, "out-of-place" animals often become so because of shifting ocean currents as the planet warms up and local climates start to change.

Many cryptozoologists believe, I imagine, in the theory of evolution. They are thus well placed to ask themselves whether or not the "survival of the fittest" instincts that served our remote ancestors so well are really the best principles on which to base a modern civilised society.

The Market Forces creed, based on these literally mindless instincts, leads to the mass dumping of car batteries in the Philippines and leaky drums of poisonous waste in Western Africa, for instance. Norway and Japan continue their assault on the whale populations. Planned obsolescence wastes energy and resources. Asthma and skin allergies are rife as car emissions soar and our food is debased with additives, all in the name of global market competition.

Yet how many of us muse over any of this as we push our shopping trolleys around the local supermarket?

The power of the Market is, of course, sometimes brought to bear on the problem - but usually a heavy social price has to be paid. For instance, old age pensioners and the unemployed are subsidising big industry when water companies clean up contaminated beaches: the cost is added to every water bill and the big users actually get discounts! It's the bosses in the boardrooms who are the parasites on society, not the New Age travellers and the widows of those who fought to deliver Europe from Nazism.

Whatever we do to the planet, it will stabilise; an

equilibrium will emerge. The question is, can that new-found state of equilibrium support the kind of life we would wish our children to enjoy - or, indeed, can it support ANY life? Venus, for instance - the planet where a runaway greenhouse effect has already occurred - is in a state of stable equilibrium. The surface temperature is always around 400 deg. C and the hurricane winds faithfully spray not just water, but sulphuric acid, around the Venusian landscape.

Environmental stability is not and never will be an issue, where Earth is concerned. Quality of life is - or should be.

The Market Forces creed is catching on: here in England all three main political parties are now fully committed to it. America has been for years - with the appalling social and environmental results we already know about. Russia is now moving from totalitarian neglect to Market Forces.

On Earth, cryptozoologists and green activists join climatologists as being among those people with the clearest view of the warning signs. These signs are not merely of academic interest; they should make us ask the question, what are we going to do about it all?

Yours faithfully

Steve Johnson
Wolverhampton

OMENS OF MISFORTUNE?
by Neil Arnold

Science is often quick to dismiss tales of unfamiliar animals; however the sightings continue. Some of these animals seem prehistoric - not so much out-of-place as out-of-time. Others are so hideous and unrecognisable that they could be participants in some obscure animal freak show. There are even those entities that seem to have the characteristics of ghosts. Although many strange reports are dismissed as hallucinations or hoaxes, or the product of attention-seeking idiots or overactive imaginations, a number of these sightings suggest some sort of spiritual entity - they are beyond the possibilities of nature as we usually understand it. The thought that we humans may be sharing our void with other forces is frightening enough, but there are many reported experiences which lead me to believe that our "dimension" is on a parallel with some other - and somewhat darker - plane. A plane where creatures from millions of years ago are still alive: a dimension where dinosaurs still roam, serpents swim the seas and giant birds soar overhead.

Through the years, tales of great winged beasts, disfigured monsters, hairy men and cyclops cats have come to the fore. The most common of phantom creatures are the Black Dogs and the sinister Mothman. It is as if these demons are created on an emotional level by certain individuals and, like ghosts, can flicker onto our everyday level - but often only being perceived by the "invoking" individual. Various folklore omens and "signs" have emerged through the ages - for instance, death is the misfortune forewarned by the cry of a banshee, the howl of a hound and the screech of a cat. However, visions which seem to foretell moments of fate and misfortune may be misidentifications of other phenomena. The Black Dogs of Devon are seen by many and not always is a death impending - they may be mistakenly-identified black leopards. Mothman may in fact be an oversized eagle. Still unlikely; but a lot more natural than two-headed tigers, griffins and horses with human faces - creatures surely beyond nature's capabilities.

This certainly seems to be the case when we flick through the thousands of sightings which link misfortune and odd spectral-type creatures. The glowing red eyes and apparent ability of Mothman and the Cornish Owlman to disappear seems to point to supernatural forces rather than the existence of undiscovered flesh and blood creatures. These people may be seeing things that have stemmed from their emotional level - as one might see a deceased relative. Many entities are perceived as signs of 'darker' things - a theme that has run throughout folklore, and is still evident today. It is obvious that something beyond our every day level is at work. It may be extraterrestrial - and this may explain the connections some people have mentioned between UFOs and Bigfoot, big cats and lake monsters. This, though, is beyond human mentality. We may find out all we need to know when we reach our destination after death; until then, we are playing guessing games. However, we may never know why such odd and dark creatures infiltrate our void. We have not even worked out our own minds - but are still trying to solve mysteries beyond our science. There is much to learn about the animal kingdom here on Earth: just look at the surprises that keep popping up.

Messengers of doom? Omens of misfortune? Symbols of destiny? - the jigsaw is nowhere near complete. And the Ark was not what it seemed!

BOOK REVIEWS

'THE UNEXPLAINED' By Dr. Karl P.N.Shuker. (Carlton £16.99 248pp)

When Karl Shuker told me that he was writing a book on 'mysteries', and providing a scientific analysis for such phenomena as crop circles, vampires and UFO's, I am perfectly prepared to admit that I was not 100% enthusiastic about the project. After all, Shuker is possibly the most important cryptozoological writer of his generation, and being somewhat of a 'purist' I was worried to think that he might be diverting his undoubted talents into new and unfamiliar territory.

I was wrong.

This book is completely superb. There are an awful lot of 'weird phenomena' books on the market, and as we approach the millennium I suspect that there will be a lot more. This is the only book that I have seen for years which deserves to be mentioned in the same breath as 'Phenomena' by Bob Rickard and John Michel, or Francis Hitching's 'Atlas of World Mysteries'. I have been collecting this sort of data for years and I have upwards of 600 books on the subjects. Even so, Karl has managed to present information that is completely new to me. Even when discussing such well known phenomena as 'The Dover Demon' he does so in a refreshing and insightful way and presents illustrations that I have never seen before.

He even credits me for 'inventing' the term "Zooform Phenomena" (although the section on 'The Mad Gasser of Mattoon' fails to mention a singularly good song on the subject recorded by a certain art-rock band from Exeter). Despite that appalling omission I have no problem at all in describing this book as 'Essential'. It is one of the best books I have read all year, and I cannot praise it highly enough.

'THE ASHOVER ZOOLOGICAL GARDEN - The True Story' by Clinton Keeling. (Clam Publications 134pp) £7.00 (including p&p from C.H.Keeling, 13 Pound Place, Shalford, Guildford, Surrey. GU4 8HH).

When he gave it to me at Zoologica this year, Clinton Keeling described this book as 'very bitter'. Not for the first time I am forced to disagree with him within the pages of 'Animals & Men'. This is not a bitter book. It is perhaps the most cathartic volume of autobiography I have read in many years but it is positive in its outlook and avoids the twin pitfalls of needless recrimination and self-pity which would indeed have marked this book down as 'bitter'.

The title is self explanatory but only gives part of the story. This is not just the story of the seventeen year history of what Maxwell Knight described as "one of the few zoos - almost certainly the only privately-owned one - to make a serious attempt to justify its existence", but it is the story of a remarkable man, who is also a remarkable zoologist, and his struggle against bureaucracy, apathy and a failing marriage. As someone whose own marriage has recently collapsed in circumstances which present a number of parallels to those of Clinton's first marriage I found this book very uplifting as well as an interesting read from a zoological point of view.

Ashover Zoo was many years before its time,

being the first zoo to see its role as primarily educational. The list of species it kept is extremely impressive, as were its achievements, especially considering the almost total lack of support that they received from the zoological establishment of the time.

It may seem pretentious of me, but I can find no parallells for this book within the canon of zoological literature. The nearest comparison that I can make is a 1974 LP by Neil Young called *'Tonights the Night'* which, like this book is catharticly autobiographical and which provides a valuable insight into the mind of its creator.

This is a wonderful, but sad book, and I have no reservations in recommending it most highly!

'Biological Anomalies: Mammals 1' by William Corliss (Ed). (Sourcebook Project). 286pp

For two decades now William Corliss' Sourcebook Project has been cataloguing scientific anomalies and presenting them in a form which provides an invaluable body of reference material for any fortean researcher. This latest volume, is perhaps the most important yet from the point of view of the fortean zoologist because it includes a wealth of material not readily available elsewhere.

The book is divided into three categories:

External appearance and morphology
Behaviour and
Unusual talents.

Each of these categories is subdivided with extreme care. This book has even surpassed the previous sourcebook *'Incredible Life'* in importance in my personal library, and I can only say that I am agog with anticipation as regards the next volumes in the series.

'Sea Head Lines' by Tony 'Doc' Shiels (Giss'On Books. 44pp £2.50).

Although best known to fortean readers as a monster hunter, magician and purveyor of naked witches, Tony Shiels is by trade a surrealist artist. This gorgeous little book of writings and drawings, which also includes a wonderful playlet from Simon Parker, is the first published fruits of 'Doc's' latest project: The Sea Heads In Elemental Locations Scheme. (Grok the acronym).

Tony weaves a marvelous spiders web of interconnecting images, invoking sea monsters, Moby Dick, Pablo Picasso and the elephant squid producing a bewitching miasma of Grotesque Sensuality (GS). As he writes:

"A Sea Head is a celebratory image. It raises the spirits and inspires uninhibited revelry. Mighty crack altogether. I raise my glass to it!"

And so say all of us.

Editor's Note: To celebrate all sorts of things, and to win a copy of the Sea Heads bookeen and a free years's subscription, be the first person to telephone me and tell me, according to Jonathan Richman what was Pablo Picasso never called?

'Quaggas and other Zebras' by David Barnaby (Bassett Pubs. £9.00).

I am a fan of David Barnaby. I like the way that he writes, with style and not a little wry humour but this latest book of his, in my opinion at least, surpasses his previous books by a considerable measure. This is a book about the Quagga, an

African wild horse extinct since the late nineteenth century. Barnaby examines the history of the species and its decline. He makes some very interesting points about its taxonomy and he lists the few surviving museum specimens. He also presents an impressive list of cultural and literary references to the beast (including my favourite one from 'King Solomon's Mines' by H.Rider-Haggard).

Furthermore he sets the entire saga over a backdrop during which he describes an exciting trip to South Africa undertaken with A&M Contributor Chris Moiser during the summer of 1995.

Most importantly, however is the current project to 'reconstitute' the species by intensive breeding experiments, and the book offers the almost fantastic hope that man can, in a small way at least, attempt to repair a little of the damage he has done to the biodiversity of this planet!

SPECIAL OFFER

'Quaggas and other Zebras' will be published on 22nd December. In conjunction with Bassett publications we can offer this remarkable book at a pre-publication price of £7.50. (Including P&P). This offer ends on the 22nd December, and Bassett Publications have asked us to state that whilst they will make every effort to do so, they cannot guarantee that the books will be delivered by Christmas.

THE CRYPTO SHOP NOVEMBER 1996

POSTAGE AND PACKING

This will now be charged at cost. If you include £0.75 per book and 25p per periodical with your order then we will either refund the balance or invoice you for any extra postage due. Payment can be in cash (UK or US currency). International Money Order, Eurocheque, or cheque drawn on a UK bank.

Please make all cheques payable to JONATHAN DOWNES. Please telephone to ensure that the goods you want are still in stock. If no telephone call is received and an order is received for something that is out of stock then a credit note will be given..' If you live outside the EC please add 10% surcharge to cover adittional post and packing.

Every effort will be taken to ensure prompt delivery within 21 days. Orders outside Europe are sent by surface mail unless aditional postage is paid.

NEW BOOKS FROM PUBLISHERS

BARNABY David *The Elephant that walked to Manchester* 66pp Pb. A wonderfully bizarre little book which tells the story of the sale of Wombwell's Menagerie in 1872 the aftermath when Maharajah the Elephant walked from Scotland to his new home in Manchester. Includes lots of interesting vignettes on 19th Century travelling menageries which are of interest to the readers of 'Animals & Men'.Autographed by the author. Usually £6 now £5.00

BARNABY David & BENNETT Clive: *The Reptiles of Belle Vue 1950-77*. 156pp A4 pb. Another excellent book of anecdotes from the author of 'The Elephant that Walked to Manchester'. It provides a wondeful insight into the workings of the reptile department at Manchester's Belle Vue Zoo, but also provides a large body of otherwise unavailable anecdotal evidence for 'out of place' exotic reptiles in the North of England over a period of 27 years. Highly reccomended and autographed by the authors £7.00

BARNABY David *"Quaggas and other zebras"*. A superb book about one of the most notorious extinct animals of all time and the desperate attempts that are being made to reconstitute the species. Usually £9 until 22.12.96 £7.50

FARRANT D *'Beyond the Highgate Vampire'* AUTOGRAPHED BY AUTHOR 43pp 1992 'Excellent personal history of the phenomenon by the one person really qualified to write about it. Reccomended' £3.95

GREEN Richard *Wild Cat Species of the world*. 163pp Pb Illustrated. The best book on the felidae since Guggisberg in the mid 1970's. Includes The Onza (in colour) Very Highly reccomended. Usually 12.50 our special offer of £10.00

THOMAS Dr Lars *'Ordbog Over europiske dyr'* Pb 180 pp 'Completely wonderful book from our Danish Representative which lists the commpon names of European mammals, birds and fish in all European languages. Essential! £7.00

SHUKER Dr K.P.N: "IN SEARCH OF PREHISTORIC SURVIVORS" 192pp hb Many Illustrations. This is arguably the most important new book on general Cryptozoology since 'On the track of Unknown Animals' was first published forty years ago. Shuker calmly and sensibly presents evidence for the survival of dozens of species of animals, presently known only from the fossil record. We cannot praise this book highly enough Price £18.99

SHUKER Dr K.P.N. "DRAGONS - A NATURAL HISTORY" 120PP PB. Gorgeously illustrated, this book must be one of the most attractive books that I have seen in many years. It is also a must for anyone with an interest in things Draconian. Shuker proves that he is not only a meticulous scientist, but a fine story teller to boot. £11.00

SHUKER Dr K.P.N. "THE LOST ARK - NEW AND REDISCOVERED SPECIES OF THE 20TH CENTURY". hb. Another lavishly illustrated book which is an essential purchase for anyone interested in the advances that zoology has made over the past ninety-five years. NOW OUT OF PRINT £18.00

SHUKER Dr. K.P.N. *The Unexplained*. Probably the best book of general forteana to have been published for many years. I cannot reccomend this book highly enough £16.99

Dr Shuker will, personally autograph his books for you at no extra cost. Telephone for details.

HEUVELMANS Dr B "ON THE TRACK OF UNKNOWN ANIMALS" Hb 677pp many illustrations. At last this classic work, which essentially defined the methodology and practise of the science of Cryptozoology is now available again. This is a reprint of the 1962 edition but has a new introduction and many new illustrations. If you have not already got a copy then you should certainly buy this. £20.00

LEVER Sir C: "They Dined on Eland" 224 pp, illus. Excellent investigation of the 19th Century Acclimatisation Society andtheir founder Frank Buckland., from the author of 'Naturalised Animals of the British isles' etc Amusing and erudite. Lever is one of the great experts on Naturalised animals and he writes with great skill and aplomb.The Publishers price was £18.50 our price is £12.00

CARTER R. Loch Ness the Tour. 22pp 1996. Useful guidebook to Loch Ness. Very Good. £1.50

STEENBURG, T.N. Sasquatch; Bigfoot - The continuing Mystery. 125pp 1993 Ed. Excellent Reviewed in A&M 10. Well worth getting for those interested in North American BHM phenomena. £10.00

NEW BOOKS FROM PUBLISHERS

BARNABY David *The Elephant that walked to Manchester*. 66pp Pb.
A wonderfully bizarre little book which tells the story of the sale of Wombwell's Menagerie in 1872 the aftermath when Maharajah the Elephant walked from Scotland to his new home in Manchester. Includes many interesting vignettes on 19th Century travelling menageries. Autographed by the author. Usually £6, now £5.00

BARNABY, David & BENNETT, Clive: *The Reptiles of Belle Vue 1950-77*. 156pp A4 pb.
A wonderful insight into the workings of the reptile department at Manchester's Belle Vue Zoo, and a large body of otherwise unavailable anecdotal evidence for 'out of place' exotic reptiles in the North of England over a period of 27 years. Autographed by the authors. £7.00

BARNABY, David: *Quaggas and other zebras*
A superb book about one of the most notorious extinct animals and the desperate attempts being made to reconstitute the species. £9.00

FARRANT, D: *Beyond the Highgate Vampire*
43pp 1992 'Excellent personal history of the phenomenon by the one person really qualified to write about it. Autographed by author. £3.95

GREEN, Richard: *Wild Cat Species of the World* 163pp Pb. Illus.
The best book on the felidae since Guggisberg in the mid 70's. Includes the Onza (colour) Usually £12.50: special offer: £10

THOMAS, Dr Lars: *Ordbog Over Europiske Dyr*. Pb 180 pp.
A wonderful book which lists the common names of European mammals, birds and fish in all European languages. Essential! £7.00

SHUKER, Dr K.P.N: *In Search of Prehistoric Survivors*. 192pp hb
Arguably the most important new book on general Cryptozoology since the 1950's book 'On the track of Unknown Animals'. Shuker presents evidence for the survival of dozens of species of animals, presently known only from the fossil record. We cannot praise this book highly enough £18.99

SHUKER, Dr K.P.N. *Dragons - A Natural History*. 120pp pb.
Gorgeously illustrated, this book must be one of the most attractive books that I have seen in many years. It is also a must for anyone with an interest in things Draconian. Shuker proves that he is not only a meticulous scientist, but a fine story teller to boot. £11.00

SHUKER Dr K.P.N. *The Lost Ark - New and Rediscovered Species of the 20th Century*. pb.
Another lavishly illustrated book which is an essential purchase for anyone interested in the advances that zoology has made over the past 95 years. NOW OUT OF PRINT £18.00

SHUKER Dr. K.P.N. *The Unexplained*
Probably the best book of general forteana to have been published for many years. I cannot recommend this book highly enough £16.99

Dr Shuker will, personally autograph his books for you at no extra cost. Telephone for details.

LEVER, Sir C: *They Dined on Eland*. 224 pp Illustrated
Excellent investigation of the 19th Century Acclimatisation Society and their founder Frank Buckland, from the author of 'Naturalised Animals of the British Isles', etc. Amusing and erudite. Lever is one of the great experts on naturalised animals and he writes with great skill and aplomb. The publishers price was £18.50; our price is £12.00

CARTER, R. *Loch Ness - the Tour*. 22pp 1996.
A good and useful guidebook to Loch Ness. £1.50

STEENBURG, T.N. *Sasquatch; Bigfoot - The Continuing Mystery*. 125pp 1993 Ed.
Excellent. Reviewed in A&M 10. Well worth getting for those interested in North American BHM phenomena. £10.00

GREEN, J. *On the track of the Sasquatch*.
Large format Pb. 64pp. Many illustrations. 1980. Excellent. £8.00

GREEN, J. *Sasquatch - The apes amongst us*. 492pp Lavishly illustrated.
Possibly the best book I have read on North American Man Beasts. This is a classic of cryptozoology and should be bought! £12.00

Animals & Men — Issue Eleven

GREEN, J. *On the track of the Sasquatch*. Large format Pb. 64pp. Many illustrations. 1980. Excellent £ 8.00

GREEN J. *Sasquatch - The apes amongst us*. 492pp Lavishly illustrated. Possibly the best book I have read on North American Man Beasts. This is a classic of cryptozoology and should be bought! £ 12.00

FULLER, E. *The Lost Birds of Paradise*. 160pp Hb. Reviewed in A&M 10. Gorgeous illustrations. Highly reccomended. £ 30.00

BILLE, M. *Rumours of Existence*. 192pp Pb 1996. Excellent book of zoological anomalies and cryptozoology. Reviewed in A&M10. £ 12.00

GARNER, B. *Monster Monster - A survey of the North American monster scene*. 190pp Pb. Excellent book reviewed in A&M10. Highly reccomended £ 10.00

BOUSFIELD E.L. and LEBLONDE P *"An account of 'Cadborosaurus willsi, New genus new species, a large aquatic reptile from the Pacific coast of North America"*. Illustrated with photographs and line drawings. Only a few copies left 32pp A4 This offer will not be repeated £3.50

BEER Trevor *The Beast of Exmoor*. The first book on the subject by the man at the centre of the investigation. Highly reccomended. 44pp with illustrations. £ 3.00

SHIELS Tony 'Doc', *Sea Headlines* The long awaited firsst fruits of the legendary Sea Heads project. Unmissable at £2.50

OUR OWN PUBLICATIONS

DOWNES, J 'Road Dreams' A month of strange goings on. The author, his wife, and a reasonably well known rock and roll band travel across England with a bunch of engaging weirdos. pb 120pp £ 5.00

ANIMALS AND MEN BACK ISSUES

ANIMALS & MEN ISSUE 1 (out of print) Photocopy only. *Relict Pine Martens/Giant Sloths/Sumatran and Javan Rhinos/Golden Frogs/ Frog Falls...and much more* £ 2.00

ANIMALS & MEN ISSUE 2 (out of print) Photocopy only. *Mystery bears in Oxford and The Atlas Mountains/ Loch Ness/Green Lizards/Woodwose/Tatzelwurm...and much more* £ 2.00

ANIMALS & MEN ISSUE 3 (out of print)Photocopy only. *Giant Worm in Eastbourne/Lake Monsters of New Guinea/Giant Lizards in Papua/Mystery Cats/Black Dogs on Dartmoor/Scorpion Mystery...and more* £ 2.00

ANIMALS & MEN ISSUE 4 (out of print)Photocopy only *Manatees of St Helena/Lake Monster of New Britain/The search for the Thylacine/much more..news/letters etc* £ 2.00

ANIMALS & MEN ISSUE 5 (out of print)Photocopy only. *Mystery cats/Loch Ness/The Migo Video/Boars and Pumas/Hairy Hands of Dartmoor/News Reviews, obituaries, HELP etc* £ 2.00

ANIMALS & MEN ISSUE 6 *Owlman of Mawnan/Humped Elephants of Nepal/Mystery Cats/news, reviews and more* £ 2.00

ANIMALS & MEN ISSUE 7 *Mystery Whales,/Strangeness in Scotland/On collecting a cryptid/Bodmin Leopard Skull/Cryptozoological Books/News, reviews and more* £ 2.00

ANIMALS & MEN ISSUE 8. Green Cats, Mystery Whales, Cryptozoological books, news etc £ 2.00

ANIMALS & MEN ISSUE 9. Hong Kong Tiger, Hoirseman in Lincolnshire, Scottish BHM, Congo Peacock, Mystery whales etc £ 2.00

STEAMSHOVEL PRESS #14

Highly regarded US magazine about conspiracy theories and the truth 'they' are not telling us. You've seen 'The Lone Gunman' on the X Files. This is the 'real' thing. UFOs/Politics/Conspiracies/ Hoaxes and counter hoaxesand all good stuff 64pp £ 3.50

LOBSTER

AN Excellent UK based conspiracy theory magazine which is highly reccomended. £2.00 an issue

MUSIC

JON DOWNES & THE AMPHIBIANS FROM OUTER SPACE - the world's only fortean rock band have the following recordings available.
'The Case' (ten tracks) CD £10.00 Casette £5.00
'Contractual Obligations' (four tracks) Cassette £3.50

SECOND HAND BOOKS

ATTENBOROUGH D *'Zoo Quest to Guiana'* 158pp 1956 Hb 185pp 'Excellent animal collecting book from the 1950's packed with zoological information unavailable elsewhere. 8 plates. !Recommended' £ 2.00

BATES, M. *'The Forest and the Sea'* (1960). Interesting book on North American ecology. Pb 216pp £2.25

BERLITZ, C. *'The Bermuda Triangle'* 1996 Ed. 201 pp Pb. Mildly entertaining quasi fortean tosh Don't believe a word! £ 3.00

BERLITZ, C. *'Mysteries from forgotten world'*. Slightly more entertaining and less tosh mostly about Atlantis. OK if you like that sort of thing! 24 plates. 219pp Pb 1996 Ed £3.00

BINNS R.J. *'The Loch Ness Mystery Solved'*. 1983. 8vo Hb 227pp. 18 photographs, numerous line drawings and text figures. An increasingly sought after book £ 5.00

BORD J and C *'The Evidence for Bigfoot and other man beasts'* 1984 pb 254 pp. Numerous photographs and illustrations. A super book on the subject of North American and other man-beasts.Increasingly highly sought after and unfortunately out of print. Near Mint Condition. We only have limited numbers of this book £ 6.00

BOTTRIEL L.G *Umbalala*. The story of the african leopard and its relationship to the wild. Largee format hb with d/w. Lavishly illustrated. From the author of 'The King Cheetah' very difficult to find and sought after 214pp £ 10.00

CARRINGTON Richard *Mermainds and Mastodons* 251pp 1961 Ed. Excellent book. Numerous illustrations. Sought after in hardback edition. £10.00

COLEMAN, Loren *'Tom Slick and the search for the Yeti'*. 1989. Pb. 176pp Excellent work on an unusual and little explored aspect to Cryptozoology. Many illustrations. Reccomended. £5.00

COX B *'Prehistoric Animals'*, pb 150pp 1969. Excellent pocket guide. £2.00

CAWSON F *The Monsters in the Mind*. Hb d/w 174pp 1995. Scholarly examination of the human need for 'monsters'. Re-evaluates much data and finds some surprising conclusions. Excellent £ 12.00

CLARK Arthur C. *Astounding Days*. 224pp pb 1989. Entertaining autobiography crammed full of fortean snippets. £ 2.00

DENIS, Michaela. *'Leopard in my Lap'*. Heartwarming true life reminiscences from veteran zoologist and film maker. Many classic pictures. 288pp Hb (Book Club Ed). 1956. £ 3.50

DINSDALE T *'The Story of The Loch Ness Monster'* 1973 pb 124pp 8 photos. Several Text Figs £ 2.00

DROSCHER V.B. *The magic of the senses*. Pb 333 1969. Fascinating look at animal behaviour and senses. Excellent bargain. Many illustrations. Reccomended £2.00

FORT, C. *'New Lands'* 1974 Pb Ed. 205pp. Essential reading £ 3.00

HAINING P.Ed *The Ancient Mysteries Reader Book 1'*. Anthology of short stories with fortean themes. Includes writings by Poe, Wells and Conan-Doyle £ 2.00

HARMSWORTH A. *'Loch Ness - The Monster'*. 32pp. Full colour booklet on Loch Ness with much useful information. Slightly tatty hence £ 2.00

HITCHING F *World Atlas of Mysteries*. 1980 book club Ed. hb d/w. 256 pp. Excellent - one of my favourite books on the subject of general forteana and mysteries. Includes much of interest to the cryptozoologist £12.00

HOLIDAY Ted. *'The Great Orm of Loch Ness'* 1970 Avon PbUS 224pp. Illustrated with plates and line drawings. Classic book on the subject. £ 4.00

HOWELLL & FORD *The true history of the Elephant Man*. Pb. 1980 ed. 223pp. 2 plates. Excellent and difficult to find. £4.50

IZZARD R. *The Abominable Snowman adventure*. 1st Ed. 1955. Hb d/w. 302pp. Quite rare and very sought after (17.00

JEFFREY, AKT. *'The Bermuda Triangle'*. 1975pb 144pp. Entertaining nonsense about a thoroughlly unbelievable subject. An esential purchase if only to compare it with Berlitz in a vain attempt to see who can write the worst rubbish. £ 2.00

JONES, Ken *'Orphans of the Sea'*. The story of the Cornish Seal Sanctuary. Autographed by author. 124pp Pb. 8 pages of plates. 1970. Nice collectors item £ 3.00

LOCKWOOD, A.P.M. *'Animal Body Fluids and their regulation'* Scholarly and interesting. Hb. 177pp. Illustrated with disgrams and line drawings.. £ 4.00

MORRIS Desmond *Catlore* 114pp hb d/w line drawings. Fascinating sequel to the acclaimed 'Catwatching' which tells you all you could want to know about moggies. £5.00

MORRIS Desmond *'Dogwatching'*. This is the essential guide to dog behaviour and is a fascinating collection of information which is hard to find elsewre. 106pp Hb d/w £5.00

SHIELS, Tony 'Doc'. *'13'*. 23pp 1967. Very rare book of magic from the Wizard of the Western World. This is the only copy of this edition we have ever seen, and it is bound to be sold almost

Animals & Men — Issue Eleven

immediately £ 7.00
SMITH D.K. *'Secrets from a Star Gazers notebook - making Astrology work for you'*. This book is perhaps the most absurd that I have ever read. Pb 492pp 1982 £ 2.50

WALKER A. *'Little One'* 1994. Pb. 291pp. Cosmic drivel from an authoress who should know better. Buy it to give to a gullible friend. The authoress claims to have been channelling a Red Indian spirit who tells her how to save the ecology of our planet! Worth buying if only for the kitsch value! £1.50

WILLIAMS J.H. *'Elephant Bill'*. 1955 Book Club Ed. Hb. Elephants in the Burmese Jungle. Illutrated with archive photographs. 245pp £ 3.00

WITCHELL N, *'The Loch Ness Story'* Revised Book Club Edition of 1979. Hb with dustwrapper 236pp illustrated throughout. *'Excellent book '*. £ 15.00

WITCHELL N, *'The Loch Ness Story'* Revised and Updated Edition of 1989. Pb 230pp 16 Photographs *'Excellent book '*. Good condition £ 5.00

WITCHELL N, *'The Loch Ness Story'* Revised and Updated Edition of 1989. Pb 230pp 16 Photographs *'Excellent book '*. Slightly tatty hence £ 3.00

WITCHELL N, *'The Loch Ness Story'* 1982. Pb edition 208pp 16 Photographs *'Excellent book '*. Slightly tatty hence £ 3.00

SPECIAL OFFER

DIMITRI BAYANOV : "IN THE FOOTSTEPS OF THE RUSSIAN SNOWMAN"

£10.00

VERY LIMITED SUPPLIES SO PLEASE ORDER QUICKLY.

GILROY R, *'Mysterious Australia'*. pb illustrated 288pp. A fascinating collection of antipodean forteana which includes several large sections on cryptozoology and allied disciplines. The Yowie; Giant Lizards; The Thylacine Marsupial 'Panthers', Bunyips, lake and River Monsters, relict dinosaurs and much more. Highly reccomended. £2 off publishers price at £ 9.00

10% OFF THE PUBLISHERS PRICE OF ALL FORTEAN TIMES PUBLICATIONS:

FORTEAN TIMES 1-15 *'Yesterdays News Tomorrow'* 400 pp pb *Includes Wolf Children, Moon Mysteries, poltergeists, Pyramids, Water Monsters, Fortean USA and lots more)* £ 17.99

FORTEAN TIMES 16-25 *'Diary of a Mad Planet'* 416 pp pb. *Includes Bleeding Pictures, Animal Attacks, Morgawr, close encounters, swarms of animals and much more. Too much to list* £ 17.99

FORTEAN TIMES 26-30 *'Seeing out the seventies'* 320 pp pb *Owlman, Mystery animals, ABC'sanimal attacks, Stigmata, Fish Falls, Fortean Phenomena in ancient pamphlets & more* £ 13.50

FORTEAN TIMES 31-36 *'Gateways to Mystery'* 416 pp pb *The touch of death, mystery blob, wildmen and hermits, mystery cats, Owlman, Doc Shiels, Morgawr, in searchg of dinosaurs, the Yeren and more* £ 17.99

FORTEAN TIMES 37-41 *'Heavens reprimands'* 416 pp pb *The Man who invented flying saucers, The Hackney Bear, Mystery Kangaroos in USA, plants growing out of peoples eyes and much more* £ 17.99

FORTEAN TIMES 42-46 *'If Pigs could Fly'* 416 pp pb *Mystery Submarines, Werewolves in Devon, horned humans, lake monsters across Europe, phantom attackers and much more* £ 17.99

FORTEAN TIMES 47-51 *'Fishy Yarns'* 416 pp pb *Australian Mystery Animals, Yeti Sightings, Lizard Man, Mystery Cats and many pages of news and much much more* £ 17.99

FORTEAN STUDIES VOLUME ONE 350pp Long research papers on many subjects includes: Karl Shuker on Mystery Bats,Michel Raynal on the Giant Octopus, The Luminous Owls of Norfolk, Mike Dash on the Great Devon Mystery, Michel Meurger on Medieval French mystery cats and much more. Essential. £ 17.99

FORTEAN STUDIES VOLUME TWO 320pp Long research papers on many subjects includes: Karl Shuker on the physical evidence of British Mystery Cats, Michel Raynal and Gary Mangiacopra on Out of Place Coelecanths, Michel Muerger on Icelandic Water Monsters, Bob Rickard on Fish Falls and much more £ 17.99

PERIODICAL REVIEWS

We welcome an exchange of periodicals with magazines of mutual interest - although because we now exchange with so many magazines we have been forced, much against our fortean methodology, to categorise them.

CRYPTOZOOLOGY AND ZOOMYTHOLOGY

DRAGON CHRONICLE, The Dragon Trust, PO Box 3369, London SW6 6JN. A fascinating collection of all things draconian which now appears four times a year. Now A4 and Glossy..how do they DO it?

THE BRITISH COLUMBIA CRYPTOZOOLOGY CLUB NEWSLETTER, 3773 West 18th Avenue, Vancouver, British Columbia, Canada. V65 1B3. Excellent and well put together, and they are now on the Internet as well!

CREATURE RESEARCH JOURNAL, Paul Johnson, 721 Old Greensburg Pike, North Versailles, PA 15137-1111 USA. New issue devoted to Pennsylvania Bigfoot reports 1994-5.

CRYPTOZOOLOGIA, Association Belge d'Etude et de Protection des Animaux Rares, Square des Latins 49/4, 1050 Bruxelles. Belgium. A French language magazine published by the Belgian society for Cryptozoology.

CRYPTOZOOLOGY REVIEW, 137 Atlas Ave, Toronto, Ontario. Canada. M6C 3P4. Excellent new publication on cryptozoology.

EXOTIC ZOOLOGY, 3405 Windjammer Drive, Colorado Springs, CO80920 USA. A free newsletter from the author of 'Rumours of existence'. Useful round-up of information on new and rediscovered species.

FRINGE SCIENCE

SCIENCE FRONTIERS, Sourcebook Project, PO Box 107, Glen Arm, MD21057, Newsletter of William Corliss' invaluable Sourcebook Project. Fascinating snippets of useful information.

NEXUS 55 Queens Rd, E. Grinstead, West Sussex RH19 1BG. Intelligent look at the fringes of science. Well put together. Very impressive.

FORTEAN/EARTH MYSTERIES/FOLKLORE

TEMS NEWS, 115 Hollybush Lane, Hampton, Middlesex, TW12 2QY. An entertaining collection of odds and sods and generally weird stuff. A magazine I always enjoy reading. Recommended.

COVER UP, David Coleman, 39 Limefield Crescent, Bathgate, West Lothian, Scotland. EH48 1RF. The magazine of the Lothian Unexplained Phenomena Research group. UFOs, animal mutilation, ghosts etc. This is a useful addition to the scene and the editor should be congratulated for his hard work.

DELVE, Gene Dyplantier, 17 Shetland St. Willowdale, Ontario. Canada.M2m 1X5. Fortean magazine. New issue includes an article on the flying snake of Namibia.

3rd STONE, PO Box 258, Cheltenham, GL53)HR. Magazine of the Gloucester Earth Mysteries Group. Wittily and intelligently put together.

DEAD OF NIGHT, 156 Bolton Road East, Newferry, Wirral, Merseyside, L62 4RY. An amusing and intelligently put together Fortean magazine. My favourite fortean journal..

deVILLE'S ADVOCATE, Mike White, 62 Goodmore Crescent, Churchdown, Glos, GL3 2DL. A highly entertaining philosophical/fortean magazine. The Spring issue includes a Dave Sivier article on the social implications of the Internet and the possibilities of its future use and misuse; snippets of weird news from the world's press; a detailed account of a UFO sighting over Cheltenham; and a cryptozoological analysis of Gloucester's Deerhurst Dragon legend. The quote on page 2, "Research, like love, requires no reasons and no excuses," sums the magazine up.

ZOOLOGY/NATURAL HISTORY

HERP LIFE, 19 Elmdale Road, Bideford, N Devon, EX39 3LF. South Western Herpetological Society 4pp newsletter Entertaining and informative newsletter from a thriving organisation. Contains some quasi-fortean oddments.

BIPEDIA, Francois de Sarre, C.E.R.B.I, 6 Avenue George V, 06000 Nice, France. Issue twelve of this scholarly magazine is now available. Written partly in French, partly in English, it explores the obscure, but fascinating theory of Initial Bipedalism, and its allied disciplines.

MILTON KEYNES HERPETOLOGICAL SOCIETY 15 Esk Way, Bletchley, Milton Keynes. Excellent A5 magazine containing handy hints, informative articles and news of what appears to be an exciting organisation.

MAINLY ABOUT ANIMALS, 13 Pound Place, Shalford, Guildford, Surrey GU4 8HH. Veteran Zoologist Clinton Keeling edits this wonderful A5 magazine which is, as the title says, mainly about animals. This is a genre of magazine that I and many others feared was lost forever and it comes with your editor's highest recommendation.

ESSEX REPTILES AND AMPHIBIANS SOCIETY, 6 Chestnut Way, Tiptree, Colchester, Essex, CO5 ONX. Another excellent and lively regional reptile society.

NATIONAL ASSOCIATION OF PRIVATE ANIMAL KEEPERS, 8 Yewlands Walk, Ifield, Crawley, West Sussex, RH11 0QE. Useful publication including a wealth of information about wild animal husbandry. This is an organisation which, especially in the present political climate needs your support.

THE MANE, Wild Equid Society, Flat 19, 119 Haverstock Hill, London NW3 4RS. Fascinating journal about wild horses and their relatives. Includes much of interest to the cryptozoologist.

MISCELLANEOUS

NETWORK NEWS, P.O BOX 2, LOSTWITHIEL, CORNWALL PL22 0YY. Anarchism. Earth mysteries, weirdness, and even a little cryptozoology. This is the sort of monumentally groovy collection which should be encouraged.

PENDRAGON, Smithy House, Newton by Frodsham, Cheshire WA6 6SX. A scholarly and massively entertaining magazine on things Arthurian. Manages to keep an entertaining balance between literature and history. Highly recommended.

animals&men
THE JOURNAL OF THE CENTRE FOR FORTEAN ZOOLOGY

Cartoon by Mark North

THE WAILING *WAL*RUS

Typeset by a Batchelor Boy   ISSN 1354 0637

ISSUE 12

DECEMBER 1996

THE JOURNAL OF THE CENTRE FOR FORTEAN ZOOLOGY

This issue contained one of my favourite articles - an in-depth investigation of the news that the Barbary lion may well have escaped an ignominious extinction. Still with the mindset that `theme` issues were a good idea, I continued the lion theme across a series of other articles.

Darren Naish did us proud in this issue as well. We produced this edition in early December 1996, as I was (not) looking forward to my first Christmas alone after the departure of Alison and Lisa. This was the beginning of the period of severe depression and near alcoholism which I recount in `The Embarrasing Years` - a chapter in my 2004 autobiography *Monster Hunter*.

I am happy to say, looking back at these issues for the first time in a decade, that despite the turmoil in my personal life, we were still producing a pretty creditable magazine...

Animals & Men
The Journal of the Centre for Fortean Zoology

The Chinese Invasion of Britain - Mitten Crabs in the Thames
More feathered Dinosaurs; The mystery animals of Germany; The Barbary Lion; New Zealand Extinctions and much more...

Issue 12   £2.00

animals & men

THE JOURNAL OF THE CENTRE FOR FORTEAN ZOOLOGY

Animals & Men # 12 — Who's who and what's what

The ever changing crew of the 'Animals & Men' mothership presently consists of:

Jonathan Downes : Editor
Graham Inglis : Newsfile Editor/Spin Doctor
Jan Williams : Associate Editor
Mark North : Artist
Richard Muirhead : Newsagent from Nowhere
Special Agent Tina Askew : Rhine Maiden

CONSULTANTS

Dr Bernard Heuvelmans
(Honorary Consulting Editor)
Dr Karl P.N. Shuker
(Cryptozoological Consultant)
C.H. Keeling
(Zoological Consultant)
Tony 'Doc' Shiels
(Surrealchemist in Residence)
Darren Naish
(Palaeontology/Cetology Consultant)
Chris Moiser
(Zoological Consultant)

REGIONAL REPRESENTATIVES

U.K
Scotland: Tom Anderson
Surrey: Nck Smith
Yorkshire: Richard Freeman
Somerset: Dave McNally
West Midlands: Dr Karl Shuker
Kent: Neil Arnold
Sussex: Sally Parsons
Hampshire: Darren Naish
Lancashire: Stuart Leadbetter
Norfolk: Justin Boote
Leicestershire: Alaistair Curzon
Cumbria: Brian Goodwin
S. Wales/Salop: Jon Matthias
London: Richard Askew (No relation)
Tyneside: Simon Elsdon

EUROPE
Switzerland: Sunila Sen-Gupta
Spain: Alberto Lopez Acha
Germany: Wolfgang Schmidt & Hermann Reichenbach
France: François de Sarre
Denmark: Lars Thomas and Eric Sorenson
Eire: The Wizard of the western world

OUTSIDE EUROPE
Mexico: Dr R.A Lara Palmeros
Canada: Ben Roesch

DISCLAIMER
The views published in articles and letters in this magazine are not necessarily those of the publisher or editorial team, who although they have taken all lengths not to print anything defamatory or which infringes anyone's copyright take no responsibility for any such statement which is inadvertantly included.

CONTENTS

3 Editorial
4 NEWSFILE
11 NEWS FROM NEW ZEALAND - Darren Naish
13 THE LION SLEEPS TONIGHT including
   13 The Barbary Lion - Chris Moiser
   14 Lost Lion Renaissance - Darren Naish
   16 Between The Lions - Jonathan Downes
19 STRANGERS IN A STRANGE LAND - Wolfgang Schmidt looks at unexpected animals in Germany.
26 OUT OF THIS WORLD - exobiology column by Graham Inglis
27 New discoveries of yet MORE bird-like dinosaurs - Darren Naish
32 North of the Border - Tom Anderson
32 Now That's What I Call Crypto - Neil Nixon.
33 THE TIM DINSDALE FILM - Richard Carter blows the dust off that old Nessie footage with nautical footnotes by Steve Johnson.
36 Family, Friends, and Out-of-place Animals - Neil Arnold.
39 Letters to the editor.
40 REVIEWS AND SALES
48 Cartoon by Mark North.

SUBSCRIPTIONS

For a Four Issue Subscription:
£8.00 UK
£9.00 EEC
£12.00 US, CANADA, OZ, NZ
(Surface Mail)
£14.00 US, CANADA, OZ, NZ
(Air Mail)
£15.00 Rest of World
(Air Mail)

'Animals & Men'

**THE CENTRE FOR FORTEAN ZOOLOGY,
15 HOLNE COURT,
EXWICK, EXETER,
DEVON, EX4 2NA**

Tel 01392 424811

Animals & Men # 12 — Editorial

The Great Days of Zoology

Dear Friends,

Welcome to a new issue. We are now in our fourth year, and are - we hope at least - getting bigger and better all the time. There have been an awful lot of, what we in the wilder fringes of the biological sciences refer to as 'Cock Ups' in the last few months, and we can only apologise for them. Although every current subscriber was sent a copy of issue 11 it seems that a whole batch became lost in the post and singularly failed to arrive. If you did not receive issue 11, then please let me know and you will have one posted to you within the next few weeks. The production of The 1997 Yearbook, my book *The Owlman and Others*, and the first issue of our 'sister' magazine *The Goblin Universe* has also been fraught with problems.

As you all know 1996 was a very difficult year for me. The events of the final few weeks, after the publication of Animals & Men #11 were on a par with those of the rest of the year. We found a new, and very cheap printer who claimed to be able to print directly from disk. The masters for the three publications listed above were delivered to them in late November. Like fools we didn't keep back-up copies because he managed to erase the disks and we had to start (almost) from scratch.

The Goblin Universe is available NOW. Subscribers should have received theirs a week or so ago. *The Owlman and Others* will - or at least SHOULD - be ready by the time that you receive this magazine and the 1997 Yearbook will be with you very soon. I can only apologise to you all for the delay.

We have now got a new printer (*Devon Design and Print* at the *Devon County Council*), and so far their work has been excellent. We decided to publish this issue before completing the Yearbook because we have had personal experience of too many organisations who have allowed their publishing schedule to go seriously awry and have ended up producing nothing for years. We were determined not to go down this path, and, indeed are reasonably convinced that we shall be back on schedule within the next month or so. As a result of the delay, however, the WWW site, and the computerised database have also been held up. We apologise to those of you who have written for lists of available data. Again, this service will be available as soon as possible.

Our publishing schedule for 1997 is, again, a heavy one. We plan to put out *The Mystery Animals of Hong Kong* (by Richard Muirhead and myself) in the summer, and a reprint of *The Cantrip Codex* by Tony 'Doc' Shiels (hopefully) in time for the UNCONVENTION in April. Graham and I are also planning a book called *Weird about the West* (to tie in with our fortnightly BBC Radio series), and there will of course be a 1998 Yearbook as well as four issues of this magazine and a similar number of *The Goblin Universe*. Thanks for your ongoing support. If it wasn't for you, we would not be able to do what we do - and I think that would be a pity!

... Are not done

Animals & Men # 12 — Newsfile

EDITOR'S NOTE: I would like to welcome Graham to the post of full time Newsfile Editor. He replaces Jan Williams who essentially co-founded this magazine with me three years ago, and without whom we would not exist - at least not in our present form. She wanted to step down as Newsfile Editor last summer, but she very kindly hung on until now because of the problems caused by my separation and forthcoming divorce from Alison. Jan is still involved with Animals & Men, and on behalf of all our readers and editorial staff I would like to thank her for all she has done over the past three years.

MYSTERY CATS

Dundonald, Ayrshire

Stephen Steiner and his brother, two night-time rabbit and hare hunters, have endorsed reports of a puma-like animal living wild near Dundonald, Ayr (south west Scotland). They and other witnesses have described sightings - and attacks on cars - by an animal the size of a Labrador dog, but more agile. *"It moved like a cat,"* Stephen said. *"I think we'll start taking a camera."* Kilmarnock Standard 18-10-96

Scottish reports of big cats are being collated by Mark Fraser, editor of "Haunted Scotland" - 01563 539509.

Hampshire

Police in Totton (near Southampton) are confident the "Beast of Basingstoke" has not moved south. A *"black panther-like creature"* seen near a local school was more likely to be a labrador, they said. Southampton Daily Echo 13-5-96.

Gavin Wright reported a jet-black creature with a *"curled tail"* seen in his headlights, east of Southampton. Daily Echo 30-5-96.

NEWSFILE
EDITED AND COMPILED BY GRAHAM INGLIS

Animals & Men # 12 — Newsfile

A black creature with a grey head and green eyes glinting in the moonlight was reported by motorist Gareth Savage to be near Southampton. *"Its tail kind of went down and then swooped upwards,"* he said. Southampton Daily Echo 18-7-96.

Maastricht, Netherlands

Dutch police mounted a hunt for a puma in the area around Maastricht, and the public were warned to avoid a local forest, after the animal was videotaped by a witness. Westfalenpost 18-10-96

Normandy, France

An 'escaped lioness' was hunted by 75 police and park rangers after being spotted by several witnesses near Dieppe. No circus is missing a lion. Westfalenpost 7-8-96

Pyrenees, France

French police looked for a *"black animal with a long tail"* in the Pyrenees Mountains after sightings by 3 witnesses. Kolner Express 23-8-96.

Some newspapers recently have been carrying adverts for "soothing music for your cat". It claims the £10 tape can induce "visions of mice", for instance. Perhaps the Fortean world is ready for the Alien Big Cat control tape - at a Big price of £100, of course.

VARIETY IS THE SPICE OF LIFE...

TWO-HEADED SNAKE

A two-headed herald snake found in South Africa was blamed by the superstitious for the then-recent floods that killed more than 100 people in the region. Columbus Dispatch via COUDI.

BLONDES HAVE MORE FUN
(don't they Special Agent Askew?)

Blonde hedgehogs...

Blonde hedgehogs have been reported in several parts of Somerset, England. Excessive uv radiation (due to ozone depletion) or changing global weather patterns have been suggested as causes. Western Daily Press 5-6-96

Blonde blackbird

A blackbird rejected from its nest for being yellowy-orange is living in a wildlife sanctuary in north Devon. Chris Mead, of the British Trust for Ornithology, said, *"In 30 years I have never heard of one... it is certainly a bird in a million."* Sanctuary staff believe the bird is male but are not sure. Daily Mail 14-8-96

EDITORIAL NOTE: Albino and partially albino blackbirds are relatively well known. I have even seen a beautiful mutation, described by the man who bred it as a Chinchilla Mosaic. This bird was speckled like a Mistle Thrush and had most peculiar striped patterns on its underbelly. This bird, however, is unique in my experience and appears to be a lutino - the mutation which has recently produced some of the most striking cage birds presently on sale - the 'yellow' African Ring-necked Parakeets.

NEW AND REDISCOVERED...

Cuban mini-frog

A frog the size of a fingernail has been discovered on a remote mountainside in Cuba. At an average length of 10 mm it is just 1 mm larger than the smallest tetrapod in the world, a frog found in Brazil in 1971. New York Times 3-12-96.

Fair-skinned jungle tribe

A previously-unknown tribe of light-skinned people has been discovered in a remote jungle region of the Indonesian province of Irian Jaya, by geologists from an American mining company. The origin of the 90-strong group of men, women and children is unknown but one theory holds that they may be descendents of Japanese soldiers. The Japanese invaded and held the area during World War 2. *Westfalenpost 17-8-96.*

WASPS

In Tanzania, Africa, entomologist George McGavin found what is possibly the tiniest winged insect in the world - a member of the wasp family about the size of this full stop. McGavin suspended sheets under the branches of an acacia tree, sprayed the area with an insecticide, and then examined his catch. Unfortunately, he then lost it - among the 11,500 other specimens. *Daily Mail 3-8-96.*

Rather more combative is *Dinocampus coccinellae*, a small parasitic wasp spreading fast in Scotland, that reproduces without mating and lays its eggs in the body of the common seven-spotted ladybird. When the grub is fully fed, it severs the main nerves to the ladybird's legs, immobilising it. Scientists have warned of a potential threat to crops, as the ladybird, important in pest-control, could be wiped out in Scotland.

The less-common two-spotted ladybird is naturally immune to such attacks, however, and some scientists are contemplating a genetic engineering intervention, on behalf of the threatened seven-spotteds.

New bird: graveteiro

A warbler-sized bird, the pink-legged graveteiro ("*twig gatherer*" in Portuguese) - *Acrobatornis fonsecai* - that occupies the Brazilian forest canopy has been described as "*the most exciting new species of bird discovered in decades.*" It occurs principally over coffee plantations, however, and, following the crash of cocoa prices in the late 1980's and the spread of a destructive fungus (witch's broom), many plantation owners have been cutting down their old trees and selling the timber to raise cash. *New York Times 19-11-96.*

I'm not a Pheasant plucker...

A species of pheasant, long thought extinct, has been rediscovered in central Vietnam. The last known capture of an Edward's pheasant - *Lophura edwardii* - was in 1928. Three expeditions between 1988 and 1994 searched for, and failed to find, this dark blue-black pheasant, which is believed to be native only to Vietnam. *The Times 5-9-96*

SLIME WAVE

Snail rustling is on the increase in Sussex at a National Trust site location that's being kept secret. Wardens protecting the Roman snails (*Helix pomatia*), which are protected under the Wildlife and Countryside Act, are catching poachers suspected of intending to sell their prey to restaurants. Top restaurants charge about £20 for a dozen escargots. *Evening Standard (London) 19-8-96.*

MORE SLIME

A horde of snails crawled onto a Moroccan railway line, halting an express train that lost its grip on the slime. *Boston Globe 16-5-96.*

WHALES

Measles in the Med

Blue whales in the Mediterranean are dying of what's thought to be a strain of measles. Four specimens had the characteristic mottled skin and a fever. Pollutants - mercury, lead and cadmium - have been cited as a possible cause, as these affect the immune system. *Independent 1-1-96.*

"Good evening, sir. Your cetacean research sample, sir..."

A Japanese whaling ship sailed into a Tokyo port with 77 Minke whales caught during a 2-month "research" mission in the Pacific.

Although ostensibly caught for scientific reasons, most of the meat ends up as an expensive delicacy in Japanese restaurants. *Aberdeen Press & Journal 18-9-96.*

I got dem ol' California Blues

Since 1991, blue whales have been entering an area of the Pacific 75 miles north west of Los Angeles. Several thousand are now thought to be there: electronic tagging may help scientists find out why they are congregating. These crustacean-eating animals have been listed as endangered since 1966. *Christian Science Monitor 23-7-96.*

Dead Right

On the eastern seaboard of the USA, six right whales, the most endangered of all the great whales, were washed up in the space of three months in 1996, compared to only two in all of 1995.

Biologists are puzzled over why the species continues to decline despite being federally protected since 1935. *USA Today 11-3-96.*

Animals & Men # 12 — Newsfile

WHALE MEET AGAIN..

An article in Soviet Weekly examines the mystery of cetacean navigation and suggests that they orient themselves using their own compass. A crystalline substance possessing magnetic properties, similar to that found in pigeons and man, has been discovered in dolphins' brains.

Dr Victor Golovko writes, "If cetaceans indeed have compasses, small wonder they have learned to use them ... in the tens of millions of years of their existence." He suggests that the unexplained beachings ("suicides") that sometimes occur might be due to anomalies - local whirlpools - in the Earth's magnetic field. If a magnetic anomaly occurs with land in the vicinity, "the herd may get into a geomagnetic deadlock. Such a situation plunges the mammals into ... a state of anxiety and fear. Those leading the way, pressed by those following, then find themselves grounded."*Soviet Weekly 13-6-91.*

OUT OF PLACE

BEARS

Two bears were spotted in Lower Saxony, Germany, trotting in the general direction of Harz. The witness informed the local police, who, after checking that it wasn't a joke, summoned a hunter to paralyse the bears. *Sie sollen in einem zoo untergebebracht werden* - they will be sent to a zoo.

The paper, *Westfalenpost (9-9-96)* offered no explanation as to their origin.

Chinese mitten crab

The Chinese furry-clawed "mitten crab" is rapidly establishing itself in the River Thames, to the detriment of the indigenous and already endangered crayfish. The crabs are not only large and aggressive, but can carry a parasite which can cause serious illness in humans. *The Richmond and Twickenham Informer, 13-9-96.*

NEWSFILE EXTRA: MITTEN CRABS

I have been fascinated by Freshwater crabs for most of my life. As a child in Hong Kong I kept all three native species - *Potamon anacutholon* (The Mountain Freshwater Crab), *P. hongkongsiensis* (The Hong Kong Freshwater Crab), and *Eriochier sinensis* (The Chinese Mitten Crab). Writing in *A Colour Guide to Hong Kong Animals* (Hong Kong Government Publications 1981), Dennis Hill and Karen Phillips note:

"This is actually a member of the family Grapsidae which are marine crabs with a few species either semi-terrestrial in Mangroves or high up on the shores and a few estuarine species. The Mitten crab is a brackish water species which can live quite happily in fresh water but is alleged to require salt or brackish water to breed. (....) The hairy 'mittens' are characteristic of the species together with the patches of body hair and the four anterior-lateral teeth on the carapace.

It is an important animal zoologically in that it is the natural host for larvae of the Oriental Lung Fluke (Paragonimus westermanni). Thus, the local practice of eating steamed mitten crab in the winter months is fraught with hazard as lung fluke infections can result."

According to Muus and Dahlstrom, writing in *"Collins Guide to the Freshwater Fishes of Britain and Europe"* (1967):

"This crab originated in China and was accidentally carried to Europe, probably in the ballast water of ships. It was first found in Europe in 1912 in the River Aller, and has since spread to much of North West Germany, Holland, Belgium and parts of Scandinavia. It was reported from London in 1932, but has not been seen since..."

The figurative 'father' of Hong Kong Zoology, Geoffrey Herklots wrote in 1951 *(The Hong Kong Countryside*, SCMP Pubs):

"These were large, hairy crabs of the same kind or very similar to those which were introduced accidentally into Europe and did so much damage by burrowing into the banks of canals and therefore making them leaky".

It is an unworthy admission for me to make as a zoologist, (and I admit to being sentimental rather than ecological as far as this matter is concerned) but I rather like these creatures and I for one would not be too worried at seeing them become a permanent part of our ecosystem. However their impact on the environment as a whole will have to be monitored carefully!

Animals & Men # 12 — Newsfile

Journal 16-11-96.

EDITOR'S NOTE: Really, Mr. Inglis. Some people will write any headline to woo the X Files audience!

Bandicoots

Australian marsupials called bandicoots have been sighted in parts of the English midlands. Looking like large rats that can hop like kangaroos, and with a gestation period of only 15 days, the potential for crop destruction has alarmed some farmers. Bandicoots can grow up to two and a half feet long. *Leicester Mercury 26-10-96*

EDITORS NOTE: This story was widely reported at the time but then seems to have disappeared. It has been suggested by the more cynical at Animals & Men that this was nothing more than a publicity stunt for a video game called 'Crash Bandicoot' which was released at about the same time. However, as regular readers will know - stranger things have happened.

Piranhas

The discovery of two piranhas in a lake in southern France triggered a fishing rush. French pet owners pay up to £12 ($20) for piranhas, which are indigenous to South America. *Westfalenpost 16-8-96.*

GREYS SEEN NEAR NEWCASTLE

Fears are growing that one of England's last bastions for native red squirrels - Northumberland - is about to fall to American invaders. Greys, artificially introduced to Britain 120 years ago, have now been seen in the area. *Aberdeen Press &*

Lake and Sea monsters
(B r o a d l y   a n d   L o o s e l y)

Visitors to Lake Tianchi, a deep volcanic lake near the Chinese - North Korean border, reportedly saw a creature resembling the Loch Ness Monster. *Kolner Express 8-8-96.*

EDITOR'S NOTE: There have been reports for some years of a 'golden' creature found in the depths of this lake, and as reported in earlier issues of Animals & Men there are even suggestions that it has been video'ed. Needless to say, at the time of going to press we haven't actually seen a copy!

The Mermaid of Weymouth.

A sighting of a mermaid-type creature off Weymouth, Dorset, was reported by Martin Ball, describing it as a creature whose top half resembled a sea horse and its silvery bottom half covered with fins. He hesitated to report the sighting until he came across an account in his local library of a similar sighting in 1757. It seems that information on that sighting was suppressed at the time, in case it deterred the high society visitors of the day, who came to enjoy the sea water.

Nowadays, of course, there's nothing like a

Animals & Men # 12 — Newsfile

'resident monster' to bring in the tourists. *Dorset Evening Echo 24-2-96.*

JELLYFISH KILL SALMON

Thousands of salmon worth an estimated £250,000 died at a Scottish fish farm when strong tides washed hundreds of Lion's Mane jellyfish against the fish cages, forcing their poisonous tentacles through the netting. Their presence is usually appreciated by fish farmers because they feed off plankton that builds up on the mesh. *Aberdeen Press & Journal 18-9-96.*

Toads

A plague of toads brought predictions of impending tragedy to central Bolivia last February. Radio reports said a main road was *"coated with a thick layer of dead toads."* Experts said drought conditions had forced the toads out of their usual habitats and into farmland in search of water. *Columbus Dispatch, via COUD1.*

ATTACKS

THIS WAS NO PICNIC

The activities of the blood-sucking organ-devouring beast known as Chupacabras have taken a sinister turn. In an incident on the Caribbean island of Puerto Rica, the beast reportedly opened a window, destroyed a stuffed teddy bear, and then departed, leaving slime on the windowsill. *Aberdeen Press & Journal, 21-11-96.*

DEER ATTACKS

A Sussex woman described a deer attack on her and her Jack Russell terrier. *"It was like a wildcat, it just pounced. It just kept coming at me."* She was unhurt. Steve Webster of the Sussex Wildlife Trust said that deer were normally placid but this one might have been defending a faun. *Evening Argus 23-5-96*

Deer in London's royal parks have become distinctly aggressive as the breeding season gets under way. Several attacks on passers-by have been reported. Recent dry weather means bracken has not grown high enough for the deer to give birth in private and hide their young.

If man or beast goes too near a hidden baby, the mothers form a circle round the intruder, rushing it in unison and crushing it. David Smith, head ranger at Richmond Park said, *"We've had quite a few dogs killed by deer."* *Evening Standard (London) 23-5-96.*

SCIENCE

MAMMOTH TASK

Researchers from Japan want to use ancient DNA from mammoths frozen in the Siberian permafrost and modern-day elephant sperm, in an effort to create half-mammoth, half-elephant offspring. Over several subsequent generations, a creature genetically close to the prehistoric woolly mammoth could then be created. *Yorkshire Post 20-8-96.*

THE STING

The British scorpion Euscorpius flavicaudus, is still believed to be alive and well.

The one-inch-long, yellow-tailed variety is harmless, but suffers from being mistaken for its tropical cousin - people encountering one are liable to stamp on it. However, scorpion expert Bernard Betts says, *"They are the ultimate survivors. They can withstand 200 times the amount of radiation we can stand, and can go without water for a year."* He once discovered a colony of them in a London Underground station but a later fire under the platform destroyed them. *Daily Telegraph 17-2-96.*

EDITOR'S NOTE: Some years ago (and I am afraid that I have mislaid the reference) Fortean Times printed a letter from a reader who claimed to have been responsible for the scorpions found at the railway station in, I think, Epping Forest.

He claimed that British Rail had been about to close the station, so in order to keep it alive he bought some scorpions from a local pet shop and released them in the station in order to have the place kept open as a site of special scientific interest. It sounds a tall story to me - but then again...

ON THE TRACK OF UNKNOWN EVOLUTIONISTS...

Patrick Matthew, a university-educated Scottish fruit farmer, published his theory of natural selection 30 years before Darwin. And, contrary to popular belief,

Darwin never even used the word "evolution", according to *The Australian.* 'On the Origin of Species' gave scant source material, ignored the influences of others, and was preferentially promoted because Darwin had the right social connections, whereas Matthew was dismissed as an "obscure writer on forest trees". *The Australian 15-4-96.*

Wriggling for longer

Scientists have isolated a gene (dubbed 'age-1') that, in its mutated form, bestows longer life on the the worms (C. elegans) that carry it. Humans have a similar gene but the implications for human lifespan adjustment are as yet unclear. Researchers say that worms with the mutated age-1 gene are unusually resistant to the chemical effects of free radicals.

EDITOR'S NOTE: Regular readers will know that amongst my other obsessions are the novels of Robert A.Heinlien. The above story is very reminiscent of a passage at the end of *Methuselah's Children* proving once again that art imitates life and vice versa. If anyone called Lazarus Long happens to be reading this magazine....

Survival of the fittest...

The power of the corporate market is being brought to bear on protecting endangered species: under a government-backed scheme, protection of the dung beetle has now won sponsorship from Lord Montagu of Beaulieu. Other species, though, like the water vole and the brown hare, have yet to clinch a deal with the private sector. *Daily Telegraph 16-5-96.*

Obviously they didn't hire a suitably high-profile advertising agency for redefining their image and niche in the wildlife market-place. Perhaps genetic engineering will one day enable hares, for instance, to be born automatically branded with industrial logos or advertisements for alcoholic drinks.

Neural codes: crickets and bullfrogs

Scientists trying to find how much information the nervous system can handle, and how reliably, have studied what has been dubbed *"the cricket's hairy eyeball"*. The organ consists of thousands of tiny hairs on two antenna-like appendages that stick out from the cricket's belly.

Air movement around a hair triggers changes in the electrical signals generated by a nerve cell at the base of the hair. These signals help the animal maintain stable flight in windy conditions.

Tests on crickets in a wind tunnel, by University of California researcher John Miller, showed that the cricket can interpret wind pulses "with astonishing *accuracy and reliability*" using only a few nerve cells.

In another study, Stanford University researchers played taped bullfrog calls mixed with white noise, to create a *'radio with bad reception'* effect. The researchers reported that bullfrogs are surprisingly efficient at distinguishing meaningless and meaningful sound.

The scientists analysed the electrical activity in bullfrog sound-handling nerve cells and concluded that unimportant information gets filtered out by sensory nerves before it reaches the brain. *St Louis Post-Dispatch 20-3-96 via COUDi.*

Animals & Men... and flowers and germs

Genetic engineering, the practice of redesigning a species by splicing genetic attributes from one species into the genetic material of another, has been utilised in producing a new type of soya bean.

Scientists seeking to make American soya beans resistant to a *herbicide* called 'Roundup' utilised the genetic storehouse of the petunia plant, a bacteria, and a virus. They incorporated genetic material from all three into the soya plant, producing a new organism.

Greenpeace Briefing, October 1996.

Rabbits are our distant cousins?

A study that examined the detailed structure of 88 proteins common to all mammals indicates that rabbits are more like primates than rodents, and very much more like tree shrews, a group that is thought to have given rise to the primates.

The function of the proteins examined is the same in all species but the molecular structures can vary as a result of mutations. The longer it is since two groups shared a common ancestor, the more genetic differences are accumulated in their proteins. *Saginaw News (MI) 11-3-96 via COUDi.*

Newsfile Contributors:

David Coleman, Richard Muirhead, COUDi, Wolfgang Schmidt, Herman Reichenbach, Lionel Beer, Sally Parsons, Michael Playfair, Tom Anderson, Ian Sherred, Ginny Ware.

NEWSFILE EXTRA: NEWS FROM NEW ZEALAND

by Darren Naish

Many aspects of New Zealand's history, both ancient and recent, remain mysterious or poorly understood. A few discoveries made in 1996 sparked off some much needed debate regarding the timing of human arrivals on the landmass, an area with implications across both the zoological and political boards. 1996 also saw New Zealand's press speculate wildly about national disaster as, on 19th June, Mount Ruapehu provided a spectacular eruption.

Exactly when humans first came to New Zealand remains somewhat contentious. Presently the only reliably dated evidence of human presence dates to 800-850 years ago - consistent with the Maori view that they are the 'tangata whenua', or original people of the land. There is in fact a treaty, the Treaty of Waitangi, that provides the Maori with preferential treatment because of this status.

However, amateur and fortean archaeologists frequently make claims about supposed sites or artifacts that pre-date the Maori.

As reviewed thoroughly in November's *Fortean Times* (92: 12-13), the latest of such to receive wide attention is a wall-like structure found in North Island's Kaimanawa State Forest. Photographs show the 'wall' to have a vertical face and a series of near parallel vertical lines that provide the impression of large, man-made blocks. The entire structure is 7 metres long, and each block structure is 2 metres long and 1 metres high.

Championing the 'wall' as an ancient, man-made structure is New Zealand author Barry Brailsford - a figure who has previously been strongly criticised for his views by both Maori and Pakeha (non-Maori) alike. According to Brailsford and his colleagues 'the wall' may be a link to ancient Egyptian and South American civilisations and must have been made by the stone-carving Waitaha people, an imaginary group, hypothesised by Brailsford to have colonised New Zealand 2000 years ago. As qualified archaeologists and geologists have been able to show, however, all 'archaeological' evidence of supposedly ancient New Zealand structures dissolves on analysis - Brailsford's 'wall' being no exception. Photographs published in *Geology Today* (12: 137-8) show one end of the wall grading nicely into a fairly standard exposure of the volcanic rock ignimbrite, and a number of reputable archaeologists and geologists have been able to pin the 'wall' down to the 330,000 year old Rangitaiki Ignimbrite. Rocks such as this regularly produce near-parallel or parallel joints and cracks as they cool, and the man-made appearance of the structure is an illusion.

Exit the Kaimanawa 'wall'.

Ironically, however, November saw the publication of some serious scientific data backing the view of a pre-Maori New Zealand invasion. The data came from rats. Dr. Richard N.Holdaway, best known for his work on the giant New Zealand eagle *Harpagornis*, undertook radiocarbon tests on the bones of the Kiore or Polynesian Rat (*Rattus exulans*). He found that rat remains from fifteen sites were more than a thousand years older than they should have been, dating at about 2,000 years before present (*Nature* 384:225-6). This suggests that Kiore were resident on New Zealand at least 1,000 before the Maori.

According to Maori legends, the first Kiore may

Animals & Men # 12

Newsfile Extra

According to Maori legends, the first Kiore may have been taken to New Zealand in the war canoe of the great chief Te Kupe, and conventional explanations have favoured introduction of Kiore by the earliest Maori settlers.

Dogs and Kiore were the only mammals that the Maori introduced to New Zealand, and the archaeology of the Pacific islands shows that Polynesians took their rats right across the region. If, however, the rats were there before the Maori, then someone else took them there (they could not have rafted in). It would seem that transient human visitors dropped by New Zealand about 2,000 years ago, and either moved on without settlement, or died out before the Maori invasion at 800-850 years before present.

Acceptable archaeological evidence, does not, as yet, indicate the presence of such pre-Maori peoples, but perhaps they are the Moriori - an ancient mythical tribe believed in by some.

With an extra 1,000 years worth of rats on both islands of New Zealand, the possible effect of these rodents on New Zealand's fauna must be re-analysed.

Kiore eat more fruit than the other rats that accompany man and are less aggressive, as shown by the elimination of them when Brown rats (*R.norveigicus*), Black rats, (*R.rattus*) and Stoats (*Mustela erminea*) were introduced between 1769 and 1884. If, however, any of New Zealand's fauna was vulnerable to Kiore predation, presumably it would have become extinct before the Maori invasion at 800-850 years before present.

This means that extinctions on New Zealand would occur in two waves - an earlier one of small animals starting at 2,000 years before present., and a later one of large animals (particularly moa) starting at 800-850 years before present. Better resolution of the archaeological record is needed before this theoretical model is supported.

If it is correct, then a second wave of Kiore brought by the Maori may have had little effect on an already Kiore-affected fauna. The European introduction of the agile, climbing Black Rat seems to have adversely affected tree nesting birds that the Kiore did not kill, so there may also be an extinction of ground or bush-nesters caused by Kiore, followed by an extinction of tree-nesters caused by the Black rat.

Extinction of tree-nesters is also correlated with deforestation, and this occurred throughout Maori and Pakeha residence. The introduction of Mustelids in 1884 may not have had adverse affects because the better part of the damage had already been done, but mustelids and Black rats together predated upon - ironically - the Kiore. Clearly, the chronology of extinction on New Zealand is anything but simple!

Newsfile Extra: The Lion Sleeps Tonight.

EDITOR'S NOTE: During the closing months of 1996 I was flicking through the news pages of the teletext when I noticed a story claiming that a small group of Barbary Lions, thought extinct, had been found starving to death amongst the effects of a bankrupt travelling circus in Mozambique. I have unfortunately lost the reference, which is annoying, especially as I have not been able to unearth any further details on this fortuitous event.

I mentioned it in passing to various cronies on the A&M Editorial team and was not at all surprised when several told me that they knew quite a lot about this elusive subspecies. Here is a selection from what they sent me:

The Barbary Lion

by Chris Moiser

The Barbary Lion, *Panthera leo leo*, is one of those enigmatic subspecies that has caused much concern and debate amongst taxonomists. Officially, it became extinct when the last wild one was shot in about 1922 in the Atlas Mountains, but news of its extinction may have been premature.

As a subspecies, its recognisable characteristics were, in the male, the long black mane which ran along the full underside of the body. This visual characteristic was shared with the other extinct subspecies, the Cape Lion, *Panthera leo melanochaita*. It is generally believed that the dark thick mane of both animals evolved as a response to the lack of dense bush, and the cool winters in their respective ranges.

Prior to its extinction in the wild, the Barbary Lion was fairly well represented in captivity. For example, John Edwards, in his book 'London Zoo from Old Photographs', shows 'Sultan', a Barbary Lion in London Zoo in 1896. As lions are one of the easiest cats to breed in captivity it is likely that Sultan, and other members of his subspecies, combined many of his genes with those of other subspecies before they died out. It is, after all, only in the last thirty or so years that zoos became sensitive about keeping their subspecies pure! It is quite possible that many of the european zoo African Lions contain genes belonging to the Barbary Lion subspecies.

The story becomes more confused because it is thought that the lions that were found in the King of Morocco's zoo were a remnant population of pure bred Barbary Lions. As a result several were moved to western zoos as part of a subspecies rescue programme. There are now over fifty in various world zoos. A pride may be seen in England at Port Lympne in Kent, one of John Aspinall's establishments.

More recently another colony of eleven long and dark maned lions was found in a zoo in Addis Ababa in Ethiopia by Dr. Hym Ebedes, a South African veterinarian with zoo experience. These appear to be descendants of the lions that were in the private Royal zoo that closed after the emperor, Haile Sellasie, ('The Lion of Judah'), was deposed in 1974. In the International Zoo Yearbook for 1959 this collection was referred to as the 'Imperial Collection of Lions'.

Although it is highly likely that these lions are descended from Barbary Lions and not Cape Lions, Dr. Ebedes seems keen that long, black maned lions, should again be seen in the Cape. The Cape Lion

Lost Lion Renaissance
by Darren Naish

Reported widely in the press during the July of 1996 was the discovery, in an Addis Ababa zoo, of a group of lions that seemed to be of a supposedly extinct subspecies. News of the eleven animals has not been forthcoming, but according to reports the three males in the group are characterised by extensive, thick black manes.

Coat and mane colour is quite variable within lions, but only two sub-species - the Atlas or Barbary Lion (*Panthera leo leo*), and the Cape Lion (*P.l.melanochaita*) [1], ordinarily have thick, extensive manes. Or ordinarily DID have - for both subspecies are generally thought to be extinct. Barbary lions - ironically the sort of lions that serve as the type for all members of their species - were caught in great numbers for use in Roman Gladiatorial contests.

became extinct probably in about 1850. It is thought that Dr. Ebedes has been in touch with the Capetown Public Works Department to see if the derelict lion enclosure in the old Cecil Rhodes Zoo could be renovated to house them. Although the zoo, at Groot Schuur, on the outskirts of Cape Town closed, and was largely dismantled in the 1980s, the lion enclosure was a very substantial structure and is largely undamaged.

References.

Denis, Armand. (1964) *Cats of the World*. Constable, London.
Edwards, John, (1996) *London Zoo from Old Photographs 1852-1914*, John Edwards, London.
Green, Richard (1991) *Wild cat Species of the World*. Bassett, Plymouth.
Guggisberg, C.A.W. (1975). *Wild Cats of the World*. David and Charles, Newton Abbot.
Richardson, Douglas. (1992). *Big Cats*. Whitter Books, London.
The Times, 13th July, 1996.

Africa's biggest, most thickly furred lions - the dark Cape lion of the south, and the light Barbary lion of the north. Picture by Darren N.

Animals & Men #12

Newsfile Extra

Hunting and encroachment on their habitat further reduced the wild population, and by the late 1880s they were in severe trouble. Officially the last one was killed in 1922. Cape lions, similarly were extensively hunted in the 18th and 19th centuries. They were first eliminated from around Cape Town, and in 1858 the last one reported to have been seen in Cape Province was killed. Some authorities quote the extinction date for this form as 'mid 1850s', but it is also widely stated that the very last one was hunted down and shot by General Bisset in 1865. This was in Natal.

An interesting fact worthy of mention here is that Dr. David Livingstone himself was mauled by a lion, and in fact was dealt an arm fracture by the animal. This occurred in Bechuanaland sometime, I presume, around 1840. [2]. By this time, Cape Lions would have been extremely rare: according to some sources, Livingstone's lion may have been one of the very last. However, there is a 1936 record of a lion shot by a Mr. Lennox Anderson near Hectorspruit in the eastern Transvaal. Lions other than *P.l. melanochaita* did, and still do, occur in this area (they have even been reintroduced to parks in Natal, perhaps their most southerly stronghold in present times), but the immense size of this individual - it was reportedly 313 kg in weight, and is thus listed in some editions of the Guinness Book of Records as the biggest lion ever - suggests that it may have been a Cape lion (see below). If it had a dark, extensive mane, there is a very good chance of it having been so. In one sense at least, either or maybe both of these lions may live on. Genes from them doubtless still occur in other lions, especially those in captivity, and some cat experts have long been of the opinion that Barbary lions may still be hidden away in zoos. Some sources report that Barbary-type lions have also been reconstituted in captivity by cross-breeding. Morocco's Rabat zoo has apparently had some success with this venture. So, how does one choose the lions that might be candidates for such a reconstitution programme?

Being at the latitudinal extremes of the African lion's range, and thereby presented with similar climates, Barbary and Cape lions shared similar traits. Both were huge animals - big males of both subspecies allegedly exceeded 227 kg and 3 m. For comparision, a really big lion of another subspecies may reach 180 kg. The biggest living cat ever recorded, a Siberian tiger (*Panthera tigris altaica*) shot near the Amur River around 1933, weighed 364 kg. As noted above, there is one record of a 313 kg lion.

The lion's mane relates intimately to its sociality: there seems to be correlation between number of pride members, male territorality, and degree of decoration in males. It may also be that mane size and extent relates to climate and vegetation cover. Whatever, both Barbary and Cape lions had very large, bulky manes that extended right under their bodies to the middle of the ribcage. Though no details have been reported on the body size of the Addis Ababa zoo lions, they are said to have extensive, black manes. This allows candidature of the Cape lion without doubt, but I am not so sure if the Barbary lion had a dark mane. In his 1908 book *The Game Animals of Africa*, Richard Lydekker described the Barbary lion as 'dusky ochre'. A meticulously researched restoration by Maurice Wilson for David Day's monumental work *The Encyclopedia of Vanished Species* similarly depicts a Barbary lion with a light brown mane. Apparently, females of this subspecies had a white inner foreleg. So: more information from Addis Ababa please!

Notes

1. *P.l. melanochaitus*, seen in many sources, is incorrect.
2. EDITOR's NOTE: I checked with my father who is presently writing a book on the history of pre-colonial Africa, and he gave me the following information, gleaned from *Livingstone*, by Tim Jeal (Book Club Assoc, 1973 ed, London). Livingstone's encounter with the lion occurred at Mabotse on the sixteenth of february 1844. His arm was splintered at once, and the long teeth of the beast made a series of "*gashes like gunshot wounds*". The wound suppurated for months and Livingstone was very ill, before being nursed back to health by Mary Moffat, whom he eventually married. (pp 58-9).

BETWEEN THE LIONS
by Jonathan Downes

According to Guggisberg (1975) lions were found across a far wider geographical range in historical times than they are now. They were found across the continent of Africa, and across parts of the Arab Peninsula, Asia Minor, and Iran to northern India. Surprisingly, although lions were wiped out in Palestine at about the time of the Crusades, the species was found in Iran as late as 1923, when one was killed south of Shiraz. Other animals were reported in 1928 and 1929 and Guggisberg wrote that *"...some authors think it possible that there may be a few Persian lions even today."* He also noted that an entomologist called Hugh Scott had presented evidence to suggest that lions were found relatively recently in the Yemen. The Asian lions of the Gir forest in northern India are well known, but the possibility of surviving lions elsewhere in Asia is a tantalising possibility.

In classical times they were found in Greece but there have been no European lions since the days of Hercules. Surprisingly, however, Guggisberg notes that the last surviving tigers and leopards in what is still technically Europe died out well within living memory.

He writes that the Anatolian Leopard P.p.tulliana was last seen in the Caucasus in 1956, and says that a few animals may still exist in Transcaucasia. The last tigers in the region were shot in about 1930. Emotionally information like this is a double-edged sword.

Whilst it is unarguably exciting to discover that such wonderful creatures were found so widely within a relatively recent time-scale, the knowledge that within the lifetime of my parents, humanity has managed to decimate the distribution of so many of the animals with whom we share this planet is indeed a heavy load to bear!

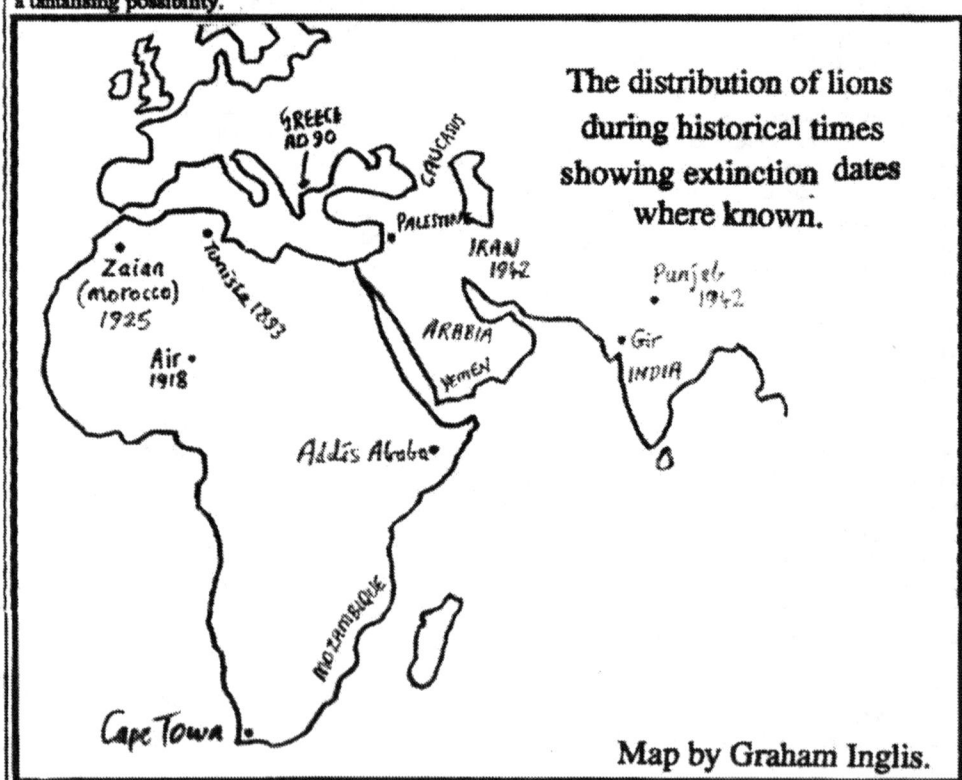

The distribution of lions during historical times showing extinction dates where known.

Map by Graham Inglis.

Animals & Men #12 — Newsfile Extra

LIFE WITH THE LIONS
(THE END BIT)

Two days before we went to press I received a telephone call. It was from Michael Playfair, a long-time Animals & Men reader and newsfile contributor. He asked whether I had heard about the Barbary Lion being found in Mozambique. I told him, essentially what I had written in the introduction to this section, and he told me to check my post the following morning. This I did, and found a cutting from the *Sunday Express*, dated 22.12.96. The Akef Egyptian Circus had been stranded in Maputo in Mozabique since its owner El Sayed Hussein Akef had absconded in January 1996. The animals had been cared for by the few members of staff at the circus who remained, but by the time rescuers from a London based animal charity called Animal Defenders arrived on the scene the creatures, which included a python, horses, dogs, tigers and eight lions - three of which; a male, now named Akef, after the circus in which he was found, and two lionesses - Hod and Nazine, are suspected to be Barbary lions.

The animals have now been moved to Hoedspruit Research and Breeding Centre for Endangered Species in South Africa.

The article also featured two photographs of the (admittedly slightly mangy looking) Akef, and these pictures go some way towards resolving the mystery about the mane colour of this subspecies. In one of the pictures he is seen sitting on the ground. The picture is taken in profile and his mane appears to be brownish in colour - not a million miles away from the 'dusky ochery' described by Lydekker. The other picture shows him, anaesthetised and lying on his side, surrounded by people, on what appears to be a vet's operating table. Here one can see that the lower part of the mane is much darker - indeed it appears to be black. The mane appears to have two 'layers' - an upper lighter one and a lower dark one. Therefore the apparent inconsistencies between the different accounts of the morphology of this subspecies appear less significant.

It appears, therefore, that there are now two, presumably unrelated, groups of what appear to be Barbary lions in South Africa.

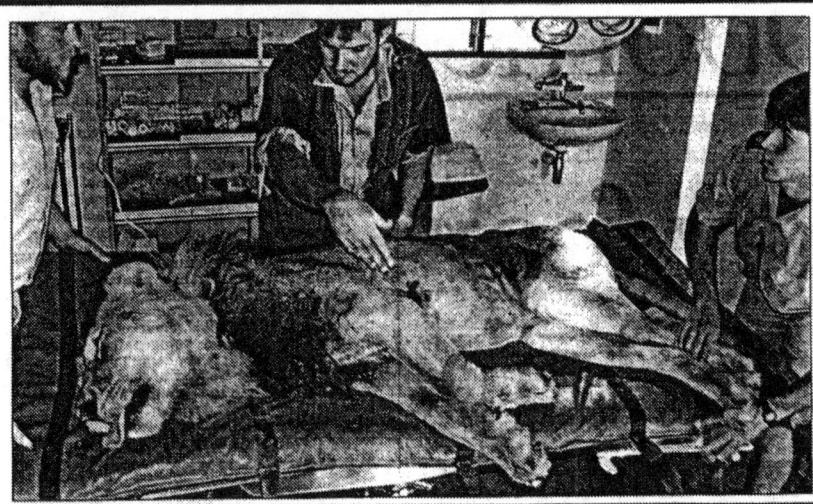

Akef the lion being examined by a vet. Note dark coloured mane on nder side of body. Picture c/o The Copyright Liberation Front.

animals&men

THE JOURNAL OF THE CENTRE FOR FORTEAN ZOOLOGY

Animals & Men #12 — Newsfile Extra

Their provenance seems uncertain, especially as the Rabat Zoo project was only carried out relatively recently, (and it appears that the lions owned by the Emperor Haile Selassie were donated by the King of Morocco many years before), and until we get more information on the three which were found in Mozambique, the true significance of this episode is uncertain. The plot thickens ..... watch this space!

EDITORIAL NOTE: Who in 1969 released a dreadful album called *'Life with the Lions'*, and where was side A. recorded? The first person who telephones me with the answer gets a free year's subscription to this wonderful magazine. The editorial team were somewhat disappointed with the response to the phone in quiz in the last issue. Several people noted that *'Glass Onion'* from The Beatles 'White' album (1968) featured the word 'walrus' in the lyrics. Darren Naish (bless him) suggested *'God'* from the *John Lennon/Plastic Ono Band* album, which also features the word. No-one, however suggested the third song *'Death Trip'* by my old Boss Steve Harley (and Cockney Rebel), and as no-one got more than one walrus ditty the prize remains unclaimed!

Akef the lion - a slightly moth-eaten 'King of the Beasts' (Picture c/o The Copyright Liberation Front.

The Editor would like to thank the following people for being on hand with their help during the research for this collection: J.T.Downes, Dr Karl Shuker, Chris Moiser, Graham Inglis, Simon Elsdon, Darren Naish, Austin Orchard, Michael Playfair and anyone else he's fogotten!

Strangers in a Strange Land

Sightings of Unexpected Animals in Germany: Where on earth did they come from?

by Wolfgang Schmidt

With a total population of nearly 80 million people living in a comparatively small country, Germany is one of the most densely populated countries of the world. The last thing that would be expected here are large, exotic animals roaming the wild. Nevertheless, individuals step forward to claim some of the most unbelievable sightings: wolves, bears, even pumas, panthers, and crocodiles. Can they all be mistaken? Or is this some kind of elaborate hoax by some pranxters to fool the population? Or is there another explanation? In the following article, which is by no means complete, the author will try to shed some light on a few of those unusual sightings.

1. Re - Imports: They're back

Some big animals that once populated Germany but were driven to extinction by habitat loss and extensive hunting. Now, after the „Fall of the Wall," they are starting to return to their former territory.

Go West - not just a popular song by the Pet Shop Boys frequently played in European Football stadiums, but also a new trend in the migration patterns all over eastern Europe. The end of the Cold War has lead to the largest migration in Europe in nearly 50 years. It's not just humans trying to improve their miserable economic situation by moving to West Europe looking for a new home; but also a number of animals long extinct in densely populated western Europe, such as wolf, bear, lynx, and wildcat, are showing a remarkable comeback in areas where they have been absent for a century or more. (1)

The most prominent (and numerous) of all returnees is the *Wolf, canis lupus*. After centuries of being relentlessly hunted, it was practically wiped out in West Europe, except for some smaller populations in Spain and the French *Ardennes* and *Pyrenees* mountains. In East Europe, particularly remote regions of the former Soviet Union (such as Siberia), it was still relatively common, numbering about 150,000. (2)

With the end of the cold war and the „Fall of the Wall," the wolf started recovering its old territory. After being extinct in Germany for nearly 150 years, eastern Germany was among the first places where the wolf returned. Moving in from western Poland on age-old paths that had been blocked for decades, wolves were seen more and more often, especially in the German state of Brandenburg. An estimated 20 - 40 wolves are now roaming that area again and, as the number of contacts with humans increase, the first fatalities have occurred - on the side of wolves! Several already lost their lives in car accidents and cases of mistaken identities (when thought to be feral dogs and shot).(3)

Mecklenburg and Niedersachsen are two more northern German states where wolf sightings are on the increase. Following the stabilization of the wolf populations in eastern Poland (about 900 animals) and Slovakia (450) in the 1970s, individual wolves kept moving westward into the northern German low plains. Since the German-German border was opened and the mine fields were swept, the numbers of wolf visitors from the east have gone up further. (4)

The *Bayrischer Wald* (Bavarian Forest), where the last recorded shooting of a wolf occurred in 1846, also reported the return of the wolf. After first sightings were reported and wolf tracks were found near the Czech border in the fall of 1992, a wolf was reported by witnesses and later run over by a train when trying to cross a railway in December of 1994. The autopsy revealed its identity to be that of a free-roaming wolf. Since subsequent checks with zoos did not show any escapees, the best explanation is it crossed over from the Czech Republic. (5) Today, there are at least six wolves confirmed to be living in the *Bayrischer Wald Nationalpark*. (6)

Even more surprising was the case of „Lupo," a wolf seen and finally shot in the *Eifel* region (close to Köln), in spring of 1994. After killing three sheep and avoiding several traps by local hunters, the wolf was finally ambushed and shot by police and finished off by a local hunter. When a closer examination of the body showed the wolf to be a female, *she* was appropriately renamed „Lupine." (7, 8)

In the fall of 1995, the area around Val Ferret in neighboring Switzerland, was haunted by a ferocious beast, described alternatively as a dog, wolf, or simply „beast," which killed more than 70 sheep. The description of the animal was usually „like a German Shepherd Dog, with longer hair and light-colored spots." (9) Half a year and more than 30 dead sheep later, the Swiss authorities decided to put an end to the reign of the „Beast of Val Ferret" and hunted it down. As it turned out, the „beast" was indeed a wolf; it was shot, but managed to escape wounded. This is the last I have heard about the creature's destiny. (10)

In Germany, as in most of Europe except for a few eastern countries (Slovakia, Romania, etc.), the Brown bear, *Ursus arctos*, has been extinct for more than a century. In the middle ages it used to inhabit most of Europe, but was decimated and hunted to extinction in most of western and central Europe several hundred years ago. Humans and bears just didn't seem to be able to get along, and, as a result, wherever human population density increased, bear populations decreased. (11)

Expecting to be perfectly safe from any dangerous predators in highly industrialized South Germany, unsuspecting tourists in the *Bayrische Alpen* (Bavarian Alps) had a scary moment when running into brown bears in spring of 1996. Fortunately, no harm was done. These bears are believed to have wandered across the border from Slovakia after the end of the cold war and have found a new home in the Bavarian Alps. (12)

These pocket populations may have a chance to survive if given proper protection by the alpine countries they have wandered into. Generally smaller and less aggressive than their North American conterparts, the European brown bears have not caused many problems when inhabing remote areas or particularly rugged terrain without much contact with humans. (13)

Another, even more surprising encounter with bears took place between the towns of Langelsheim and Seesen in Niedersachsen. In September of 1995, a local hunter spotted two brown bears slowly crossing a highway and called the police. An animal hunter was summoned and managed to catch the couple, which was then sent to a nearby zoo. No information was given where the two bears may have come from. (14)

Although the European lynx, *Felis lynx*, never posed any real threat to man or his livelihood, it was never-the-less hunted out of existence in Germany and most of central Europe. Now, after more than 100 years of extinction, the lynx has returned to Germany. (15).

In other parts of central Europe, such as the *Karpathes* and *Pyrenees* mountains, pocket populations managed to survive. Experts disagree, where the first specimens showing up in the *Bayrischer Wald* forests came from: whether they were secretly or illegally reintroduced into the wild or had crossed over from the Czech Republic. Increased environmental awareness has lead hunters and farmers to give the lynx another chance, and there are now between 10 and 15 lynx living in the *Bayrischer Wald* forest.. (16)

Another region that has reported the return of the lynx is the *Pfalz* region in South Germany. With a total size of 163,000 hectar, the *Pfälzer Wald* is the largest uninterrupted forest remaining in Germany. A number of sightings took place near the towns of Pirmasens and Landau. The total population of lynx, which are believed to have wandered over from the French *Vogesen* mountains, is estimated to be between 8 and 10 strong. (17)

2. Winter Visitors:

The unusually long and strong winter of 95/96 forced a lot of birds from cold regions to leave their home territories and move - at least temporarily - to more moderate climates. As a result, the citizens of Germany had the rare opportunity to see birds usually found only in the Scandinavia or the arcic circle.

When a Golden Eagle, *Aquila chrysaetos*, was spotted over the island of Fehmarn (Baltic sea) this caused quite a stir among ornithologists and nature

Animals & Men #12

Feature

lovers. After all, the bird has been extinct in Northern Germany for several hundred years and exists in central Europe only in a few secluded pockets in the southern Alps. (18)

Perhaps even more of a surprise were sightings of several **Snowy Owls**, *Nyctea scandiaca*, in northern Germany. Indigenous to the tundras of eastern Europe, northern Scandinavia, and the Arctic, this huge, near-white owl (wingspan up to 1.6 m or 5' 4") must have left quite a n impression on the witnesses who saw it.

The Glaucous Gull, *Larus hyperborus*, with a wingspan of up to 1.8 meters (6 feet) among the largest of all gulls, usually spends its time patrolling the waters of the northern Scandinavia looking for fish. Only occasionally, during the coldest of winters, does it come down to the continental European shores. This winter, however, it was seen as far south as Hannover (Niedersachsen), where unsuspecting witnesses probably didn't believe their eyes.

Even more surprising was the arrival of the **Iceland Gull**, *Larus glaucoidus*, in northern Germany. The German name („Polarmöve") indicates that it is usually indigenous to the arctic circle and the northernmost regions of Iceland and Greenland, but this particularly strong and long-lasting winter forced birds well-adapted to the cold, such as the Iceland gull, to look for a different place to spend the winter.

Another unexpected visitor from the arctic north to Germany was the **Pink-footed Goose**, *Anser fabalis brachyrhynchus*, which usually lives in Iceland and winters in Great Britain. No less surprising were sightings of **Great Northern Divers**, *gavia immer*, better known as the Common loon which calls the arctic waters of North America, Greenland, and Iceland its home.

Due to its size, beauty, and rarity, the **Gyrfalcon**, *Falco rusticulos*, belongs to the most highly valued birds of prey in the world: Arabic falconers pay up to 100,000 DM (US-$ 65,000) for a single bird. Normally, its territory is restricted to the arctic and some coasts of northern Scandinavia, but this winter there were a number of sightings in northern Germany near Hamburg. (19,20,21,22)

3. Cryptids: Exotic Intruders

Then, there are sightings of exotic animals that were never known to exist in Germany, weren't missing from any private or public menageries, but continued to be seen in different places before - usually - disappearing without a trace.

In the late 1980s, Germany experienced a wave of *Alien Big Cats* (ABC) sightings. Ulrich Magin, one of Germany's leading Fortean researchers, summarized the in a Fortean Times article and found that most of the sigtings took place in South Germany (Hessen, Saarland, Bayern). After mounting several large-scale, but unsuccessful, searches, the German police - ill-equipped to deal with Fortean phenomena - tried to solve the problem by claiming that there never was a panther to start with. (23)

Animals & Men #12 — Feature

Despite the police's claims to the contrary, the „panther" sightings didn't go away. In the early 90s, a new wave of sightings occured in the same general area as the previous one: *Saarland, Pfalz,* and *Bavaria*.

In June of 1992, several German newspapers reported sightings of big black cats in Germany. One particular incident made the headlines: In an open field near Saarbrücken, two policemen observed what they described as a „large, black feline," about 1 meter (40") long and what they declared to be a panther. (24)

Obviously, this big cat sighting was taken very seriously, since the professional football team of Bayern München (Munich), preparing for an upcoming game, switched to a different practice field.

Besides this newswire report, there were more sightings of „black panthers," mainly in South Germany. Either the cat(s) moved considerable distances in a short time, or the sighting „triggered" more encounters.

The following year, southern Germany got its next big cat scare. This time, it was „just" a cougar, that was spotted several times in the *Pfälzer Wald* near Landshut and Homburg. As usual in these kinds of sightings, nobody was attacked and the police mounted a large-scale - but unsuccessful - search for the beast. Later, they declared that it had to have been an escaped pet, although they couldn't find anybody missing one. (25)

In 1994, we went back to the black panthers, again. Two anglers, fishing in the „Schwarze Laabe" river in the *Bayrischer Wald*, reported seeing a black panther. Those big cats sure like southern Germany! No need to mention that nobody was attacked, the police did not manage to catch the cat ... (26)

There may be yet another explanation other than mass hallucinations, hoaxes, escaped pets, or hordes of exotic cats invading Germany: The *Bayrischer Wald Nationalpark* is one of the last big natural habitats left in Germany. Due to its size and physical proximity to the Czech border, wolves, moose, and even bears haved already crossed the border into Germany. Because of its appeal to wildlife, the Bavarian forest was selected as a site for the reintroduction of the lynx, who now thrives in small but steady numbers. Could it be that young adults, wandering off and looking for their own territory got a little „carried away" and ended up in other areas of southern Germany (where most of the big cat sightings occurred)? You couldn't blame Germans, not used to seeing either cougar or lynx, or any other big cats for that matter, for confusing the two during a surprise encounter in the wild!

While all those „dangerous" big cats minded their own business and didn't hurt anybody, a seemingly harmless Nandu (a south American relative of the African ostrich) „created havoc" in Germany. For two days, the big bird scared and even attacked pedestrians near Hellental (*Eifel* region) until the police finally took the reports seriously ans sent a patrol car to investigate. The two officers tried to subdue the bird, but it resisted arrest and tried to escape. One of the two officers chased the bird and actually managed to catch it (!!!). (I don't ever want to hear jokes about the physical fitness of our lawmen anymore.) Despite being scratched in his face by the powerful 1.5 meter animal and unable to get his handcuffs, the officer used his uniform tie to tie up the big bird. Apparently, this was too much for the Nandu, since it died shortly afterwards in an animal shelter where the police had brought it. A police spokesman later suggested that the bird had escaped from a private compound ... (27)

There were other, equally unbelievable reports of strange animal sightings in Germany, such as piranhas downriver from a nuclear power plant near Düsseldorf, but let me finish with a mystery that never was: crocodilians in Germany:

In 1994, news of a crocodile hunt made the headlines in German newspapers. *Sammy*, who was actually a caiman, had escaped when his owner, Jörg Zars, took him for a swim in a lake. The 4-foot reptile caused such a „panic" that the police was called to shoot the beast to protect the public. Although the police claimed they succeeded, several days later *Sammy* was caught unharmed and now spends his time in Falkenstein zoo, while his owner released a Rap CD to raise money for his legal bills: resort hotels had sued him for losses due to the „terror" that *Sammy* had caused around the lake.

A few days later, a dead caiman was found in a different, far-away lake; probably the casualty of some „copy cat," who had tried to get rid of his overgrown pet when the „crocodile hype" was still on.

In april of 1996, a Düsseldorf(Nordrhein-Westfalen) judge ruled the caiman ro be returned to Jörg Zars, where he will join *Sammy 2*, who was purchased as a substitute Zars (28)

At first, police in my hometown of Lennestadt (Nordrhein-Westfalen) thought it was a practical joke: „Help me, there is a crocodile in front of my apartment door." Again, the „crocodile" turned out to be a caiman - a scary twelve-incher! After interrogating the caller, it turned out he was actually the owner of the small reptile: his former girl friend had simply returned the little critter after they broke up and left it in a basket in front of his apartment. (29)

No wild crocodiles (or caimans) in Germany, it seems. Big surprise! But how about those sightings just west of Germany, in Belgium and Holland?

In Belgium, an angler fishing a lake for pike, caught what he thought was the biggest fish of his life: Pulling in the line, it turned out to be a 3-foot caiman instead. (30) Talking about „pulling:" maybe, somebody was pulling our leg here.

In Holland, *at least there really was a caiman*. After it escaped from its terrarium, the owner called the police, who started a large-scale search for the reptile. In the meantime, the Dutch media got wind of this, and turned the 3-foot caiman into a „ferocious beast coming straight from the jungle and viciously attacking and devouring anybody coming its way." The search was finally called off, and the caiman found - quietly and peacefully sitting in a closet of its owner's house. (31)

OUT OF THIS WORLD: EXOBIOLOGY

by Graham Inglis

OBITUARY: CARL SAGAN 1934-1996

Carl Sagan, longtime professor of astronomy and space science at Cornell University in New York, died in December aged 62. He was most widely known for his efforts in explaining astronomy and science to the public - particularly in his tv series Cosmos.

His entheusiasm for astronomy spanned virtually the whole of his life. His first major research was on the atmosphere and surface conditions of Venus, at a time when our knowledge of the planet was skimpy: this was before the days of the Venera spaceprobe landings. His conclusions were a major contribution to Mankind's newly-emerging awareness of a phenomenon later dubbed "the greenhouse effect".

Sagan was not only an astronomer. He'd gained an MSc in physics (1956) and studied biology and genetics, and, in 1983, together with biologist Paul Erlich, first proposed the "nuclear winter" theory - dust from enough nuclear strikes could reduce sunlight and put all life on Earth at risk. This was politically "hot stuff" at the height of the cold war.

Sagan's forays into anthropology led him to explain much of human behaviour in terms of the 'R-complex' - that part of the brain we have inherited from our remote ancestors, the reptiles. While humans have evolved extra 'layers' of the brain, such as the cerebral cortex, the R-complex remains responsible for base instincts like territoriality, greed and agression - still very much part of human behaviour today.

Despite this, as millions know from *Cosmos*, Sagan was essentially an optimist. Most astronomers automatically dismissed Jupiter as a possible life-bearing planet. Not Sagan. He postulated an entire ecosystem in the ammonia/helium cloud deck, including floating animals operating like hot-air balloons. When assessing the liklihood of life elsewhere in the universe, Sagan, unlike many scientists, tenaciously believed it thrives in abundance.

OBITUARY : CLYDE TOMBAUGH 1907-1997

Clyde Tombaugh, the discoverer of Pluto in 1930, died in January aged 90. Tombaugh conducted a long and systematic search in an area of the sky calculated to be the likliest area for Planet X - a planet gravitationally tugging at Neptune and Uranus. However, it was soon realised that Pluto was not the Neptune-perturbing Planet X after all: it's too small. Tombaugh's discovery was, though, an important link in the ongoing investigation of the mysterious outer edge of our Solar System.

Sources:

1. *Rotkäppchen und der Marderhund.* Bild der Wissenschaft, p. 14-19, 3/96
2. *The wolf packs are back.* Fortean Times 76:16
3. *Hilfe, die wilden Tiere kommen.* Bild am Sonntag, January 21, 1996
4. *Wieder Wölfe in Deutschland,* Die Pirsch, 20/96.
5. *Aufregung um getöteten Bayernwolf,* Die Pirsch, May 1995
6. N-TV, April 6, 1996, 6:30 pm
7. *Wolfs-Jagd in der Eifel.* Bild. March 25., 1995
8. *Eifel-Wolf erschossen: Plötzlich sind alle traurig.* Bild, April 10, 1995
9. *Wilde Bestie zerreißt 70 Schafe.* Westfalenpost. August 24, 1995
10. *Schweizer jagen Wolf und Büffel.* Westfalenpost. February 8, 1996
11. SIELMANN, HEINZ (Hrsg.): 1995, *Das große Buch der Tierwelt*, Weltbild Verlag, Augsburg, p. 128 ff
12. *Rotkäppchen und der Marderhund.* Bild der Wissenschaft, p. 15, 3/1996
13. REICHHOLF, JOSEF: 1995, *Die farbigen Naturführer - Säugetiere*, Time Life Bücher, Amsterdam, p. 136 ff
14. *Braunbär-Pärchen geht spazieren.* Westfalenpost, September 9, 1996
15. REICHHOLF, JOSEF: 1995, *Die farbigen Naturführer - Säugetiere*, Time Life Bücher, Amsterdam, p. 176 ff
16. *"Pinselohr" kehrt zurück.* Die Pirsch, 22/96
17. *Luchs aus Vogesen zugewandert.* Die Pirsch, 20/1996
18. *Steinadler kehren zurück.* Bild, February 8, 1996
19. *Für Vögel vom Polarkreis ist auch das winterliche Deutschland mild.* Welt am Sonntag, February 18, 1996
20. PERRINS, CHRISTOPHER: 1995, *Encyclopedia of Birds*,
21. STEINBACH, GUNTHER: 1995, *Die farbigen Naturführer - Wasservögel*, Time Life Bücher, Amsterdam
22. STEINBACH, GUNTERH: 1995, *Die farbigen Naturführer - Landvögel*, Time Life Bücher, Amsterdam
23. MAGIN, ULRICH: *The Odenwald Beast.* Fortean Times 55:30-3
24. Westfalenpost, June 28, 1992
25. *Puma streift durch Pfälzer Wald.* Westfalenpost, November 30, 1994
26. *Schwarzer Panther.* Kölner Express, October 6, 1994
27. *Eifelsafari: Polizist fing Vogel mit Schlips.* Kölner Express, May 23, 1995
28. *"Sammy darf nach Hause.* Kölner Express, April 24, 1996
29. *Krokodil wartete vor der Tür.* Bild, December 7, 1995
30. *Krokodil geangelt - und das in Belgien.* Bild, June 2, 1995
31. *Krokodiljagd in den Niederlanden.* Westfalenpost, August 23, 1995.

New Discoveries of yet MORE bird-like Dinosaurs.

by Darren Naish

Few non-specialists realise just how many types of dinosaur are presently known (about 800 genera), or how often new kinds are discovered. Presently, about one new species is discovered every two months - a very high rate considering that, just a few decades ago, some people thought that there were no new dinosaurs to be found. Furthermore, this rate of discovery is not slowing down, but speeding up! Clearly we have a long way to go before we can view Mesozoic ecosystems in their entirety. In recent months, a surprisingly high number of smaller, bird-like dinosaurs have come to light. These provide yet more irrefutable evidence for the notion that birds are dinosaurs, and part of that bipedal, mostly carnivorous group; the Theropoda.

Theropods first appeared in the late Triassic, (approx. 225 million years before present): already they were superficially bird-like with short, stiff bodies, long curved necks and long legs. It was not until the late Jurassic (approximately 150 m.y.b.p), however, that small climbing forms began to experiment with true flight: at this time the first member of Aves, Archaeopteryx, appeared. Previously it was thought that Archaeopteryx lithographica and the strata in which it was found were of Tithonian age, but recent dating studies suggest instead that they are Kimmeridgian. (Figure One) - thus the appearance of birds is pushed back a few million years. Also, in 1995, Andrea Weigert reported Archaeopteryx-like teeth from the late Oxfordian-early Kimmeridgian of Portugal. A second Archaeopteryx species, A.bavarica, described by Peter Wellnhofer in 1993 is younger than A.lithographica and is probably uppermost Kimmeridgian in age, while a new specimen from Korea, reported in 1994, dubbed the 'Korean Archaeopteryx' may extend the range of these birds into the earliest Cretaceous. (Fig. 2).

Fragmentary bits and pieces from the late Jurassic of Wyoming and the early Cretaceous of Romania have also been referred to Archaeopteryx, but in most cases they have turned out to be either indeterminate or pterosaurian (pterosaur bone fragments often look very bird-like). There has also long been debate over whether Archaeopteryx lithographica itself contains two additional taxa: Archaeornis siemensi (Dames, 1897), for the Berlin specimen, and Jurapteryx recurva (Howgate, 1984), for the tiny Eichstatt specimen [1]. Further complexity is added by the fact that the name Archaeopteryx lithographica (Von Meyer, 1861), seems to have been first created for an individual feather found in 1860 [2], rather than for a complete skeleton. However, yet again a dinosaur group known initially from a single taxon is found to be more diverse than first imagined.

Considerable debate still exists over whether the Archaeopterygids were tree or ground dwellers, though it is now generally agreed that they were capable of competant flight. Those who support the ground-dwelling theory point out that the theropods closest to the archaeopterygids were ground-bound too. However, a minority opinion holds that all dinosaurs are descendents of small, tree-dwelling forms which were the direct ancestors of birds, and if this is correct we are simply missing small, tree-dwelling dinosaurs that predate birds from the fossil record.

Evidence that bird relatives were primitively arboreal may come from a new specimen, as yet undescribed, but now displayed on the www. Discovered accidently by the Linster family of Stephensville, Montana, this small theropod, one of

Animals & Men #12 — Feature

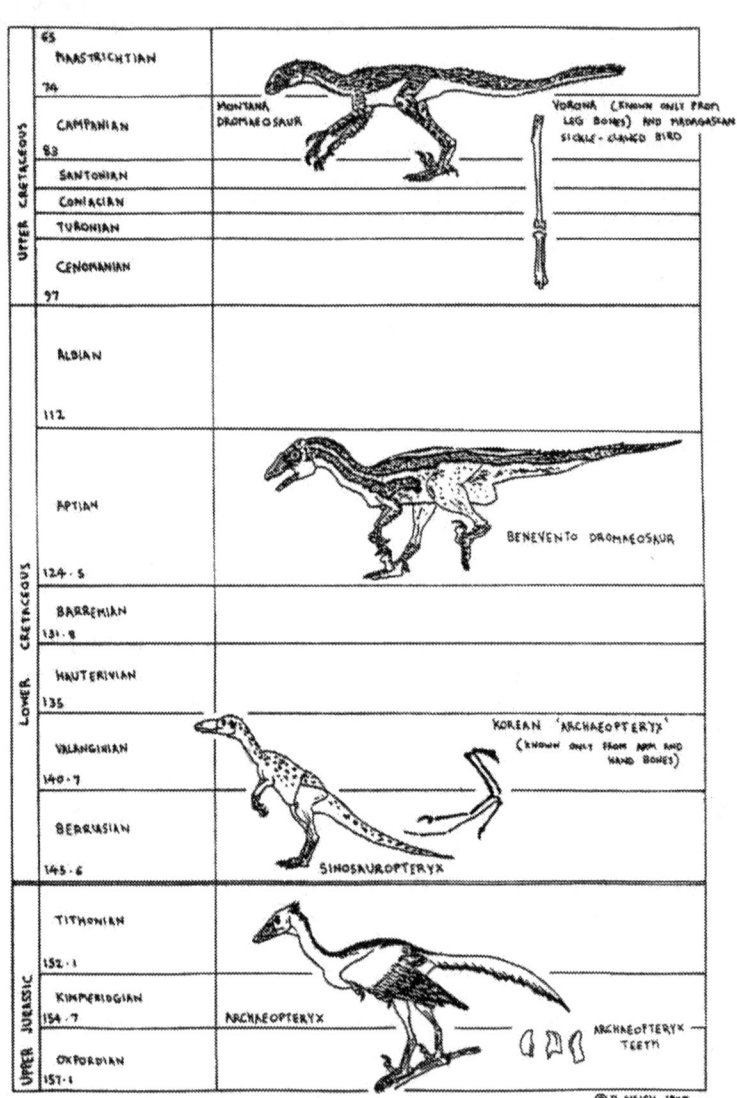

Figure 1.
Chronological distribution of the animals mentioned in this article.
The numbers represent millions of years before present.

the dromaeosaurids - the most bird-like of all non-birds - is represented by a virtually complete specimen. It is only about 25cm tall at the hips, and has extraordinarily long arms and legs, a relatively small skull, and a large fercula (wish bone). Its teeth and other features suggest that it belongs to the genus *Saurornitholestes*, named by Hans-Dieter Sues in 1978.

Despite this animal's small size it does not appear to be a juvenile. Juvenile vertebrates, including dinosaurs, have proportionally large tall skulls with huge eyes. This dinosaur does not. Also, in juvenile dinosaurs, elements like the fercula tend not to be ossified, whereas in this specimen they are. For full details we shall have to wait for the complete description: this should be published early in 1997.

This animal's foot is especially important. Like all other dromaeosaurids (and primitive birds including *Archaeopteryx*), it has a raised second toe which supports a 'sickle claw'. In large dromaeosaurids, this was a devastating and awesome weapon with which they could slash open the bellies of prey dinosaurs, but in these small forms it was relatively feeble. Perhaps it first appeared in arboreal forms as an aid to climbing. More importantly, the first toe (the hallux), of this new specimen is very low down on the foot and may have been opposable to the other toes. In birds, such an adaptation allows them to grasp branches. If this feature evolved before birds did, it may be further evidence that the theropod ancestors of birds were already in the trees.

Another tiny dromaeosaurid, but this time from the Aptian of Benevento, Italy, was described by Leonardi and Teruzzi in 1993 (it also got front-cover billing in the Italian magazine *Oggi*). Unlike the Montana dromaeosaurid, it has rather short limbs and its proportionally large skull (shaped very much like that of *Archaeopteryx*) indicates its juvenile status. It represents an as-yet-unnamed new genus. Its feet are missing, so we cannot tell whether it had an opposable hallux like the Montana specimen, but it is exceptional in that it has a soft tissue outline preserved.

This reveals that dromaeosaurids did not have wing membranes (called patagia) as one authority, David Peters, has suggested.

Though telling in that regard, the preserved outline of the Benevento dromaeosaur does not preserve any integuement.

As announced in A&M 11 (pp. 10-11), a small, apparently feathered theropod, informally named *Sinosauropteryx* was announced in October 1996. News articles also appeared in the *New York Times*, Science 274: 720-1 and the January 1997 *BBC Wildlife*). A second, larger specimen was reported soon after in early December - this one is apparently 'covered' in small feathers. As of writing, no study has conclusively proved that these theropods are definitely feathered. If they are, we can safely assume that dromaeosaurids - being even more closely related to birds than *Sinosauropteryx* - were clothed in feathers too.

With newly discovered dromaeosaurids being more bird-like, it should not come as a surprise that newly discovered birds are more dromaeosaurid-like. David Krause, Katherine Forster and their colleagues have recently discovered a rich Campanian dinosaur fauna in the Mahajanga Basin, Madagascar. By Campanian times, Madagascar was an island - it separated from Africa late in the Jurassic, but remained linked with India until later on (finally becoming isolate about 88 m.y.b.p). It was home to unusual endemic dinosaurs.

Among these are three new birds. One, *Vorona berivotrensis*, was described in 1996 but is known only from a few leg bones. The other two birds remain presently undescribed, but one of them is known from a virtually complete, albeit headless specimen that has been displayed at several palaentological conventions. Like dromaeosaurids it has a sickle-claw - an unexpected feature in a bird of this age. Presumably, it is therefore a kind of anachronistic relict that survived in the isolation of Madagascar, and its sickle-claw may be further evidence for the close relationship between dromaeosaurids and birds (it is, however a

Animals & Men #12 — Feature

possibility that the sickle-claw in this form is convergently evolved: the seriemas (Cariamidae), a relatively recently evolved group clearly not closely related to dromaeosaurids, also have a sickle-claw).

And another new theropod, an Upper Cretaceous dromaeosaurid from Patagonia has hips virtually indistinguishable from the Madagascan bird. Fernando Novas, the palaentologist who will describe this South American theropod, further reports that it has very bird-like arms and shoulders: it could tightly fold its arms as birds do their wings.

The fact that many of these very bird-like dromaeosaurids appear after the earliest birds has led some workers to suggest that they are actually descendants of early birds like *Archaeopteryx*, rather than late-surviving descendants of bird ancestors.

This former view, first suggested by Greg Paul in 1988, was recently (1995) supported by Mongolian theropod expert Andrzej Elzanowski who showed that the bizarre Upper Cretaceous oviraptorids, another group of bird-like theropods, were actually more like modern birds than *Archaeopteryx*[3].

However, *Archaeopteryx* and other birds have been shown to posess features that other theropods do not.

Also, the idea that bird-like theropods did not debut until after birds is not true: a number of recent discoveries (mostly of little teeth), have now shown that they were around before birds after all.

Overall, disputes are clearly far from over. - as highlighted by another group of newly discovered theropods, the mononykids. More about them next issue.

* * * * *

Notes.

1. Both of these are almost certainly not distinct enough to be representative of different taxa, though some niggling features suggest to me that *Jurapteryx* may perhaps be different enough to rank as a different species at least (in which case it would be *Archaeopteryx recurva*).

2. The skeletal remains we associate today with the name *Archaeopteryx* first turned up in 1861 (the London specimen), and though referred by von Meyer to the same taxon as the feather, it is possible that the feather came from another animal - in which case, technically the name *Archaeopteryx* does not belong to the animal that we know today as *Archaeopteryx*.

No doubt, if this turned out to be the case, the London specimen would be strictly defined as the type for *Archaeopteryx*. How pure and simple taxonomy is.

3. Also worthy of note in the context of theropod-bird relations is the discovery of several oviraptorid skeletons from the Upper Cretaceous of Mongolia preserved lying on top of nest-filled eggs.

One such find was described in 1995, and another in 1996. Embryonic remains described in 1994 had already shown that certain fossil remains from the Upper Cretaceous of Mongolia, previously thought to have been laid by the primitive horned dinosaur *Protoceratops*, actually belonged to an oviraptorid. More about oviraptorids in a future issue.

Figure Two: Some of archaeoptrygid diversity. The life restoration is of *Archaeopteryx lithogaphica*: in life about 50cm in length and about 300g in weight. The uppermost skeleton is of the only known skeleton of *Archaeopteryx bavarica* (about 40cms total length). It is characterised by an ossified sternum and especially long legs. The skeleton on the left is the Eichstatt specimen named as *Jurapteryx recurva* by Howgate (1984). It is thirty cm in total length and in life would have weighed about 69g. The illustration to the right represents all that is currently known of the so-called 'Korean *Archaeopteryx*'. Three fingers and the arm bones are preserved: the relative proportions of them show that this bird is not the same as *Archaeopteryx*. *A. bavarica* after Wellenhofer 1993, Eichstatt and Korean specimens from photographs.

NORTH OF THE BORDER

(home of the free range haggis and the tastefully boxed souvenir prophylactic in Balmoral Tartan for our visitors from Japan)

with

Tom Anderson

'He sure as hell doesn't hide his light under a bushel' (The Undertaker's Review 1922).

The first known carnivore fossil found in Scotland has been unearthed on the Isle of Skye. It is believed to be the vertebra of a theropod of the Jurassic period. The 175 million-year-old bone is reported to have belonged to a 10m upright carnivore similar to Tyrannosaurus Rex and a further search is underway to locate the skull and teeth.

The Black Grouse shooting season opened on August 20th last summer. In the last twenty years their breeding grounds have shrunk by 28%. Their numbers have diminished over the last century from around 500,000 to around 35,000, the vast majority of them living in Scotland whereas once they lived as far south as the heathlands of Surrey. Their much rarer cousin the Capercaillie, has diminished by 64% over the last 25 years and now numbers only about 10,000. Apart from being shot out of the skies, the major hazard to both species are the high, wire deer fences which claim around 10,000 birds a year. These are essential to protect newly planted trees and areas of natural regeneration, but are a 'wall of death' for black grouse and Capercaillie acording to the British Trust for Ornithology who claim that they are seriously jeopardising the survival of both species.

The 'glorious twelfth' last year for Red Grouse was most certainly not! Stocks are so low that some shooting estates never opened, and one holding larger stocks stipulated that only flintlock guns were to be used to make it more of a 'challenge'. Another way of redressing the balance would be if every fifth bird had been fed Semtex H with their seed - the resulting conflagration would not only add a piquancy to this dreary pastime but create employment in new fields,. Asbestos roofs for Range Rovers, fireproof tweed, and non-combustible spaniels to name but three!

NOW THAT'S WHAT I CALL CRYPTO

Another excerpt from that legendary compilation album of zoomythological rock'n'roll mayhem..

by Neil Nixon

Nestled somewhere in the grey area in which dance, ambient and avant garde collide are the idiosyncratic talents of Ultramarine. Generally a duo, the band have been augmented from time to time by others. On their Bel Air album, released in 1995, they allowed a similarly idiosyncratic duo - known as Pooka - to contribute a few guest vocals and lyrics.

The four wayward talents fashioned a handful of hypnotic soundscapes with ethereal voices, one of which - Mutant - stands as a celebration of strange life. Marine's strength is the development of bubbling walls of sound created with the latest dance technology but ripe with acoustic elements and samples. The floating vocals on 'Mutant' overlay the incessant sound with a series of cliches.

"He's a tall and ugly stranger/but I like the danger".

Elsewhere, we get a series of potted facts that fire the imagination without spelling out too many specifics. His body is stretching all over the floor, his genitals are ping-pong balls, he smells like the dentist, he's naked etc.

Ultimately it explores the idea of 'Mutant' and leaves open every possibility from the spawn of some Sellafield saddo to a primeval throwback mocking evolutionary theory and scientific order as he oozes slime from the tips of his throbbing tentacles.

Then again, maybe they got the inspiration from a picture of Andrew Lloyd-Webber!

Next Time: The Revolutionary Bigfoot Diet.

EDITOR'S NOTE: Pooka are not only a fine band with a name oozing in cryptozoological and fortean symbolism but their last two CDs were the victim of the ritual "sharing out the CD collection" that took place when Alison and I parted company. If anyone comes across CHEAP copies for sale grab 'em for me please. Thanx.

TIM DINSDALE'S FILM - ANOTHER LOOK
by Richard Carter.

For thirty years now a piece of 16mm Cine film has been acclaimed as the proof for the existence of a large creature living in the depths of Loch Ness. The film was taken by Tim Dinsdale who, after reading about the Loch Ness mystery made a trip to the Loch in the hope of filming the monster.

On the morning of 23rd April 1960, his last day at the Loch, he was making his way back to the Foyers Hotel from Fort Augustus when he saw an object moving across the loch. It then turned left and continued down the loch. He filmed it with his 16mm Bolex camera.

This piece of film became the centre of many debates until in 1966 the Joint Air Reconnaissance Centre (JARIC), a unit of the Royal Air Force carried out their own examination of the film, and then published a report.

This report convinced most people that the film showed an unknown object swimming in the loch. After reading Tim's book and studying the JARIC report it is my contention that the film actually shows a local fishing boat filmed under dull weather conditions. The JARIC report starts by stating that the film was taken from a point three hundred feet above the loch, but if this measurement is wrong and the height was actually only 290 feet then all other measurements would be in error by +10% I have reached this figure because the vertical angle over one mile works out at only 0.1 feet per 100 feet of height which would not make a difference to the size of the object.

Their next observation is that between frame 700 and frame 1700 the speed of the object was 10 m.p.h and it covered 630 feet. Both the speed of the object and the distance covered MUST be incorrect. The camera was set at 24 frames per second. The gap between frame 700 and frame 1700 is 42 seconds of film time. No mention was made of the fact that the camera was clockwork driven and needed to be rewound every twenty seconds - twice during the duration of the sequence. Each rewinding takes around 12 seconds This adds 24 seconds to the running time which would reduce the speed of the object to six and a half miles per hour - within the speed obtained by local fishing boats.

The report then goes on to state the difficulty of measuring the object whilst it is travelling away from the camera. It also says that between frames 1-384 the object was travelling away from the camera at a speed of ten miles per hour and covered 240 feet.

They then discuss the size of the object describing it as black, triangular in shape with a base of 5.5 feet and a height of 3.7 feet. Because of the angle it was filmed from this would be very near to the true size of the object. This corresponds with the dimensions of a fifteen foot fishing boat with a pilot, seen moving away from the observer.

Just to confuse things further their next calculation is taken from frames 816 to 1440 when the object is travelling parallel to the opposite shore. They give a speed of 7 m.p.h - again a speed obtainable by a local fishing boat, but JARIC say that the speed might be as fast as 10 m.p.h even though the object was travelling across the film, because of the difficulty of estimations at that distance. They were, however, happy with their own calculations made whilst the object was moving away from the camera.

When JARIC start to interpret the film, their first mistake was assuming that the 'marker' boat was filmed under exactly the same weather conditions. It is obvious to anyone viewing the film that it was made on a much brighter day.

Animals & Men #12   Feature

The main objection they have against the object being a fishing boat with a non-planing hull is that a fishing boat of fifteen feet could only achieve a speed of between 6.5 and 7 m.p.h. Apart from the one measurement between frames 1 and 384 they give no reason to suppose that the object was travelling any faster than that. Their only argument against the object being a boat with a planing hull is that most speed boats are painted so as to be photo visible at all times. Tim Dinsdale, however, said that the object was reddish-brown at a distance of nearly a mile. The JARIC report ends by trying to estimate the size of the object beneath the water. They don't even suggest the possibility that it could be a floating object!

When Tim Dinsdale arrived at the loch on the evening of the 17th April 1960 he had two cine cameras with him. The 16mm Bolex loaded with black and white film and an 8mm cine camera loaded with colour film. He also had a 35mm still camera loaded with colour film. He never used either.

Whilst driving from Dores to Foyers he saw an object floating in the Loch. He swung his car across the road and prepared to film it. At the last moment he looked at the object through his binoculars and noticed a single leaf blowing from what turned out to be a floating log. If he had not used his binoculars would this too have become a film of 'The Loch Ness Monster'?

On the evening of the 21st April he set up his camera on the hill behind Foyers Bay. Suddenly he saw a disturbance at the mouth of the river. He focussed his 16mm Bolex camera and filmed the object for about 30 seconds. He then decided to go closer and drove down the zig-zag road to the river mouth. He then stated that he had got film of the monster 'in the bag' but little mention is made of this film.

This does suggest that he was 'hyped up', because from that vantage point he could have studied the object through his binoculars, or even used one of his back-up cameras to capture the creature in colour.

At 9.00 on the morning of the 9th April Tim was cruising towards the Foyers Hotel when he saw an object cruising across the Loch. With his naked eye he could see that it was reddish-brown in colour and through his binoculars he could see a dark mark on its side. When he first saw it the object was sideways-on but it turned away from him although he could still see it from end to end. This would, of course be impossible from this angle at a distance of nearly a mile. As the object started to move he turned his 16mm camera on it and started to film in short bursts like a machine gunner. The object then turned across the loch, then turned left and moved parallel to the far shore. It was at this point that he saw foam breaking the water, which he later likened to paddle strokes. When the film is examined, however, these look more like the bow of a small boat being bumped up and down in the water as it travels at top speed.

He then decided to try and get closer and 'shot off down the zig-zag road like a rocket', leaving a cloud of dust behind him and sounding his horn. He missed his turning and had to circle a group of houses. This he did with his tyres squealing but when he reached the water there was nothing to be seen.

Was it because the monster had returned to its lair? Or was it because a small fishing boat would be very hard to pick out at water level? He later sent out a marker boat for comparison, but it was very much later and the sky was much brighter. He 'clocked' the speed of the marker boat as 7 mph as it returned to its base. On his way home, before he developed the film he stated that he had "caught it by its tail and that no power on earth" would make him let go.

I think that it is quite clear that what has been taken as our greatest piece of evidence for the existence of the Loch Ness Monster might not, after all, be anything more exciting than someone out after his early morning fish. I hope to return to the Loch this year and reproduce this film with both cine and video cameras. I hope that this will help to solve the mystery once and for all.

As an active Loch Ness investigator I feel that it is important to analyse all that is taken as evidence so that we know what is truth and what is not, and only then can we hope to try and solve the mystery!

EDITOR'S NOTE: Richard Carter and Richard Askew are planning an expedition to Loch Ness in July. They are looking for people to go with them and they can be contacted c/o the editorial address.

On a personal note, I have come under quite a lot of criticism vis-a-vis my attitude towards the Loch Ness Monster. On the Weird CD ROM I state my opinions that it is not possible to explain the sightings at Loch Ness within a purely zoological framework. I stand by this view. When Doc Shiels said, "*I don't believe in the Loch Ness Monster even though I've seen the shagger and photographed it.*" on my last CD ('Invocation of my Daemon Brother') he was stating what I truly believe to be the case.

Whilst it is certain that, as Neil Arnold stated in the 1996 Yearbook, there are some outsized specimens of known species within the loch, and whereas I am also convinced that there are more sightings of seals than is generally credited, the main body of 'Nessie' sightings can only, I feel, be explained within zooform terms of reference. Richard Carter's masterly analysis of the Dinsdale film can only, I feel, support this hypothesis.

The editorial team are aware, however that this is a contentious viewpoint and would welcome letters on the subject from people on both sides of the metaphorical fence. The irony is that whereas, theoretically, at least - if there is a living creature of an unknown species in Loch Ness this COULD be proven if a specimen is secured - if my theory is correct then the case will never be settles to anyone's satisfaction!

A ZOOLOGIST-FRIENDLY GUIDE TO BOAT HULLS

Planing, in the boating context of the previous article, is similar to road "aqua-planing". Driving a car too fast in wet conditions causes the tyres to lift up from the tarmac and onto the water surface - and grip is immediately lost.

Non-planing boat hulls stay submerged; planing hulls rise to the water surface when the revs are on.

- - -

Non-planing hulls are found on many small boats, on barges, log canoes, and on passenger liners.

A typical non-planing hull

Fairly flat, or a shallow U shape, such hulls trudge through the water and have a well-defined limit to the speed at which they can be "rammed" through the water ahead. If an over-powerful engine were fitted, the boat would rear up and could even somersault backwards.

Planing hulls, more V-shaped at the base, are found on most speedboats. Such a hull lifts almost horizontally up to the water surface as speed increases.

A typical planing hull

Skilful driving of a boat with a planing hull requires the pilot to maximise submersion of the propeller(s). If too much thrust is applied, the boat keeps taking off, like a skimmed pebble, with the propeller at the back over-revving every time it emerges into the air. It's a waste of petrol - but a great way of pulling the chicks, though.

Family, Friends and Out of Place Animals

by Neil Arnold

She opened the door of her flat and was horrified at the sight before her. The 'thing' was like something from a horror film. She stepped back into the doorway and almost shrieked as the wings fluttered and the form wriggled. It was roughly six inches in length and appeared to be similar to a centipede. However, it was certainly unlike anything she had ever seen before. It was a browny-green colour, had many legs and 'feelers' and appeared to have no fear. It was too big to be a normal British insect. Perhaps it had come from another country, either imported by mistake or an escapee from a private collection. Whatever the case, she could only describe it as 'weird' because it appeared to be too hideous for normality. It was a vile little thing that squirmed and fluttered but it didn't go anywhere - simply staying by the doorstep.

The flat was quite high up and the balcony overlooked a pile of rubble as well as mattresses and other rubbish. This heap of trash was to be the maze in which the creature would escape.

Very reluctantly she struck out at it. She noted that it appeared to be solid and seemed to be unflustered by the contact. Each time she moved towards it, she cringed and flinched, but eventually succeeded in moving it. She pushed the creature over the balcony and it fell between two of the old mattresses and disappeared into the dark cavity.

Just after this experience she rang her father who came over to her flat to try and find the creature. It was not to be found but just by looking at his daughter's frightened face he could see that this odd encounter had really happened.

This may seem like an excerpt from a horror novel but it is in fact a true case that happened to a woman I know. She is the daughter of a man I visit regularly and although they have had many strange experiences, this one in particular rates highly on their list of natural, though peculiar occurrences.

It seems that the creature was a type of insect, but certainly not one that is at home in the United Kingdom. The woman concerned is certain that this was no dragon-fly, wasp, centipede, spider or beetle. Even if she was mistaken it would appear that the animal was very oversized. The woman was horrified at what she saw and to this day has not changed her story.

EDITOR'S NOTE: *There is one other story that I know of within the annals of forteana which has several startling parallels to this one. It is taken mostly from 'Mystery Animals of Britain and Ireland' by Graham McEwan (Robert Hale, London, 1985). Borley Rectory in Suffolk, popularly known as the 'most haunted house in England' burned down, under mysterious circumstances in early 1939. The summer before it was destroyed, a lady was painting in the garden when she had a curious experience:*

"Looking up I saw the queerest object with impelling eyes advancing toward me at about eye level. It seemed to be coming out of a mist. It was accompanied by a wasp on its left..."

Alarmed, she hit out at the creature, knocking it to the ground. Worried, both that she might have injured it, and that it might attack her, she searched around for it, but it seemed to have disappeared. She later described it as being "quite three inches in length, its body entirely black and composed of sections enabling it to expand and bend at ease.

Animals & Men #12

Feature

"Its eyes were large and the colour of bloomy black grapes"...

The picture she drew appears to be of a long serpentine insect with huge eyes. It appears to be anatomically correct; the wasp in the same picture is perfect, but like the animal described by Neil Arnold, it does not correspond to any recognisable insect species!

The next case is even more spectacular and is a lot stranger than the last. For some it may even border on the ridiculous! This is probably a paranormal occurrence rather than a known, though out-of-place animal. This account concerns the father of the woman described above and suggests the existence of 'little folk'.

One night the fellow in question was sitting at home talking to two other people, one of whom may have been his daughter, although his memory is a little frail where names and dates are concerned. The two guests were sitting on a couch facing the mantlepiece whilst my friend, whom I shall call Mr X in order to protect his identity was sitting in his favourite armchair facing in the same direction. During the conversation Mr X noticed a movement on the mantlepiece. He stood up to get a better view and was amazed at what he saw. Of course, by this time the two women on the sofa had spotted the thing and were watching it!

Mr X said that the creature was a pale colour and appeared to be lying down. It then stood up in the way that a person would stand a cigarette lighter on its end. He spotted no legs or arms but he did see wings which flickered quickly. He also saw a tiny face, similar to that of a human. As the creature stood up Mr X sought his camera. The creature began to hover away from the mantel and Mr X took a photograph. The creature hovered as it flew and made its way towards the two women on the sofa, but when it got there it simply vanished.

This may seem an unlikely story, but three witnesses enjoying a cup of tea seem to have no reason to make it up!

The photograph came out but the 'creature' is just a white blur against a background of flowery wallpaper.

Mr X said that it was quite tiny, about the size of his thumb, but he laughs whenever people bring up the subject of fairy phenomena. However the creature did appear to have a human face although other characteristics appear more insect like. the fact that it vanished suggests that it was psychical in nature, and it must be said that most people who witnessed such a thing would fear for their sanity! There is no doubt, however, that SOMETHING was seen. A fairy? A ghost? An insect? Who knows? This was no figment of the imagination and I don't think that Mr X and his companions were drinking 'herbal' tea!

The next few anecdotes will seem comparatively normal when considered alongside the first two. I feel, however, that they are worthy additions to the files of Fortean Zoology.

Everyone has heard of the monster of Loch Ness and the following account is one of many encounters with the animal. A friend of mine was privileged enough to travel the waters of the lake because he knew someone who owned a boat. Although he had been camping at the side of the loch for some time he was unlucky and had not seen the animal. His friend who owned the boat was, however, fortunate enough to see something which certainly seems to have been the elusive 'Nessie'.

One night, whilst fishing the waters, he noticed a disturbance - something so great that it could not have been caused by a normal fish. The commotion was caused by a long neck which rose out of the water before crashing back down and disappearing into the depths. Whatever this thing was it terrified the boatman who quickly came ashore. He now thinks that he had a lucky escape because his boat would have turned over quite easily had the creature been any closer.

The legend lives on - despite the hoaxes and the

sceptical viewpoints, which are slowly erased by the continual stream of sightings.

The final two incidents involve a certain man who encountered what may have been Alien Big Cats. Unfortunately he never caught a decent view of the things and tends to attribute them to ghosts rather than to mysterious feline prowlers.

The first encounter occurred in Richmond Park during the early 1950s. He was in the army at the time and his camp was within the park which was a large one inhabited by deer.

One night he was late returning from London and had to climb over the wall or fence surrounding the park. When inside he noticed that the deer were agitated. Suddenly several deer came running past him pursued by what appeared to be a great white hound. This horrific creature had huge teeth but he could see no other distinguishing characteristics.

The weird thing noticed by the alarmed man was that the creature appeared to be a foot above the ground as it ran. To this day he is convinced that whatever it was, it was spectral in nature because of its hideous appearance and huge size. The deer, however saw it as well!

EDITOR'S NOTE: For a similar story concerning spectral hounds 'walking on air' see my article 'Hellhound on my Trail', in issue four of 'The Goblin Universe'.

This spectral beast may have some sort of connection with a famous ghost which apparently haunts Richmond or Kingston Hill.

The ghost is reputedly of a ghostly knight mounted on a white horse. I don't think that the knight caused that disturbance but connections between haunted sites and mysterious animals are common.

The final story took place in the back garden of the man who saw the giant hound in Richmond Park. Here, again, the 'creature' may have been supernatural in origin, but we cannot be sure because whatever it was - it was simply too fast to be seen.

The man was strolling down his garden path when he was frightened to hear what appeared to be two whip cracks behind him. These sounds were extremely loud but he continued his walk. He was looking for his pet cat but what he heard and felt was certainly no normal creature. Further down the path an invisible 'presence' seemed to hurtle past him and he heard an unearthly scream like that of a big cat.

He described it as a 'banshee wail' and he felt something very large brush past him. Soon after he found his cat which was hiding in a shrub. The cat was wide eyed and would not come out from its shelter.

There is a fine line between natural and supernatural. Out-of-place creatures are extraordinary and can't all be explained as normal animals. Some appear to defy nature herself with the ability to vanish or other weird characteristics. More natural creatures such as huge hounds or mystery felids seem to have unnatural connections which make it difficult for observers to determine exactly what they are.

The Loch Ness Monster may be a living dinosaur but then again despite the fact that it appears solid enough, it could be a spectral creature!

I have no doubt that Big Cats roam parts of Britain and that other creatures are metaphorically in 'the wrong part of town', but we must not close our eyes and ignore the more wondrous possibilities simply because many of the alien animal encounters seem a little beyond belief. Maybe some people would rather remain sceptical and cast aspersions upon a person's sanity rather than accepting those sightings with more supernatural aspects.

Not all these encounters are the result of the mind playing tricks, but then again not all hounds 'run on air'.

Animals & Men #12 — Feature

LETTERS

Disclaimer:

The Editor welcomes letters on any subject of interest to readers of this magazine. However, he reserves the right to edit and/or omit letters as appropriate. He would also like to stress that opinions expressed by correspondents are not necessarily those of the editor or the magazine staff, and would also like to stress that whilst every care has been taken not to infringe anybody's copyright or to libel them, any such offences are the responsibility of the individual author rather than of the editor, the editorial team or The Centre for Fortean Zoology.

Dear Mr Downes,

The article reproduced below was published in the Daily Mail on the 29th May 1990. Since then I have never heard any more news about the discovery. I would be very grateful if you, or any of your readers could tell me if the skeletons have been identified.

Yours sincerely,

Roderick Moore
Liverpool.

Riddle of sea cave monsters

SYDNEY: Scuba divers have discovered the remains of creatures unknown to scientists.

Experts are baffled by initial examination of photos of the monsters from an underwater cave off the Fijian island of Matagi in the Pacific. 'They bear no resemblance to any marine creature I know,' said diver Kevin Deacon, who has videoed the remains of four creatures amid coral and seaweed. The two largest are 30ft overall with yard-long skulls. 'They look more like a land animal or an amphibian,' he said.

STEVE MOORE ALSO KNOWS THE SCORE

Dear Jon,

Just a quick note to follow up Richard Freeman's letter in Animals & Men #10 about the yeti comic strip for Dr Who Weekly. The story was called 'Yonder The Yeti', and ran in four two page installments from the 14th May to 5th June 1980. I am afraid, though, that it wasn't drawn by the wonderful Dave Gibbons, but by the equally wonderful David Lloyd. And the script wasn't written by the god-like Alan Moore, either, but by the wizened and unsymmetrical bag-of-spiders who pens this letter. It's not the sort of story that I'd usually go around boasting about, you understand, but I don't see why Alan should take the blame for the wretched thing!

Seeya,

Steve Moore,
London.

BOOK REVIEWS

DIMITRI BAYANOV: In the Footsteps of the Russian Snowman.

Publisher: Crypto-Logos, Moscow.
ISBN 5-900229-18-1. 239pp. Pb.

Bayanov has compiled the first English-language record of sightings and investigations into relict hominids in the lands of the former Soviet Union.

Bayanov presents descriptions of archaeological and cryptozoological expeditions, investigations into the folklore and historical accounts of nomadic snowmen, and examples of discussions within academic circles on the subject.

"...And there, right at my feet, an unknown creature lay asleep," a testimony from a geologist surveying a mountainous area of Tajikistan runs. "It was lying fully stretched out ... about a metre or so in length. The whole body of the animal was covered with shaggy hair. I looked at my Tajik guide ... he pulled me silently by the sleeve and indicated that we must run at once..."

According to the inhabitants of the valleys there were several families of those 'devs' living in the mountains. The creatures were considered to be of the animal kingdom, not supernatural beings, but it was considered to be an evil omen to meet one.

The account goes on to suggest that the 'dev' was a young one and discusses the implications of the food remains seen beside the sleeping entity.

Many of the accounts have been subsequently followed up by Bayanov, in exploration of the question, is the Russian snowman a reality? This book presents a convincing body of evidence to suggest that the answer is a resounding 'YES'!

GI.

KEITH HOWMAN: Pheasants of the World.

ISBN 0-88839-280-X
Publ: Hancock House, USA & Canada, 1993. Large format. 184pp.

A revised and updated edition, this well-laid-out reference book, with many colour illustrations, covers pheasant breeding and captivity management (including housing, dietary needs, stock selection, disease identification and incubation techniques) and a detailed guide to the principle 49 pheasant species.

The Edwards' Pheasant, recently rediscovered in Vietnam, is described (with suitable caution on its extinct status) as follows:

Lophura edwardsi. CITES classification: 1.

"Originally discovered in 1895, nothing more was heard of them until 1923 when Dr Delacour went on an expedition to central Annam province in Vietnam. A number of pairs were trapped and a total of 15 birds were successfully shipped to Cleres where four cocks and three hens were retained and successfully bred in 1925. Since then they have been widely distributed, although inbreeding has lead to stock degeneration and infertility... Surveys in Vietnam in 1988, 1989 and 1990 found no trace of the species survival and ICBP and World Pheasant Association now consider that there may be none remaining in the wild. The importance of maintaining the genetic viability of the Edwards' Pheasants in captivity is, therefore, vital."

"Minimum aviary size: 150 sq ft (14 m2)
Status in captivity: vulnerable to inbreeding.
Full adult plumage: First year
Egg clutch size: 4-7 eggs
Incubation period: 22 days
Feeding habits: nonvegetarian and grain products"

This is an excellent book and one which will make a valuable addition to the library of the zoologist, and the bird-keeper alike. GI.

Animals & Men #12 — Book Reviews

Malcolm Penny & Caroline Brett:
PREDATORS - Great Hunters of the Natural World.

ISBN 0 09 180749 2. Large format. Ebury Press, London, 1995. 224pp.

Based upon the Anglia TV series, *Survival*, this book is not so much a reference book as partly expedition anecdote and partly documentary in style. It has no index.

Predators looks at lions, snakes, hunting dogs, grizzly bears, killer whales and leopards and cheetahs, within the settings of their natural habitats - or as natural as Man has allowed them to remain.

The authors describe the animal's behaviour and senses, their methods of survival and feeding and behaviour patterns, in customary *Survival* style.

Gaby Roslin (of *The Big Breakfast* fame) accompanied the tv crew and gives her view of things in each section of the book. In the snakes chapter, for instance, she's pictured with a yellow and white boa constrictor around her neck. "It felt heavy ... after 10 minutes my neck and back were killing me..." (I too remember being surprised by the weight of one that was draped around me at the last Zoologica exhibition.)

Predators is superbly illustrated - there are very few pages without a colour photograph. It's a visual treat and an entertaining read.

GI.

FORTEAN STUDIES VOLUME 3

Edited by STEVE MOORE.

John Brown Publ, London, 1996.

This volume is the best and most informative yet (and I am not saying this because it includes my long paper on singing mice.) The highlights for me are Karl Shuker's erudite paper on giant mystery birds, and Mike Dash's excellent overview of the legend of Spring Heeled Jack.

While the re-evaluation of Bernard Heuvelman's sea serpent theories contains much convincing data, it does seem at the moment that it's 'open season' on him.

There have been a number of articles in various fortean magazines recently which seem overly antagonistic towards him. I think that it is very important that his massive contribution to the science of cryptozoology is not ignored, in what appears to be an over-eager 'rush' to re-evaluate his theories. He has justly been named 'the father of cryptozoology' (he formulated its methodology) and cryptozoology in its current state would not exist without him.

That apart, this book is excellent and comes with my highest seal of recommendation. JD

GreenScene

Currently lurking in that administrative hellhole known as "the pipeline" is GreenScene, the cousin magazine to Animals & Men and the A & M sister magazine Goblin Universe.

Edited by Graham Inglis, GreenScene, the magazine of Devon Greenpeace, will mainly focus on worldwide environmental and zoological news and current affairs.

GreenScene is looking for correspondents in all parts of the world who can send news items or clippings for GreenScene Newsfile. If you would like to participate, then please contact Graham at the GreenScene editorial address:

7 Queens Terrace, Exeter, Devon, EX4 4HR.

GreenScene supports the aims and objectives of the Greenpeace Devon local support groups and of Greenpeace nationally and internationally.

Animals & Men #12

Totally unbiased advert

Three years after the debut of Animals & Men comes our long awaited 'sister' magazine.. The price is the same as A&M, and though we say it ourselves it is equally groovy. This is a fortean mag of the sort that you didn't think existed anymore. Why not give us a try? The multiverse is stranger than you think!

The Gobin Universe — The Parish Magazine of the Outer Edge

WE'RE BACK - AND THIS TIME IT'S LUDICROUS!

Issue Four: Essential Reading for the Incurably Sane

£2.00

IN THIS ISSUE:

News from Nowhere

Strange stories from around the universe.

St Neot: Weirdest Village in the West?

Sightings and sketches of ABCs; ectoplasmic earthlights; UFOs; and a wolf pack - all in one area!

Exeter Strange Phenomena Group

Psychic phenomena, biorhythm mind control, The Beast of Haldon, crop circles and ghostly apparitions are all investigated by the E.S.P. research team..

Hellhound On My Trail

Current psychic investigation into the case of The Naked Witches And The Black Dogs of Buckfastleigh.

We Are What We Eat

Unfortunately most of us eat a load of chemically-altered rubbish. This article tells you all you ever needed to know about butalated hydroxy-toluene

PLUS:

Eyewitness reports, stories from West Africa, animal mutilations, books and music reviews, a call for no cover-ups on the next Mars mission - and more!

Animals & Men #12 — Sales

NEW BOOKS FROM PUBLISHERS

BARNABY David *The Elephant that walked to Manchester*. 66pp Pb.
A wonderfully bizarre little book which tells the story of the sale of Wombwell's Menagerie in 1872 the aftermath when Maharajah the Elephant walked from Scotland to his new home in Manchester. Includes many interesting vignettes on 19th Century travelling menageries. Autographed by the author. Usually £6, now £ 5.00

BARNABY, David & BENNETT, Clive: *The Reptiles of Belle Vue 1950-77*. 156pp A4 pb.
A wondeful insight into the workings of the reptile department at Manchester's Belle Vue Zoo, and a large body of otherwise unavailable anecdotal evidence for 'out of place' exotic reptiles in the North of England over a period of 27 years. Autographed by the authors. £7.00

BARNABY, David: *Quaggas and other zebras*
A superb book about one of the most notorious extinct animals and the desperate attempts being made to reconstitute the species. £9.00

FARRANT, D: *Beyond the Highgate Vampire*
43pp 1992 'Excellent personal history of the phenomenon by the one person really qualified to write about it. Autographed by author. £ 3.95

GREEN, Richard: *Wild Cat Species of the World* 163pp Pb. Illus.
The best book on the felidae since Guggisberg in the mid 70's. Includes the Onza (colour) Usually £12.50: special offer: £10

THOMAS, Dr Lars: *Ordbog Over Europiske Dyr*. Pb 180 pp.
A wnderful book which lists the common names of European mammals, birds and fish in all European languages. Essential! £ 7.00

SHUKER, Dr K.P.N: *In Search of Prehistoric Survivors*. 192pp hb
Arguably the most important new book on general Cryptozoology since the 1950's book 'On the track of Unknown Animals'. Shuker presents evidence for the survival of dozens of species of animals, presently known only from the fossil record. We cannot praise this book highly enough. £ 18.99

SHUKER, Dr K.P.N. *Dragons - A Natural History*. 120pp pb.
Gorgeously illustrated, this book must be one of the most attractive books that I have seen in many years. It is also a must for anyone with an interest in things Draconian. Shuker proves that he is not only a meticulous scientist, but a fine story teller to boot. £ 11.00

SHUKER Dr K.P.N. *The Lost Ark - New and Rediscovered Species of the 20th Century*. pb.
Another lavishly illustrated book which is an essential purchase for anyone interested in the advances that zoology has made over the past 95 years. NOW OUT OF PRINT £ 18.00

SHUKER Dr. K.P.N. *The Unexplained*
Probably the best book of general forteana to have been published for many years. I cannot recommend this book highly enough £ 16.99

Dr Shuker will, personally autograph his books for you at no extra cost. Telephone for details.

LEVER, Sir C: *They Dined on Eland*. 224 pp Illustrated
Excellent investigation of the 19th Century Acclimatisation Society and their founder Frank Buckland, from the author of 'Naturalised Animals of the British Isles', etc. Amusing and erudite. Lever is one of the great experts on naturalised animals and he writes with great skill and aplomb. The publishers price was £18.50; our price is £ 12.00

CARTER, R. *Loch Ness - the Tour*. 22pp 1996. A good and useful guidebook to Loch Ness. £1.50

STEENBURG, T.N. *Sasquatch; Bigfoot - The Continuing Mystery*. 125pp 1993 Ed.
Excellent. Reviewed in A&M 10. Well worth getting for those interested in North American BHM phenomena. £ 10.00

GREEN, J. *On the track of the Sasquatch*. Large format Pb. 64pp. Many illustrations. 1980. Excellent. £ 8.00

GREEN, J. *Sasquatch - The apes amongst us*. 492pp Lavishly illustrated.
Possibly the best book I have read on North American Man Beasts. This is a classic of cryptozoology and should be bought! £ 12.00

Animals & Men #12 — Sales

GREEN, J. *On the track of the Sasquatch.* Large format Pb. 64pp. Many illustrations. 1980. Excellent £ 8.00

GREEN J. *Sasquatch - The apes amongst us.* 492pp Lavishly illustrated. Possibly the best book I have read on North American Man Beasts. This is a classic of cryptozoology and should be bought! £ 12.00

FULLER, E. *The Lost Birds of Paradise.* 160pp Hb. Reviewed in A&M 10. Gorgeous illustrations. Highly reccomended. £ 30.00

BILLE, M. *Rumours of Existence.* 192pp Pb 1996. Excellent book of zoological anomalies and cryptozoology. Reviewed in A&M10. £ 12.00

GARNER, B. *Monster Monster - A survey of the North American monster scene.* 190pp Pb. Excellent book reviewed in A&M10. Highly reccomended £ 10.00

BOUSFIELD E.L and LEBLONDE P *"An account of 'Cadborosaurus willsi, New genus new species, a large aquatic reptile from the Pacific coast of North America".* Illustrated with photographs and line drawings. Only a few copies left 32pp A4 This offer will not be repeated £3.50

BEER Trevor *The Beast of Exmoor.* The first book on the subject by the man at the centre of the investigation. Highly reccomended. 44pp with illustrations. £ 3.00

SHIELS Tony 'Doc'. *Sea Headlines* The long awaited firsst fruits of the legendary Sea Heads project. Unmissable at £2.50

OUR OWN PUBLICATIONS

DOWNES, J 'Road Dreams' A month of strange goings on. The author, his wife, and a reasonably well known rock and roll band travel across England with a bunch of engaging weirdos. pb 120pp £ 5.00

ANIMALS AND MEN BACK ISSUES

ANIMALS & MEN ISSUE 1 (out of print) Photocopy only. Relict Pine Martens/Giant Sloths/Sumatran and Javan Rhinos/Golden Frogs/ Frog Falls...and much more £ 2.00

ANIMALS & MEN ISSUE 2 (out of print) Photocopy only. Mystery bears in Oxford and The Atlas Mountains/ Loch Ness/Green Lizards/Woodwose/Tatzelwurm...and much more £ 2.00

ANIMALS & MEN ISSUE 3 (out of print)Photocopy only. Giant Worm in Eastbourne/Lake Monsters of New Guinea/Giant Lizards in Papua/Mystery Cats/Black Dogs on Dartmoor/Scorpion Mystery...and more £ 2.00

ANIMALS & MEN ISSUE 4 (out of print)Photocopy only *Manatees of St Helena/Lake Monster of New Britain/The search for the Thylacine/much more..news/letters etc* £ 2.00

ANIMALS & MEN ISSUE 5 (out of print)Photocopy only. Mystery cats/Loch Ness/The Migo Video/Boars and Pumas/Hairy Hands of Dartmoor/News Reviews, obituaries, HELP etc £ 2.00

ANIMALS & MEN ISSUE 6 Owlman of Mawnan/Humped Elephants of Nepal/Mystery Cats/news, reviews and more £ 2.00

ANIMALS & MEN ISSUE 7 Mystery Whales,/Strangeness in Scotland/On collecting a cryptid/Bodmin Leopard Skull/Cryptozoological Books/News, reviews and more £ 2.00

ANIMALS & MEN ISSUE 8. Green Cats, Mystery Whales, Cryptozoological books, news etc £ 2.00

ANIMALS & MEN ISSUE 9. Hong Kong Tiger, Hoirseman in Lincolnshire, Scottish BHM. Congo Peacock, Mystery whales etc £ 2.00

ANIMALS & MEN ISSUE 10 Mystery moth of Madagascar, Bengal Leopard Cats, The Derry, Wild Boars in Kent, a new Irish lake monster, mystery whales and the truth about the Essex Beach Corpses etc

ANIMALS & MEN ISSUE 11. Mystery Walruses, Feathered Dinosaurs, Groound Sloth Survival in North America, Mystery Whales, Initial Bipedalism and much more. £ 2.00

STEAMSHOVEL PRESS #14

Highly regarded US magazine about conspiracy theories and the truth 'they' are not telling us. You've seen 'The Lone Gunman' on the X Files. This is the 'real' thing. UFOs/Politics/Conspiracies/ Hoaxes and counter hoaxes and all good stuff
64pp £ 3.50

SECOND HAND BOOKS

ATTENBOROUGH D *'Zoo Quest to Guiana'* 158pp 1956 Hb 185pp 'Excellent animal collecting book from the 1950's packed with zoological information unavailable elsewhere. 8 plates. !Recommended' £2.00

BATES, M. *'The Forest and the Sea'* (1960). Interesting book on North American ecology. Pb 216pp £2.25

BAYANOV D. *'In the Footsteps of the Russian Snowman'*. pb Russian Ed in English 239pp. Very rare £10.00

BERLITZ, C. *'The Bermuda Triangle'* 1996 Ed. 201 pp Pb. Mildly entertaining quasi fortean tosh Don't believe a word! £3.00

BINNS R.J. *'The Loch Ness Mystery Solved'*. 1983. 8vo Hb 227pp. 18 photographs, numerous line drawings and text figures. *An increasingly sought after book.* £5.00

BORD J and C *'The Evidence for Bigfoot and other man beasts'* 1984 pb 254 pp. Numerous photographs and illustrations. *A super book on the subject of North American and other man-beasts.increasingly highly sought after* and unfortunately out of print. Near Mint Condition. We only have limited numbers of this book £6.00

BOTTRIEL L.G *Umbalala*. The story of the african leopard and its relationship to the wild. Largee format hb with d/w. Lavishly illustrated. From the author of 'The King Cheetah' very difficult to find and sought after 214pp £10.00

CARRINGTON Richard *Mermainds and Mastodons* 251pp 1961 Ed. Excellent book. Numerous illustrations. Sought after in hardback edition. £10.00

COLEMAN, Loren *Tom Slick and the search for the Yeti'*. 1989. Pb. 176pp Excellent work on an unusual and little explored aspect to Cryptozoology. Many illustrations. Reccomended. £5.00

COX B *'Prehistoric Animals'*, pb 150pp 1969. Excellent pocket guide. £2.00

DENIS, Michaela. *'Leopard in my Lap'*. Heartwarming true life reminiscences from veteran zoologist and film maker. Many classic pictures. 288pp Hb (Book Club Ed) 1956. £3.50

DINSDALE T *The Story of The Loch Ness Monster'* 1973 pb 124pp 8 photos. Several Text Figs £2.00

DROSCHER V.B. *The magic of the senses*. Pb 333 1969. Fascinating look at animal behaviour and senses. Excellent bargain. Many illustrations. Reccomended £2.00

FORT, C. *'New Lands'* 1974 Pb Ed. 205pp. Essential reading £3.00

HAINING P.Ed *'The Ancient Mysteries Reader Book 1'*. Anthology of short stories with fortean themes. Includes writings by Poe, Wells and Conan-Doyle £2.00

HARMSWORTH A. *'Loch Ness - The Monster'*. 32pp. Full colour booklet on Loch Ness with much useful information. Slightly tatty hence £2.00

HITCHING F *World Atlas of Mysteries*. 1980 book club Ed. hb d/w. 256 pp. Excellent - one of my favourite books on the subject of general forteana and mysteries. Includes much of interest to the cryptozoologist £12.00

IZZARD R. *The Abominable Snowman adventure*. 1st Ed. 1955. Hb d/w. 302pp. Quite rare and very sought after (17.00

JONES, Ken *'Orphans of the Sea'*. The story of the Cornish Seal Sanctuary. Autographed by author. 124pp Pb. 8 pages of plates. 1970. Nice collectors item £3.00

MORRIS Desmond *Catlore* 114pp hb d/w line drawings. Fascinating sequel to the acclaimed 'Catwatching' which tells you all you could want to know about moggies. £5.00

MORRIS Desmond *'Dogwatching'*. This is the essential guide to dog behaviour and is a fascinating collection of information which is hard to find elsewre. 106pp Hb d/w £5.00

SHIELS, Tony 'Doc'. *'13'*. 23pp 1967. Very rare book of magic from the Wizard of the Western World. This is the only copy of this edition we have ever seen, and it is bound to be sold almost immediately £7.00

SMITH D.K. *'Secrets from a Star Gazers notebook - making Astrology work for you'*. This book is perhaps the most absurd that I have ever read. Pb 492pp 1982 £2.50

WILLIAMS J.H. *'Elephant Bill'*. 1955 Book Club Ed. Hb. Elephants in the Burmese Jungle. Illutrated with archive photographs. . 245pp £3.00

WITCHELL N, *'The Loch Ness Story'* Revised Book Club Edition of 1979. Hb with dustwrapper 236pp illustrated throughout. 'Excellent book '. £13.00

WITCHELL N, *'The Loch Ness Story'* Revised and Updated Edition of 1989. Pb 230pp 16 Photographs 'Excellent book '. Good condition £5.00

WITCHELL N, *'The Loch Ness Story'* Revised and Updated Edition of 1989. Pb 230pp 16 Photographs 'Excellent book '. Slightly tatty hence £3.00

WITCHELL N, *'The Loch Ness Story'* 1982. Pb edition 208pp 16 Photographs 'Excellent book '. Slightly tatty hence £3.00

THE CRYPTO SHOP JANUARY 1997

POSTAGE AND PACKING

This will now be charged at cost. If you include £0.75 per book and 25p per periodical with your order then we will either refund the balance or invoice you for any extra postage due. Payment can be in cash (UK or US currency), International Money Order, Eurocheque, or a cheque drawn on a UK bank.

*Please make all cheques payable to **JONATHAN DOWNES**. Please telephone to ensure that the goods you want are still in stock. If no telephone call is received and an order is received for something that is out of stock then a credit note will be given.' If you live outside the EC please add a 10% surcharge to cover additional post and packing.*

Every effort will be taken to ensure prompt delivery within 21 days. Orders outside Europe are sent by surface mail unless aditional postage is paid.

OUR OWN PUBLICATIONS

The Owlman and Others, by Jonathan Downes.

Two decades of Owlman evidence including sightings - mostly by girls and young women - in Cornwall. This book comes about as close to the truth as anyone ever will... Many illus.   £10.00

The CFZ Yearbook 1996

The first yearbook, with nearly 200 pages of research papers and longer articles including Sky Beasts (Karl Shuker), mystery eagles (Jon Downes), Namibia's Flying Snake (Richard Muirhead), the Nnidnidification of Ness (Tony Shiels), African Man Beasts (Francois de Sarre), and much more. Many illustrations.   £12.00

The CFZ Yearbook 1997

Karl Shuker hunts anomolous aardvarks, Darren Naish figuratively shoots the Lake Dakataua Monster and Francois de Sarre asks if humans are descended from bipedal fish.

Also articles on : the pros and cons of reintroducing extinct mammal species to Scotland, Shakesperean cryptozoology, a list of cryptozoological movies, Mexican cattle mutilation and the Chupacabras - and much more. Many illustrations.   £12.00

Morgawr: The Monster of Falmouth Bay, by A. Mawnan-Peller.

The classic 1976 booklet reissued with a new introduction by Tony 'Doc' Shiels.   £1.50

The Smaller Mystery Carnivores of the Westcountry, by Jonathan Downes

Three species thought extinct; hints of several species apparently unknown to science; and a revolutionary suggestion that a species of mammal known from mainland Europe exists in England. Over 100 pages. Illustrated.   £7.50

PERIODICAL REVIEWS

We welcome an exchange of periodicals with magazines of mutual interest. Those ones where no new issue has reached us in the last 6 months or so have been given a brief mention only. Because we now exchange with so many magazines we have been forced, much against our fortean methodology, to categorise them.

CRYPTOZOOLOGY AND ZOOMYTHOLOGY

DRAGON CHRONICLE, The Dragon's Head Press, PO Box 3369, London SW6 6JN. Jan, May, Sep. A fascinating collection of all things draconian - including ads and services.

THE BRITISH COLUMBIA CRYPTOZOOLOGY CLUB NEWSLETTER, 3773 West 18th Avenue, Vancouver, British Columbia, V65 1B3, Canada.

CREATURE RESEARCH JOURNAL, P. Johnson, 721 Old Greensburg Pike, North Versailles, PA 15137-1111 USA.

CRYPTOZOOLOGIA. Association Belge d'Etude et de Protection des Animaux Rares, Square des Latins 49/4, 1050 Bruxelles, Belgium. The Dec 96 issue includes an article by Francois de Sarre on (if my French translation's accurate) mystery fishes thought to be coelacanths.

CRYPTOZOOLOGY REVIEW, 137 Atlas Ave, Toronto, Ontario, M6C 3P4, Canada.

EXOTIC ZOOLOGY, 3405 Windjammer Drive, Colorado Springs, CO 80920, USA. A free newsletter from the author of 'Rumours of existence'.

FRINGE SCIENCE

SCIENCE FRONTIERS, Sourcebook POroject, PO Box 107, Glen Arm, MD21057. Newsletter of William Corliss' invaluable Sourcebook Project.

NEXUS 55 Queens Rd, E. Grinstead, West Sussex RH19 1BG. An impressive look at the fringes of science

FORTEAN / EARTH MYSTERIES / FOLKLORE

TEMS NEWS, 115 Hollybush Lane, Hampton, Middlesex, TW12 2QY. An entertaining collection of odds and sods and generally weird stuff. Recommended.

HAUNTED SCOTLAND. Mark Fraser, 35 South Dean Rd, Kilmarnock, Ayrshire, Scotland, KA3 7RD. An enjoyable mag covering weirdness from north of the border. Bi-monthly. The Dec issue includes an interview with a tutor of the College of Past Life Healing, a crossword (crossweird?!) and tales of visitations and encounters.

COVER UP. David Colman, 39 Limefield Cres., Bathgate, West Lothian, Scotland, EH48 1RF. The magazine of the Scottish Unexplained Phenomena Research group. Dec issue mainly has well-presented and interesting UFO sighting reports.

DELVE, G. Duplantier, 17 Shetland St. Willowdale, Ontario, M2M 1X5. Canada. Fortean.

3rd STONE. PO Box 961, Devizes, Wilts, SN10 2TS. £2.50. Once the magazine of the Gloucester Earth Mysteries Group, 3rd Stone has now moved towards *Fortean Times* type big-time: A4 glossy format. Let's hope they don't forget their roots too much, as things progress.

The spring 1997 issue stays close to home, geographically speaking, for a look at the standing stones of the Mendip hills (near Bristol) and a stone circle near Avebury in Wiltshire, and a road between Bristol and Clevedon which seems to have had weird or ritualistic things happening along it for centuries. An interesting blend of archaeology and mythology.

DEAD OF NIGHT, 156 Bolton Road East, Newferry, Wirral, Merseyside, L62 4RY. £2. An amusing, intelligently put together Fortean magazine. My favourite fortean journal. The Nov/Dec issue included bits about life on the Martian meteor, UFO reports, lots of weird news stories, and various ghostly accounts.

ZOOLOGY/NATURAL HISTORY

MAINLY ABOUT ANIMALS, 13 Pound Place, Shalford, Guildford, Surrey GU4 8HH. Veteran zoologist Clinton Keeling edits this wonderful A5 quarterly magazine.

In issue 32, Jan 1997, Darren Naish (taking a well-earned break from writing about whales and dinosaurs for *Animals & Men*) has an article on the diversity and history of New Zealand's giant flightless birds. There's also a look at rhinos, the lineolated parakeet - and news from the UK's zoological gardens.

BIPEDIA, Francois de Sarre, C.E.R.B.I, 6 Avenue George V, 06000 Nice, France. French / English magazine for Initial Bipedalism Theorists.

MILTON KEYNES HERPETOLOGICAL SOCIETY. 15 Esk Way, Bletchley, Milton Keynes. Excellent A5 magazine containing handy hints and informative articles.

ESSEX REPTILES AND AMPHIBIANS SOCIETY, 6 Chestnut Way, Tiptree, Colchester, Essex, CO5 ONX. Ads and events of the regional reptile society.

NATIONAL ASSOCIATION OF PRIVATE ANIMAL KEEPERS, 8 Yewlands Walk, Ifield, Crawley, West Sussex. RH11 QE. Useful publication including a wealth of information about wild animal husbandry. This is an organisation which, especially in the present political climate needs your support.

THE MANE, Wild Equid Society, Flat 19, 119 Haverstock Hill, London NW3 4RS. Wild horses and their relatives.

MISCELLANEOUS

NETWORK NEWS, P.O Box 2, Lostwithiel, Cornwall, PL22 0YY. Anarchism, Earth mysteries, weirdness, and even a little crypto... This is the sort of groovy collection which should be encouraged. Issue 10 is the 'Sex Magic Sacrifice'... Is there an issue 11, yet, folks?

FOAFTALE NEWS, Dept of Folklore, Memorial University of Newfoundland, St Johns, Newfoundland, A1B 3X8, CANADA.

An academic approach to contemporary folklore; a "participation" magazine with input from many many sources, and has good follow-up on issues from one newsletter to the next. Long live the ongoing investigation!.

PENDRAGON, Smithy House, Newton by Frodsham, Cheshire, WA6 6SX. A scholarly and massively entertaining magazine on things Arthurian. Keeps an entertaining balance between literature and history.

LOBSTER, 214 Westbourne Avenue, Hull, HU5 3JB.It is nice to see a conspiracy theory magazine with a UK bias which means that it is not overly obsessed with JFK.

Issue 32, Dec 1996, asks, is Libya still the prime suspect in the murder of police officer Yvonne Fletcher. It also looks at CIA and radiation experiments on humans; the connections bewteen the drugs trade and the authorities - in Wales and Mexico ; and gives a detailed guide to various web sites covering military secrets, civil liberties, etc.

Cartoon by Mark North

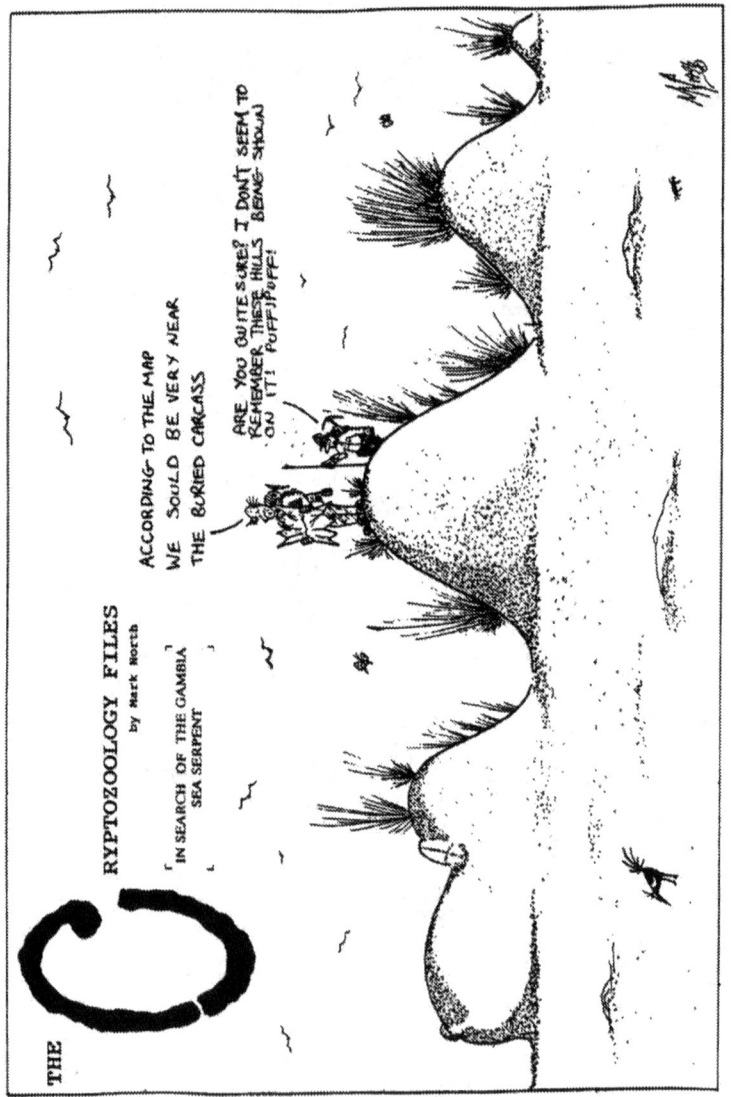

ISSN 1354 0637 Typeset by an Electrician

"I'm drilling through the Spiritus Sanctus Tonight"

ISSUE 13
APRIL 1997

This issue saw the arrival of the final piece of the CFZ jigsaw which has remained a constant ever since. Richard Freeman then a young gothic student at Leeds University made his debut appearance in this issue. Graham and I liked him a lot, and during the long summer and autumn of 1997 he visited us at the CFZ HQ in Exeter on a number of occasions.

Ten years later, he still lives there.

This also features a remarkably bizarre article by Dr Karl Shuker, sporting, what I proudly believe, are his two worst puns. Ten years later we were proud to reprint this article in the expanded version of his classic book *Extraordinary Animals Worldwide*.

animals & men
THE JOURNAL OF THE CENTRE FOR FORTEAN ZOOLOGY

Animals & Men
The Journal of the Centre for Fortean Zoology

MOLLY THE SINGING OYSTER; A PLETHORA OF PANGOLINS; MYSTERY CATS IN YORKSHIRE; THE BARKING BEAST OF BATH; MOBY THE GREAT SPERM WHALE AND MORE...

Issue 13   £2.00

animals & men

THE JOURNAL OF THE CENTRE FOR FORTEAN ZOOLOGY

Animals & Men # 13 — Who's who and what's what

The ever changing crew of the 'Animals & Men' mothership presently consists of:

Jonathan Downes : Editor
Graham Inglis : Newsfile Editor/Spin Doctor
Jan Williams : Associate Editor
Mark North : Artist
Richard Muirhead : Newsagent from Nowhere
Special Agent Tina Askew : Rhine Maiden
Alyson Diffey: Editorial Assistant and Medium

CONSULTANTS

Dr Bernard Heuvelmans
(Honorary Consulting Editor)
Dr Karl P.N.Shuker
(Cryptozoological Consultant)
C.H.Keeling
(Zoological Consultant)
Tony 'Doc' Shiels
(Surrealchemist in Residence)
Darren Naish
(Palaeontology/Cetology Consultant)
Chris Moiser
(Zoological Consultant)

REGIONAL REPRESENTATIVES

U.K

Scotland: Tom Anderson
Surrey: Nick Smith
Yorkshire: Richard Freeman
Somerset: Dave McNally
West Midlands: Dr Karl Shuker
Kent: Neil Arnold
Sussex: Sally Parsons
Hampshire: Darren Naish
Lancashire: Stuart Leadbetter
Norfolk: Justin Boote
Leicestershire: Alaistair Curzon
Cumbria: Brian Goodwin
S.Wales/Salop: Jon Matthias
London: Richard Askew (No relation)
Tyneside: Simon Elsdon

EUROPE

Switzerland: Sunila Sen-Gupta
Spain: Alberto Lopez Acha
Germany: Wolfgang Schmidt & Hermann Reichenbach
France: François de Sarre
Denmark: Lars Thomas and Eric Sorenson
Eire: The Wizard of the western world.

OUTSIDE EUROPE

Mexico: Dr R.A Lara Palmeros
Canada: Ben Roesch

DISCLAIMER

The views published in articles and letters in this magazine are not necessarily those of the publisher or editorial team, who although they have taken all lengths not to print anything defamatory or which infringes anyone's copyright take no responsibility for any such statement which is inadvertantly included.

CONTENTS

3 Editorial
4 NEWSFILE
13 Two articles on Moby The Sperm Whale by Tony 'Doc' Shiels and Tom Anderson
15 A Plethora of Pangolins including articles by Tom Anderson, Richard Freeman, Clinton Keeling and Jon Downes
21 The Barking Beast of Bath by Terry Hooper
26 Two articles on Yorkshire Alien Big Cats by Richard Freeman and Henry Moncrieff
28 Leatherback Turtles in Denmark by Lars Thomas
29 North of the Border - Tom Anderson
30 Molly the Singing Oyster by Karl Shuker
31 The Singing Mouse Contest of 1937 by Nick Smith
33 Return of the Space Walrus by Darren Naish
35 Pinnepedophobia by Clinton Keeling
36 Letters
40. Book Reviews
43. Sales
48. Cartoon by Mark North

SUBSCRIPTIONS

For a Four Issue Subscription:
£8.00 UK
£9.00 EEC
£12.00 US, CANADA, OZ, NZ
(Surface Mail)
£14.00 US, CANADA, OZ, NZ
(Air Mail)
£15.00 Rest of World
(Air Mail)

'Animals & Men'

THE CENTRE FOR FORTEAN ZOOLOGY,
15 HOLNE COURT,
EXWICK, EXETER,
DEVON, EX4 2NA

Tel 01392 424811

THE GREAT DAYS OF ZOOLOGY ARE NOT DONE...

Dear Friends,

Welcome to another issue of *'Animals & Men'*. We are now well into our fourth year and despire what Graham misquotes (via a Billy Bunter story) as *'The wings and Sparrows of Outrageous Fortune'* we are still going relatively strong.

Finally both the 1997 Yearbook and my book *'The Owlman and Others'* are now available. We apologise profusely to everyone who has been waiting for their copies to arrive. However, with the release of this issue our time-line is now back on schedule for the first time since Alison left last July.

We also now have a fortnightly radio show, however unless you can get BBC Radio Devon you are going to be unlikely to receive it. We broadcast every other tuesday afternoon from 2.00 to 2.30 and the show is called *'Weird about The West'*. Various names familiar to regular readers of this magazine have appeared on the show to date, including Darren Naish, Trevor Beer, Chris Moiser, Tony Shiels, and Richard Freeman as well as other folk from non cryptozoological areas of the fortean universe.

The *Animals & Men* posse has also been visible recently on television, most notably when I, together with Sally Parsons, Richard Muirhead, Richard Freeman, Al Pringle and Darren Naish appeared on a rather fine little discussion programme called *'For the love of ....Cryptozoology'*. In answer to all those who asked --- no I wasn't out of my proverbial tree, I just had flu (and a slight hangover).

I would like to take this opportunity to thank someone who, above all, has not only got me through the trials and tribulations of the last horrible twelve months, but whose tireless efforts have managed to save the CFZ and convert it into something with a slight semblance of efficiency about it. Graham Inglis is not only a valued colleague but a very dear friend, and he does what he does for very little reward. I would also like to congratulate him for having been elected as the Chairman of the Exeter Greenpeace Support Group. What with this and his editorship of the nascent *Greenscene* newsletter as well as his activities with the CFZ and with our sister magazine *'The Goblin Universe'* (which he edits with me), I feel sure that his days of media obscurity are fast coming to a close and that we are approaching a new era in which he shall get the recognition that he undoubtedly deserves.

We shall be out and about at several events this summer and look forward to meeting as many of you as we can. We have a large number of exciting new projects just about to start, and now we are finally up to date we shall be announcing our next generation of activities over the next few months..

Until then,
best wishes,

Jon D

Animals & Men # 13   Newsfile

NEWSFILE

EDITED AND COMPILED BY
GRAHAM INGLIS WITH THE
OCCASIONAL
INTERJECTION BY THE
EDITOR.

MYSTERY CATS

Lossiemouth, N.E. Scotland

Architect Stuart Matthews and his wife reported a *'black panther'* as they drove on a village road to Garmouth. The 5 ft (1.5 m) animal appeared from trees 200 yards ahead and bounded across the road. The couple described the tail as long and curling. *Aberdeen Evening Express 24 Mar 97*

West Lothian, Scotland

Margeret McCord saw an animal resembling a large black cat whilst she was out with her pony. She described the animal as being *'definitely not a fox'* and was apparently frightened that it would maul her pony. *Scotland on Sunday 12.1.97*

Animals & Men # 13 — Newsfile

London.

What was described as a 'mountain lion' has been reported by various witnesses living wild along a railway track in West London. Doug Richardson from Londoon Zoo was reported as saying: "Railway land would suit it perfectly. It would be undisturbed and have a plentiful food supply". In the wake of the recent US Government denials about the Roswell Incident, however we tend to look with somewhat a jaundiced eye at official denials or endorsements! *Daily Sport July 17 1996.*

West Yorkshire.

A mystery animal dubbed *'wildcat of the wolds'* has been spotted over the past few years as far afield as Goole, Beeford, Bilton, Driffield and Holme-on-Spalding-Moor. During the summer of 1996 the animals started to appear in the Walkington area where a couple who, for some reason, remain nameless were frightened by an animal the size of a great dane, as their car turned off the motorway at North Cave. A corpse was reported as having been found with *"the head and ears of a cat, the body of a dog, and fearsome teeth and claws"*. *Hull Daily Mail 27.8.96*

EDITORIAL COMMENT: What happened to the corpse is unrecorded, although we feel sure that if it had been anything of significance we should have heard about it.

At least two other corpses of supposedly mystery cats were found during the past twelve months. The first, from Cornwall early last summer turned out to be that of a dog of unspecified breed, and the second, which according to some sources, including the North Devon naturalist Nigel Brierley is supposed to be that of an African wildcat (F.lybica). No evidence has been presented to us in support of this claim, and we feel unable to aceept this record until we receive concrete proof.

LAKE AND SEA MONSTERS

Central Russia has its own version of the Loch Ness Monster, according to the news agency Itar-TASS. Sightings from Twer of a 5 m long creature with an elongated neck, in Lake Brosno, have been attributed to a prehistoric reptile. *Kolner Express 12 Dec 96*

MAN BEASTS AND BHM HAIR TODAY...

Wes Summerlin, a self-styled Bigfoot hunter from California claims to have secured hair samples from the mystery hominid. They were given to Ohio State University researchers for DNA analysis. So far there is no news.

Animals & Men # 13

Newsfile

ATTACKS

An 80-year-old Lincolnshire woman needed surgery after being attacked by a 15-stone seal. It injured her leg as she walked along a coastal path through a nature reserve near Mablethorpe. It is thought the pensioner disturbed the seal with a pup. *Southampton Daily Echo 28 Nov 96*

SIMIAN STUPIDITY

The Mayor of Boston and other dignitaries holding an impromptu news conference at the city's Franklin Park Zoo were bombarded by missiles made from straw and droppings and thrown at them by a young gorilla who obviously didn't think much of politicians. *Aberdeen Press and Journal 30.1.97*

SIMIAN SADNESS (OR SHOULD THAT BE SLAUGHTER)

Vets in the Philippines killed over six hundred monkeys at a breeding farm because of a scare over Ebola virus. *Aberdeen Press and Journal 30.1.97*

I THOUGHT I SAW A PUDDY CAT

Finnish farmer Erkki Turrunen, of Eno in western Finland, tried to stop a wolf grabbing his pet cat. He grabbed the cat's tail and entered into a bizarre tug of war situation with the ravening wolf who, (as was pretty obvious) won the competition and ate the cat. *Aberdeen Press and Journal 30.1.97*

SEE YOU LATER ALLIGATOR

A Brazilian boy was saved by his parents from the jaws of an alligator after he fell into the water at the Shark Valley nature trail in Florida. They waded into the water to beat off the creature that had attacked their child. It was reportedly the first attack on a human since the park first opened in 1947. *St Louis Post-Dispatch via COUDi 17.7.96; USA Today 16.7.96 via COUDi.*

... IN A WHILE CROCODILE

In a bizarre case, which both mirrors the last story and proves that the American legal system is fundamentally unsound the mother of a nine year old girl who was attacked by a Florida alligator, is to sue the dead beast (killed by her husband) for a million dollars.

The little girl was badly scared in the attack. *"Courtney had nightmares for a while"*, her mother said.

"She was scared, even later that day she was still shaking. Any of my kids will now tell you that alligators are bad!"

Her husband shot the beast and is now being sued for having unlawfully killed an animal of an endangered species. Mrs Novacs, citing an obscure

Animals & Men # 13 — Newsfile

endangered species. Mrs Novacs, citing an obscure piece of litigation involving Loggerhead Turtles claims that the state Game and Freshwater Fish Commission is responsible for injuries and traumas caused by endangered species under their jurisdiction.

Jim Anista, the general counsel for the state agency claims that they are not responsible for the actions of wild beasts. *"Either she has to sue us directly"* he said, *"or else the alligator has to go out and get itself a lawyer"*

The alligator is dead! (Are dead reptiles eligible for legal aid?) Atlanta Journal Oct 20 1996.

Who Killed Bambi?

A rare white stag, one of only a few in Scotland, has been shot and then decapitated, probably by a trophy hunter. The stag was a favourite attraction at Glengoulandie Deer Park. *Aberdeen Press & Journal 23 Jan 97*

WHO'S AFRAID OF THE BIG BAD...

A student playing the part of the Big Bad Wolf was shot and injured on stage after no-one bothered to check whether the gun used by the hunter who comes to the rescue of Little Red Riding Hood, in a Brazilian performance of the pantomime, was loaded. *Aberdeen Press and Journal 18.9.96*

TALKING TURKEY

A wild turkey crashed through the windscreen of a truck as it drove down Interstate 79. "It scared the hell out of me" said the driver, *"He was still alive. He must have had a rush of adrenalin. He was looking around and he stared at me for a while".* The driver climbed into the back of his truck to try and dislodge the wounded turkey which began to hiss and struggle before it finally fled through the hole in the broken windscreen. *"He's on the loose in Pittsburgh someplace"* said the driver, who swore that he would never eat Turkey at Thanksgiving again! *St Louis Post Dispatch via COUDi 19.12.96*

BEES MEAN BUSINESS

A swarm of killer bees attacked a group of farmers in central Costa Rica killing one man and injuring several others. In a scenario somewhat reminiscent of a particularly unpleasant 1980s B Movie, the creatures were reported to be a fearsome African species which escaped from a laboratory in Brazil some twenty years ago and have now spread across much of South America and some of the southern states of the United States. *St Louis Post Dispatch via COUDi 24.3.96*

The steady spread northwards of these bees was predicted in Arthur Herzog's book "The Swarm", published in, I believe, 1979.

Later in the year it was reported that similar creatures had been terrorising parts of Los Angeles (this is beginning to sound more like a B Movie than ever), and the newspaper reports said that the LAPD were receiving special training. (Cue a joke about 'honey traps'). They also note that the species first arrived in the USA in Texas during 1990. *USA Today Via COUDi 6.6.96*

NO FACETIOUS PUN FROM THE EDITOR COULD POSSIBLY DO JUSTICE TO THIS RIDICULOUS STORY!!!

According to the ITAR-Tass news agency, a man and some friends were ice fishing on a reservoir some sixty miles from Moscow when he caught a 28 inch pike. Showing off he raised the fish high and kissed it on the mouth. The pike clamped down hard on the fisherman's nose.

The pike's jaws were clamped hard on the man's nose even after his companions beheaded it. Doctors at a local hospital set him free. *Boston Globe Dec 13th 1995.*

WHALE MEAT AGAIN

Two southern right whales charged at a shark net and ripped it apart in order to save their calf who was entangled in it in the sea off Durban in South Africa. Mike Manning (no relation to Zodiac Mindwarp) said *"I was a few yards away. The young one had been caught by its tail and had been making a lot of noise, blowing in and out of its blow-hole. The two adults were trying all sorts of things but eventually both of them charged the net, ripping it apart and freeing the youngster".* People watching from the beach wanted to intervene to save the calf but in the end its parents did it themselves. *Aberdeen Press and Journal 4.11.96*

IF YOU'RE GOING TO SAN FRANSISCO

A baby alligator or perhaps caiman which was residing in San Fransisco's Mountain Lake became a figurative pawn in the never-ending publicity war between the two rival newspapers of the city. *The Examiner* ran a *'Name that Gator'* competition, and so in the next round of the ratings war *the Chronicle* hired a professional alligator hunter from Florida to hunt the beast. The Examiner retaliated by claiming that the hunter was a veritable reptile 'hit man' and hired a scuba diver to protect the beast. The police turned him away because of a local bye law forbidding swimming in the lake.

The San Fransisco Boys Chorus got in on the act and turned up on the muddy banks if the lake to croon *'Puff The Magic Dragon'* in a vain attempt to lure the rogue reptile to the surface. *The Chronicle* started a *'Gator Watch'* and the Examiner announced that they were consulting a psychic.

The San Fransisco Zoo who were officially hunting the animal withdrew their capture team in order to avoid the stupid publicity, and announced that they were stopping searching "until the commotion subsides". Meanwhile a photograph showing something looking remarkably like an iguana swimming behind a duck, and captioned *"Alligator - or possibly Caiman - pursues lunch in Mountain Lake"* proves that the missing creature - whatever it is, seems happy, healthy and unconcerned at his new found fame. *L.A.Times 9.3.96*

EURO-ROO (1)

After spending two years in the wild in forests in Belgium, a kangaroo is now recovering from frostbite at Antwerp zoo. The animal tried hard to avoid capture. How an animal indigenous to Australia came to be living in Belgium remains a mystery. *Westfalenpost 18 Jan 97*

EURO-ROO (2)

An escaped pet described as a *'litle kangaroo'* was caught by police near Hamburg. The species is not noted but it is corroborative evidence for the burgeoning populations of 'wild' red necked wallabies being found across Britain and northern Europe. *Westfalenpost January 18th 1997.*

WHITE WHALE

A Beluga whale was in danger of becoming stranded at Thurso in the Orkney Islands during September 1996.

Animals & Men # 13

Newsfile

The fears proved groundless as the animal eventually made its way towards the open sea. *Aberdeen Press and Journal 10.9.97*

SEAL'D WITH A KISS (1)

A lost seal pup was discovered basking amongst fishing boats in a Dorset harbour - 2,000 miles from its Arctic home in northern Canada. The one year old silver harp seal was netted by fishermen after pulling himself onto a pontoon in Poole. The seal was taken to the National Seal Sanctuary at Gweek in Cornwall from where staff hoped that they would be able to release him into the wild. *Teletext on 3 (Meridian News Pages) p.337 27th Jan 1997.*

SEAL'D WITH A KISS (2)

A baby seal from the North Atlantic turned up on a beach in the American Virgin Islands. The seven month old Hooded Seal, usually a denizen of cold northern waters was then sent to a sanctuary in Puerto Rico. *Aberdeen Press and Journal 9.9.96*

EDITOR'S NOTE: It is interesting how many Arctic pinnipeds, and indeed cetaceans, are now turning up in British and southern European waters. This is at least the second Harp Seal in the last twelve months and there was also a record of a Hooded Seal last year. Recent sightings of what is almost certainly a humpback whale off St. Ives are also, we believe, significant. Whether these changes in distribution are the result of climactic change or other environmental conditions, or whether (to use a totally non-scientific but probably very fortean expression) they are just 'one of those things' remains to be seen. The 'real' mystery is WHY DO ALL THESE SIGHTINGS GET REPORTED IN THE *ABERDEEN PRESS AND JOURNAL* AND APPARENTLY NOWHERE ELSE?

The return of the Multicoloured Frogs

Unusually-coloured frogs in southern England seem to be flourishing, despite their inherent inability to blend in with their surroundings. Colours ranging from orange to yellow have been seen, as has cream and pale pink. The frogs are thought to be albinos that lack their normal dark green pigmentation, and display whatever 'secondary' colour happens to be in their genetic make-up.

With no dark pigments to help them absorb the sun's energy, their development is slowed. However, more than a third of these colourful frogs are being seen in the sunny south-west of England. It has been suggested that global warming may be assisting the survival of what otherwise would be an unviable mutation. *Daily Mail 12 Mar 97.*

A LAD IN CRANE

A northern European crane was spotted at Sandhaven in Scotland last autumn. This particular specimen which was on its own's sex is unknown and was seen at a small pond near the village of Buchan. *Aberdeen Press and Journal 18.9.1996.*

Animals & Men # 13

Newsfile

THE ADVENT OF THE EUROMOTHS

A colony of rare moths has been discovered in southern England. Southern chestnut moths, normally found in warmer countries such as Italy and North Africa, have been found in the New Forest, Hampshire. The area contains lowland heather, which the larvae feed upon. Central France is the closest the moths normally come to Britain. *Southampton Daily Echo 7 Dec 96*

Researchers in China say they have found the remains of a specimen of the first bird to have a beak and true feathers. The remains have been dated to about 140 m.y.a. - the Jurassic era. The bird, standing 28 cm (12 in) tall, has been named Confuciusornis sanctus. *The Toronto Star 19 Oct 96*.

The phalarope, a wading bird common in the USA, has an unusual method of obtaining food during times of shortage. It spins around, once per second, creating a vortex in the water that lifts food particles from the bottom, for the bird to peck from the water surface. Each revolution requires 7 or 8 kicks. *The Boston Globe 19 Nov 96*

Hong Kong's pink dolphins are under threat from pollution and could be extinct in a few years, experts have warned. It is believed that only 80 are left, falling victim to pollution from rapid industrial expansion and the development of coastal schemes such as the new airport, new road and rail schemes, and an oil pipeline.

The World Wide Fund for Nature (WWF) has warned that drastic measures are needed to save the pink dolphin. With the handover of Hong Kong to the Chinese about to occur, any such rescue will primarily be down to the Chinese. *Daily Mail 31 Dec 96*

EDIOTOR'S NOTE: The new development at *Chep Lap Kok*, which will not be completed until well after the British leave the territory is a controversial one from an ecological point of view. An entire island has been levelled and a massive amount of land reclamation has taken place.

When one remembers that the channel connecting Stonecutter's Island to the mainland has also been filled in the effects on the water circulation in Hong Kong harbour must be nothing short of catastrophic. It will be unlikely to our minds at least, if the 'pink dolphin' is not the only marine casualty of this so-called progress.

When one remembers the furore surrounding the extension of Kai Tak aerodrome some three decades or more ago, one hates to think what this new development will do to the feng shui of the region. JD

Animals & Men # 13 — Newsfile

SEAL OF APPROVAL

A community of landlocked harbour seals living in northern Quebec has been studied by Professor Dave Lavigne of the University of Guelph. No-one is sure how this saltwater species came too live separetely in *Lacs des Loups Marin*. The most popular theory is that they were trapped during the last ice age, 8,000 years ago. *Toronto (?) Globe and Mail 4.12.96*

COLOSSAL FOSSIL

A gardener in Australia who took delivery of some sandstone boulders for his rockery was hosing them down when he noticed the outline of a fossil emerging from one. He contacted Dr Stephen Godfrey, a Canadian palaeontologist who was visiting Sydney for a dinosaur exhibition. The Triassic fossil is of a 6.5 ft (2 m) amphibian that looks like a cross between a crocodile and a worm, possibly a brachyopid, and dates from about 230 m.y.a.

Dr Godfrey commented after his local call-out. "I've had to go to the far ends of the earth to examine some fossils, and here I was in a friendly old fellow's garden with his wife bringing out tea and scones." *Daily Mail 15 Feb 97*

A FISHY STORY

An enormous fish weighing 303lbs and with a total length of 96 inches was caught in the Gambia River. The fish was described as a 'toppen'. According to the report in an unnamed English language Gambian newspaper (Feb/Mar 1997) 'toppen' are an Atlantic ocean species which makes its presence in a river somewhat unlikely. One can only speculate that 'toppen' is a misnomer for 'tarpon', but even so its presence in Gambian fresh or even estuarine waters would seem to be somewhat of a zoological mystery.

EVERYBODY'S GOT SOMETHING TO HIDE...

Scientists from Hanover University, Germany, have rediscovered a monkey believed to be extinct in Madagascar, off the east coast of Africa. With a mass of only 80 grams, the hairy-eared dwarf lemur (*Allocebu trichotis*) is one of the smallest monkeys in the world. Mainly nocturnal, they inhabit the still-largely-untouched rainforests. Hovever, humans are making their mark: when Europeans first arrived on Madagascar, there were 45 native species of monkey there. Now there are less than 35. *Die Welt 7 Jan 97*

NEW CRAB

A new species of crab with a multicoloured shell has been discovered in rivers near the town of Chiang Mai in northern Thailand. Because of its attractive carapace it has been named the 'elegant mountain crab'. *Westfälenpost November 14th 1996.*

WEIRD SCIENCE

TRIPLE RAMMY

A sheep has given birth to triplets - each by a different father. One lamb has the white face of its Dutch Texel sire, another is a pink-complexioned Charollais, while the third is a black-faced Suffolk. They were born in Norfolk. *Aberdeen Evening Express 24 Jan 97*

Oh no.. not those dinosaurs again!

The sudden extinction of the dinosaurs and many plant species 65 million years ago - the so-called Cretaceous Event - is widely believed to have been caused by a meteor, an asteroid larger than Mount Everest, slamming into the Earth's crust. Now,

Israeli scientists have suggested that it was cosmic radiation that bombarded the Earth - after 2 neutron stars (pulsars) collided, creating a deadly wave of radiation that destroyed the protective layers of the Earth's atmosphere. *The Toronto Star 3 Dec 96*

Bees can fly

It has almost been considered axiomic in physics that bees shouldn't be able to fly - their wings are too small. The wings of aeroplanes and birds are shaped so that the air takes longer to pass over the curved top surface than to pass underneath: this differential airflow 'sucks' the wing upwards, i.e generates lift. A bee's wings, though, are virtually flat, and shouldn't be able to lift the mass of the body. Now a research group at Cambridge University has shown that insect wings generate turbulence - small vortices - on the top surfaces. These vortices effectively increase the wing thickness, so far as the main air motion is concerned - to the extent that the radically differential airflow generates enough lift. *The Daily Mail 19 Dec 96*

STRANGE STORIES

RETURN OF THE NATIVE

Various British animals once regarded as in need of protection are reportedly re-emerging as pests. Otter, badger and raven populations have recovered sufficiently in some areas for some observers to now describe them as 'the scourge of the countryside'. Inflexibility of British conservation laws is being blamed: once protection is bestowed, it is very hard to revoke. A species no longer in danger can multiply unchecked and, say some landowners, get out of control. Official culls have been suggested as a solution.

Reports suggest that many farmers undertake illegal raids against protected animals that mount raids on their farms. And some of the racing pigeon fraternity reportedly feed their ageing pigeons poison capsules and then send them out when birds of prey are around.

Their prey will come...

Some more obscure direct action was undertaken by some pigeon enthusiasts against a peregrine that was preying on their prize birds: an ageing pigeon was sent aloft with a small explosive device attached, which went off when the peregrine struck. *The Times 11 Oct 96*

OTTER THAN JULY

In an interesting correlation to the above story, it appears that as the native British otter recovers its population numbers in the north of England, the numbers of American mink are declining. It also appears that the resurgence of the native species is also helping the recovery of another native species - the water vole. A survey of otter numbers in Northumberland showed signs of the presence of the animals at 70% of the sites visited as opposed to 23% only four years earlier. *The Times 27.12.96*

STOP THE PIGEON

A homing pigeon has turned up in Moscow after flying 1500 miles in the wrong direction. It was released from Nantes, France, and was expected to fly the 250 miles to its loft in Hampshire. Instead, it arrived safe and well in Russia more than 2 years later. *Meridian Teletext On 3, 29 Jan 97*

NEWSFILE CORRESPONDENTS

Tom Anderson, Wolfgang Schmidt, Gypsy Sherred, Sally Parsons, The Cryptic Clipper of Mannic Publications, Gene Duplantier, Chris Moiser.

Animals & Men #13

Feature

NEWSFILE EXTRA
'MOBY' THE SPERM WHALE

EDITOR'S NOTE: It is one of the truisms of fortean thinking that a single event can have an infinite number of interpretations. A recent example is the arrival and subsequent demise of a hapless sperm whale in Scotland's Firth of Forth. Here we present two differing viewpoints of the affair. Firstly from Tony 'Doc' Shiels of the Sea Head Artists Gang (S.H.A.G)

"Funny, wouldn't you say, that a sperm whale turns up in the Firth of Forth (plus his pals) just a week or so after S.H.A.G.ster Sinbad was in those Celtic coastal parts? Also S.H.A.Gster Kevin McGlue was making and shaping Sea Head stuff in the Firth of Forth just a few months (more or less) back.

EDITOR'S NOTE: See Fortean Times May 1997.

And they call it 'Moby'! AND the last time a great sperm was seen in those waters was 1977, around the time that .... (check the records). Consider these things good shipmate.

*Thar she blows,
Tony"*

The second interpretation comes from our intrepid Caledonian Correspondent, Tom Anderson:

"March 20th Firth of Forth.

A fifteen metre sperm whale beached itself at South Queensferry beside the Forth Road Bridge, the main north/south artery to Edinburgh.

By the morning of the 21st it had freed itself but, obviously in some distress, was moving upstream instead of seawards. A further three were seen downstream and it was thought that the spring tides had confused them on their migration from the arctic to the Azores. It was monitored all day by scientists in a boat, and sonar readings suggested that it was not yet in a critical stage of anxiety.

Animals & Men # 13

Feature

All attempts to drive it back under the bridge failed as traffic noise apparently frightened it. As the Forth is a busy industrial waterway it was thought that tanker traffic contributed to its disorientation. By now it was in water depth of ten to twenty metres.

B.P postponed a tanker's departure from a nearby terminal and it was proposed to limit vehicle speeds on the bridge to light noise levels. By now a flotilla of eleven craft were attempting to shepherd the whale downstream. Finally, late on the 22nd it made sonar contact with a member of the pod a mile away. It immediately turned and made for the mouth of the estuary but again ran aground for a further six hours before finally reaching open water. It then returned on the 29th, again stranded itself, and died of suffocation.

The reasons why this pod migrated down the east coast instead of the usual west are as yet unknown.

A growing body of opinion blame the oil industry's seismic explorations west of Shetland, citing previous strandings on the Belgian coast (the first since the 1700's).

This could explain the influx of previously rare cetaceans seen during the last year off the northern isles.

Further displacement down the east coast and an inability to migrate westwards south of the mainland would inevitably lead to a nil population growth for the forseeable future.

The North Sea oil industry has a minimum projected lifespan of twenty-five years.

EDITOR's NOTE: For further records of Sperm Whales stranded on the east coast of Scotland see A&M9 and for records from Norfolk during the last fifteen years see the 1997 Yearbook!

PICTURE © COPYRIGHT LIBERATION FRONT

A PLETHORA OF PANGOLINS

EDITORIAL NOTE: As has become our custom we are devoting a section of this issue to a selection of short articles about the same animal. This issue we look at pangolins. This was started by Richard Freeman who whilst staying with me at New Year mentioned that his grandparents had heard a rumour of a pangolin on the loose in the West Midlands. This spurred my editorial interest into action, and although Richard's pangolin turned out to be something completely different, this didn't stop, Tom Anderson, Karl Shuker, Clinton Keeling and even your humble editor getting in on the act...

The Pangolin that never was
by Richard Freeman

(Yorkshire representative of the CFZ investigating events in Warwickshire!)

My home town of Nuneaton is possibly the most un-mysterious and non-fortean place in England. If places like Warminster or Dyfed are 'hotspots' or 'window areas', Nuneaton is the centre for sheer blandness.

It has had but a single Crypto-incident. In 1976 a 'Black Panther' was seen prowling around the Bermuda area of the town. Police with loud-hailers warned people to stay in their homes. Of course, the whole populace of Bermuda went out immediately and prodded every bush in site with sticks. Needless to say the 'panther' (if it ever existed) disappeared without a trace. Apart from the area becoming known as the Bermuda triangle' for a while that was Nuneaton's only brush with Cryptozoology, until now.

Upon arriving home from Leeds for the Christmas holidays (1996) I was somewhat surprised to hear from my grandparents that a bizarre animal had been spotted on the outskirts of Dullsville. Apparently a local paper (they could not remember which one) had run a story about a month ago stating that a woman had seen a ant-eater like animal. My grandparents were sure it was not an ant-eater though, but something like one.

I ran past them the names of all the animals that could possibly be mistaken for anteaters. Aardvark?, Armadillo? Pangolin? To my extreme surprise "Pangolin" met with the affirmative!.

Pangolins must vie with Echidnas for the title of the weirdest mammals in the world. Once believed to be Edentates (toothless mammals) - the group that contains, ardvarks, armadillos, sloths and ant-eaters. Pangolins have now been reclassified into a group of their own and zoological arguments rage over their relationships to other animals.

Visually they look more like reptiles than mammals, with their covering of scales. One could also mistake them for giant animated pine cones. Seven species are found in tropical Africa and Asia. I have never seen one in the flesh, either in captivity or in my time in East Africa.

I suppose that they could be raised on a diet similar to that of ant-eaters in captivity. This consists of minced meat, egg, formic acid and insect eater supplement.

In the wild, pangolins feed on termites and ants, so the frozen winter ground would be a hinderance to any running free in Britain. Apparently, however, Pangolins can endure extreme cold. I was talking to the proprietor of 'The Pangolin Bookshop' in Hebdon Bridge, Yorkshire last year and enquired how the shop got its name.

She told me that when she had lived in Africa (Zimbabwe, I think), she knew a taxidermist there who wanted to stuff pangolins. Unfortunately he found them impossible to kill due to their armoured hide. In desperation he tried freezing them alive by leaving them overnight in a freezer! Luckily (for the pangolins) they were all still alive the next day. The taxidermist's cruel plan didn't work. Pangolins, it seems, are very hardy beasts.

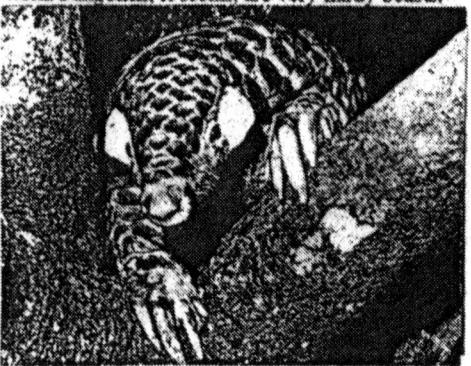

So, it was theoretically possible for pangolins to survive in Britain. But if it was a pangolin, where in God's name had it come from? No local zoos kept them. (As far as I know, they are not kept in ANY British zoo). It seemed very unlikely that they were being sold as pets by unscrupulous animal dealers. Perhaps they had teleported?

I spent the next few days in the local newspaper offices 'trawling' through their back issues. After sifting through a mountain of twee local 'news' (SHOCK! Woman loses Pencil!), I finally came upon the story.

The animal in question was not a pangolin, nor was it an anteater, nor anything even remotely like one. In fact, it turned out not to be any kind of placental mammal - it was a marsupial called a bandicoot.

(EDITOR'S NOTE: See A&M12).

Bandicoots are marsupials of the family *Peramblidae* and *Thylacomyidae* (Rabbit Bandicoots). They are insect eating animals which resemble the hypothetical outcome of a night of passionate miscegnation between a rabbit, a rat and an enormous shrew! There are eighteen types, including Gunn's Bandicoot, The Brown Bandicoot and the Spiny Bandicoot.

The story (in all its sparse detail) runs thus:

A local woman saw the animal in Hartshill Hayes, a large stretch of woodland that supports quite a variety of wildlife. The later identified it as a Bandicoot, and reported it to a radio programme. The newspapers then picked up the trail! It was printed on November 15th 1996 in the 'HARTLAND EVENING NEWS'. Unfortunately she failed to leave either a name or telephone number with either the radio station or the newspaper. The reporter who wrote up the story, John Ellis, believed that she was frightened of ridicule. The whole thing was frustratingly vague.

The newspaper folk saw fit to ask the opinion of an 'expert'. They decided to approach Molly Badham, the director of Twycross Zoo. In her 'learned opinion' Marsupials could not survive in Britain and Bandicoots had never been imported into the country. Anyone thinking that Britain is too cold for marsupials should remember these are warm blooded animals and that there are several breeding colonies of Red Necked Wallabies (and possibly other related species) in Britain. Although true bandicoots are insectivorous, and winter would therefore be a hard time for them, they are also known to eat small mammals which are plentiful in the woods at Hartshill Hayes. These woods back onto a number of gardens where these animals would be able to scavenge for domestic rubbish from dustbins. Bandicoots are formidable little mammals and could 'hold their own' against domestic cats.

As for none ever being imported into Britain, I would be very surprised if the Charles Clore pavilion at Regent's Park had not included them in its collection at some point.

Most species are not endangered and I can see no reason why there should not be any in captivity in Britain. None of the zoos within a thirty mile radius of Nuneaton (Twycross, Dudley, Drayton Manor or the West Midlands Safari Park) have bandicoots, but it is not impossible that it was an exotic pet that had escaped or been 'liberated'.

EDITOR'S NOTE: We contacted Clinton Keeling who conformed that although there are no species of Bandicoot currently in British Zoos, at least three species were kept by Regent's Park at various times. These are:

The Greater Rabbit Eared Bandicoot (*Macrotis lagotis*). The first specimen was purchased on April 28 1839 and it lived until 20 Dec 1840. The Short Nosed Bandicoot (*Isoodon obesulus*) which was first exhibited on the 15th April 1880 Long Nosed Bandicoot (*Perameles nasuta*). The first specimen in the collection was presented on the 9th June 1908 by the Government of New South Wales.

It is almost certain that specimens of all three species were exhibited subsequently, both at Regent's park and probably at other UK Collections.

At the end of the day there's not much of a case here. One woman saw something she THINKS was a bandicoot. So far, no other witnesses have come forward. As I write this several months later it seems unlikely that this strange little vignette of forteana will be resolved. However, Badgers, muntjac, tawny owls and stoats hide in this, Nuneaton's largest tract of woodland, so if Bandicoots COULD hide anywhere in the area it would be in Hartshill Hayes.

No Pangolin however!!

EDITOR'S NOTE: As noted in A&M12 the general feeling in the CFZ Offices is that the whole thing was a hoax. According to Clinton Keeling, someone set up a spurious 'Bandicoot Hotline' (this was confirmed in the May 1997 issue of *Fortean Times*) which when anyone telephoned the number given they got hold of a furious member of the public tired of being plagued by nuisance telephone calls. It seems possible that the culprits were none other than the Japanese Owned multinational Electronics company Sony. Late last year they released a 'platform video game' for their machine *The Sony Playstation*'. The game was called '*Crash Bandicoot*', and it seems not unlikely that someone involved in their publicity department dreamed up an amusing publicity stunt. Conversely it might be that the lady in Nuneaton actually saw some other species of animal, but that because the word 'Bandicoot' had been bandied (no pun intended) so widely across the media at the time as a result of the (perfectly legitimate) activities of the SONY PSX Publicity Department that she jumped to a perfectly understandable conclusion.

THE PUZZLING PANGOLINS

by Clinton Keeling.

Always of interest to the naturalist is the situation known as discontinuous distribution, or as the trendies prefer to call it 'relict distribution'.

The most famous, indeed classic instance is that of the Azure Winged Magpie, a member of the crow family, that's found in southern China, and Spain, near the border with Portugal - and without a single other one in between.

There are many other - although perhaps less extreme examples - such as the Pika (looking something like a short eared rabbit, which is found in the American Rockies, the Caucasus and the Himalayas - again with no points in between), the Monk Seal (from the Mediterranean, the Caribbean and far out in the Pacific round Hawaii), the White

Rhinoceros (down in Zululand and up by the White Nile), and the Waldrapp or Bald Ibis which is found only on two cliffs. One in Turkey and one in Morocco.

However, if you really want to see discontinuous distribution at its best, although why the state of *affairs has contrived to evolve is a complete* mystery, you'll have to travel to the dense forests of West Africa - then on to the even thicker ones of South-East Asia. In both areas you'll find (if you are observant, patient and very fortunate), the strange and highly specialised fishing owls, the chevrotains or mouse deer, (which are in fact placed scientifically on their own but are vaguely related to the camels), and the astonishing pangolins, which are amongst the most remarkable of all mammals. Let's take a closer look at them.

There appear to be seven species of these strange creatures which, being completely covered by protective, horny scales, look decidedly un-mammal-like, indeed un-animal-like as the general effect is of a grotesque moving fir cone.

I shall always remember handling an example of the giant pangolin at the Antwerp Zoological Garden and being momentarily astonished at how delightfully warm it was to the touch; for a second or so I had not equated it with a mammal.

At one time the Pangolins were classified as Edentates - the so-called toothless mammals, but now they are placed in a small order of their own - the *Pholidota*.

Strictly nocturnal, they sleep in holes that they dig for themselves by means of their long, curved and extremely powerful claws, or they may make use of the holes of other species; one authority stated that he'd often found them curled up with quite large pythons - which doesn't really surprise me as neither mammal preys on the other. Their diet consists largely of invertebrates such as ants and termites, which they obtain by tearing down their concrete like nests by means of their claws, then scooping them up with their sticky tongues. It's safe to say that the Pangolin's tongue is the most extraordinary part of an extraordinary creature, as it extends from the mouth to near the pelvic girdle, down towards the tail. In other words it's about half the owner's total length!

Obviously, the prey is swallowed alive, so things such as ants could be very uncomfortable things to have in one's stomach, but not if you're a pangolin, whose stomach is lined with horny knobs which grind up its indignant prey before it can make its displeasure known.

They have long snouts, with damp noses, so here is proof positive that they have a keen sense of smell, but as their eyes are so close to the ground, their vision must be poor to say the least! Surprisingly, the asiatic species have external ears, whereas those from Africa do not. All have prehensile tails, and some species are adept tree-climbers. The giant species from Africa, is said to be a good swimmer that readily takes to water if it feels threatened.

Pangolins are but rarely seen in zoological gardens, partly because of the difficulties attendant on obtaining them, but mainly on account of their highly-specialised diet, which we have yet to 'crack' in confinement.

Some degree of success has been obtained by feeding minced meat to which folic acid has been added to emulate the ants that would normally be consumed. However, the giant species has successfully bred in the aforementioned Antwerp Zoological Garden.

The London Zoological Garden's first specimen, a Small-scaled Tree pangolin, was purchased on May 24th 1877, but was obviously in a moribund state as it only lived three days!

A Temminck's Pangolin was deposited on June 29th 1896, and Chinese Pangolin was presented on July 3rd 1925. I recall two Giants in the old Rodent House c.1960, but don't know when they arrived or how long they lived.

PANGOLINS
by Tom Anderson

(Its a bird, its a plane... no it's Aberdeen's Mr Entertainment)

Worldwide there are seven species of Pangolin. Three Asiatic; The Chinese *(Manis pentadactyla)*, Indian *(M. crassicaudata)* and the Malayan *(M. javanica)*. The four African species have a different nomenclature and include the giant pangolin, and the tree pangolin *(M.tricuspis)*.

All are ant/termite eaters whose only defence is 'ball curling' protected by overlapping scales. These are a form of hair produced by a layer of epidermis known as the Malphagian which grows from fleshy protuberances called papillae. In addition to rolling, the tree pangolin by definition is arboreal, and is also less common near people making it the least threatened of the African species.

All pangolins are predated by man and animals, their scales are used as charms and the flesh highly prized. The tree species ability to emit noxious anal secretions similar to those of the mustelidae would appear to have some effect as a deterrent.

Prior to colonisation, the giant species was totemic to the Boganda people, but is now the most vulnerable of all the seven and is classed in the Red Data Book as endangered. Pangolins are also vulnerable to bush fires and, in cattle country, electric fencing.

Possibly due to less intensive farming methods, the Asian species are under less territorial pressure and are reputedly 'holding their own'.

Reports of species interaction are normally of the frustrated canine variety, where a jackal is thwarted by a rolled-up pangolin, but there is a record of a female tiger whose cub has been taken by hunters, engaging in manic displacement activity by literally shredding an empty pangolin carapace!

THE PANGOLIN IN HONG KONG
by Jonathan Downes.

Regular readers cannot fail to have noted that I have a particular fascination with the zoology of Hong Kong, that tiny fragment of south western China which has been a British 'possession' since 1841, and which, on the first of July this year reverts to Chinese rule amidst scare stories of Communist oppression. As both a zoologist and a lover of the colony and its wildlife, one of the most worrying aspects of the handover is the undoubted affect that will take place upon its fragile ecosystem.

Hong Kong is an area somewhat similar in size to the Isle of Wight and is inhabited by about the same number of people as live in Scotland. It is also home to a rich and varied wildlife which in recent times has included tigers, leopards and which even today includes wild cats, two species of civets and other interesting creatures including the chinese pangolin.

Writing in 1951, Herklots noted (*'The Hong Kong Countryside'*. SCMP pubs. p.90) that:

"This harmless and very useful animal is persecuted by the Chinese who believe that the scales have remarkable medicinal properties; the animal is also eaten.."

and also noted that its Chinese name ch'uen shaan kaap which literally translated means 'digging mountain scales', and also notes that kaap also means 'finger nails'.

Even in 1951 the animal was rare in the colony, but Marshall (1967) in *'Mammals of Hong Kong'* (HK Govt Pubs), recorded it both from the island of Hong Kong and the New Territories. When

Dennis Hill and Karen Phillips wrote *'Hong Kong Animals in Colour'* ((HK Govt Pubs 1981), they noted that:

"It is found in small numbers on Hong Kong Island and in scattered localities in the New Territories"..

... but over the last sixteen years its numbers seem to have declined dramatically. In issue 15 of the excellent *'Porcupine!'* (the Journal of the University of Hong Kong Ecology and Biodiversity Survey) published late in 1996 only four recent sightings were noted although the distinctive burrows had been found in several other localities on the mainland.

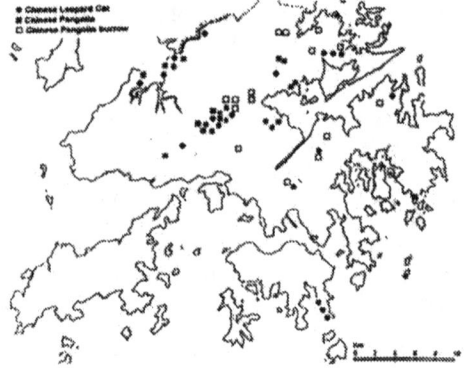

It seems that this primitive and distinctive mammal is almost acting as a 'symbol' for the decline of the native wildlife of the territory. Let us hope that as the ci-devant Crown Colony of Hong Kong enters a new era of political growth that this charming and fascinating animal will not prove to be another victim of 'progress'.

In Pursuit of Pangolins

(This is a short excerpt from Dr Shuker's forthcoming book 'From Flying Toads to Snakes with Wings' and is included with the kind permission of the author.)

Equally as incongruous ... is a scaly anteater (pangolin) abounding in England, but how else can we explain the baroque beast encountered in Dumpton Park, Ramsgate, Kent, on April 16th 1954 by Police Constable S Bishop, and described by him as a 'walking fir cone'? This is an excellent description of a pangolin, those insectivorous mammals covered in huge scales remarkably similar to a fir cone or pine cone. Pangolins, however, are wholly restricted to the tropics of Africa and Asia. Also, they are so difficult to maintain in captivity that they are seldom exhibited in zoos, and hardly ever kept as pets. So even if we do identify P.C.Bishop's beast as a pangolin, how can we explain its presence in a Kent park? We have simply exchanged one mystery for another and emerged none the wiser.

Not only are pangolins non-British, they are also non-aquatic. On January 8th 1973, the Hindu, a Madras newspaper, reported that a scaly, orange-coloured beast measuring almost three feet long, with a sharp, anteater-like nose, and able to walk and to run on land, had been caught by fishermen in the sea off Tranquebar, Madras. True, it has certain pangolin similarities, but pangolins are neither orange coloured nor amphibious. Needless to say, no further news has appeared.

THE BARKING BEAST OF BATH

by Terry Hooper

At the time off this case I was the director of UFO International as well as running a UK Branch of the late Ivan T.Sanderson's Society for the Investigation of the Unexplained (SITU). It was in my capacity as a SITU member that I looked into the affair, although I had been alerted to it by the British UFO Research Association (BUFORA) - and the Press desperately tried to get a UFO group invovled.

The BUFORA National Investigations Co-ordinator, Maureen Hall, sent me three newspaper clippings. I was told that BUFORA wanted me to look into the case and would pass me further information later. I was mystified, but with a national UFO Group seemingly possessing more information, and with the Press trying to get me involved, I assumed that there was more here than met the eye.

The Bristol Evening Post of the 12th August 1980 reported:

"Beware of the Beast! Anyone stalking the mystery beast of Brassknocker Hill, Bath, could be in for a nasty shock, RSPCA Inspector Peter Meyer warned today. Renewed hunts are being made for the creature after a policeman and a taxi driver saw a monkey about three feet tall near the woods behind the hill at the weekend. The beast first appeared last summer, damaging trees and frightening wildlife. Efforts to track it down failed. Today Mr Meyer said: 'If it is a chimp or a monkey and it has been living in the wild for so long it could be extremely dangerous...'"

The item stated that Mr Meyer's search on the 11th had been unsuccessful but that he planned another search that day. Mr Ron Harper, a retired cabinet maker, who lived on the edge of the woods was convinced that this beast was a monkey. He told reporters:

"It has been here in the wood all the time but it comes out in August when it gets warmer and the new shoots appear on the trees. We think that it was let loose from a car, probably by a foreigner who didn't want to report the loss".

I could find nothing that indicated a UFO angle. The Daily Mirror of the same date had a field day. It reported that a 'strange furry creature' was first seen in the august of 1979. The article informed us that:

".... shaggy shapes and glaring eyes made some of the locals think twice about venturing out at night".

Certainly neither Mr or Mrs Harper were afraid of the creature. John Elphinstone, a taxi driver, was driving along when the beast 'hailed him' from the roadside. A policeman dispatched to the scene was in time to see the creature lope off. Inspector Mike Price, of Bath police said:

"We were sure that this mystery creature would turn out to be a monkey of some sort. After all, men from Mars aren't hairy are they?"

I made a few preliminary checks. The RSPCA in Bath and Mr Meyer's office would not return calls or answer letters of enquiry. I tried the Bristol RSPCA who told me that I ought to try the Primate Protection League.... they never answered either! A reporter told me that strange lights had been seen. I visited the area but never found Mr Meyer. The police referred me to the Daily Mirror article and that was it. I telephoned BUFORA and explained that I thought that it was just an escaped chimpanzee. I was told that I should stick at it and I'd be passed on 'certain information' BUFORA had. I tried to get even the vaguest idea of what this 'information' was, but was unlucky.

Animals & Men # 13 — Feature

Map of area involved in the Barking Beast of Bath case.

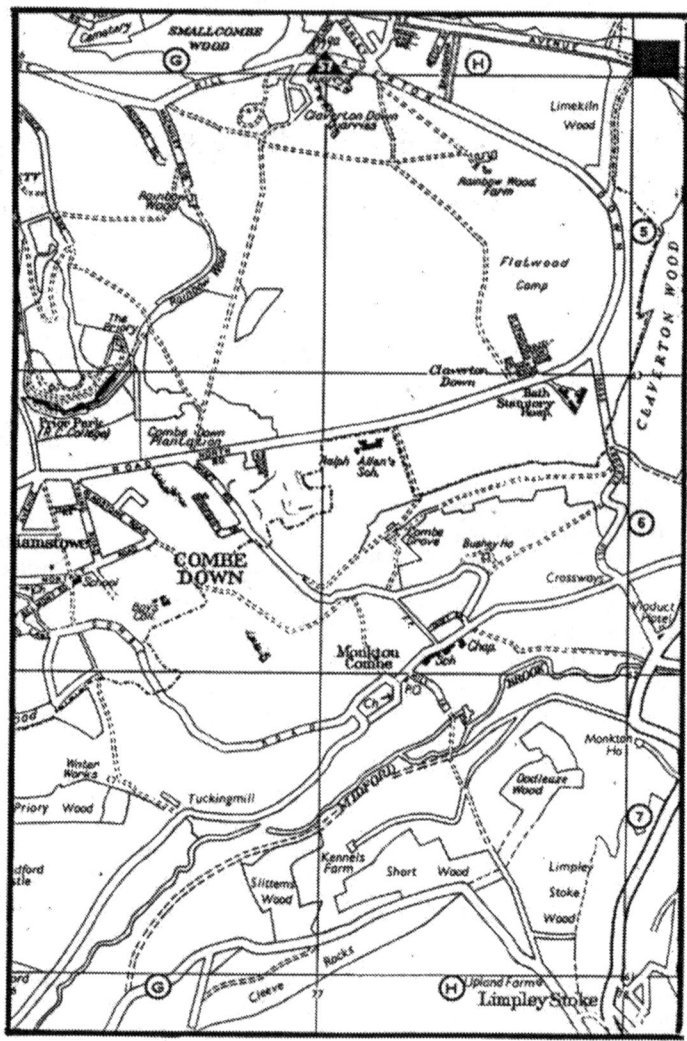

The map clearly shows how sparsely populated the area involved is and the extent of woodland; square H6 marks the Brass Knocker Hill area.

1995 T.Rooper/CITU

I suppose that the article which appeared in The Guardian newspaper on 23rd August 1979 had much to do with this mystery. Though, when I contacted the reporter involved, Dennis Barker, he couldn't recall much other than there was 'something' to do with UFOs and all his notes were gone anyway. It was, oddly enough, Mr Barker who tried to rope in UFO International in 1979 - a year before BUFORA had contacted me.

Barker's article reported that the four foot (1.2m) tall beast had shattered the peace of the little village; it gave the impression of wearing white spectacles. Pigeons, magpies and jackdaws had vanished from the area and bark had been stripped off the trees as far up as twenty feet. (6m). The theory was put forward that the creature must be able to hang upside-down and lean over to do the damage. The beast always did its work at night it seems.

In his article Mr Harper reported that it was some kind of rodent. He wrote that a man from the Bath Park's Committee had paid a visit and stated:

"You know, Mr Harper, if I was not talking silly, I would say that you have a squirrel ten times bigger than normal".

Harper also wrote that the teeth marks found were:

"Ten to twenty times the size of a squirrel's".

The Harper's pet goat would not go near to the tree that the 'beast' had attacked. At least fifty other trees in the area had been similarly damaged.

At this time no-one had actually seen the beast, until allegedly at least, Mr. Christopher Morris and a friend were driving through Monkton Combe at around 12.30 a.m. There, in the middle of the road, stood the creature illuminated by the vehicle's headlights. It was 3-4 feet (90cms to 1.2m) tall, and had bright white rings around the eyes as though it were wearing spectacles. Mr Morris thought that it looked like a baboon, although his friend claimed that it more nearly resembled a chimpanzee.

EDITOR'S NOTE: For more details on this sighting and other background information on this case please consult *Fortean Times* #30 p.10.

I decided to look around the area and for a week I lived in a tree, getting dirtier and smellier by the day. I saw no RSPCA Inspector. No member of the Primate Protection League. Not even one of the reporters who seemed to prefer to telephone around for clues but do no actual searching in the field. I did try to ignore a rather amorously involved couple one night. Nothing else though - well, nothing actually SEEN that is.

There was a strong smell of fresh monkey urine. Having spent many hours watching the great apes and other primates at Bristol Zoo I am very familiar with the odour which is unmistakeable. During the mid 1970s I was visiting Westbury Wildlife Park and achieved the dubious honour of being urinated on by a chimpanzee which left an unforgettable smell! I could find no faeces but the smell convinced me.

I contacted BUFORA and was told that they hadn't thought that there was anything to this case. The strange light had turned out to be a meteorite. I told BUFORA where to go; I'd wasted time and money for nothing. The case was closed, or so I thought.

In 1996 I received word that between one and three chimpanzees might be loose in the area.

There was nothing new after my appeal to the press, but I did get a letter from John Elphinstone who is now living on the island of Benbecula in the North Atlantic. Yes, he HAD seen a chimp and a policeman had seen it also. Mrs Harper, the widow of Ron, wrote to say that he had seen the 'chimp' on several ocasions. Primates on the loose seem to becoming common; a Baboon was sighted in Shropshire in 1996.

But does something else live in the woods now? I

Animals & Men # 13 — Feature

received a letter in November 1996 from Martha Wakelin. In August 1995 she was staying with a friend whose house is in the valley just below Brassknocker Hill.

They were preparing some jumps for pony riding when Martha turned and saw a black cat sitting on the top of the hill next to a log. They later measured the log and found it to be 2m long and 30cm high and wide. The cat seemed large in comparison.

Martha alerted her friend who was rather bemused. Both decided to approach the cat and see whether it was tame (something that I would NOT recommend that anyone do). As they got to within a hundred metres of the cat it turned and ran off into the woods. Its size was estimated as three times that of a domestic cat, probably 60cms high, 30-45 cm wide and 60 cm from the front paws to the tail.

The next day, Martha's friend's father was getting out of his car when he called out; "Look at that cat!" All three saw it in the same spot as in the previous day. The incident was not reported.

So, are there large cats also living in the woods at Brassknocker Hill? It is interesting to speculate upon the situation. No doubt, some 'foreigner' was driving past when his pet baboon/chimp and cat escaped and he just never reported it.

EDITOR'S NOTE: Methinks Senor Hooper is being facetious!

Whatever the explanation, Britain's fauna is getting some strange new additions.

EDITOR'S NOTE: There are some interesting parallels between this case and other reports from the Westcountry. As I have noted elsewhere there are a growing number of reports of what are described as 'monkeys', 'apes' and even 'bigfoot' from the United Kingdom.

The Devon folklorist Theo Brown collected a number of such stories including one chilling recollection by a friend of hers who had been walking alone at dusk near the neolithic earthworks at the top of Lustleigh Cleave on the extreme eastern side of Dartmoor.

Lustleigh Cleave is an extraordinarily strange place, and it appears to be one of those 'window areas' where an inordinate number of unexplained incidents and anomalous phenomena seem to take place on an almost monotonous basis. I have got reports of sightings of a ghostly Tudor hunting party, mysterious lights in the sky, and even the apparitions of a pair of Roman Centurions, but Theo Brown's friend saw, clearly, a family of 'cave men', either naked and covered in hair or wrapped in the shaggy pelts of some wild animal, shambling around the stone circle at the top of the cleave just a few yards away from the place where in 1978 two schoolboys found the skull of what is certainly a large predatory felid, a discovery which to this day remains a mystery.

Another report, also from the South West concerns a man who later in Africa had a reputation as a big game hunter who saw a creature at the Hangley Cleeve barrows in Somerset which he described long after the sighting as the most terrifying thing he had ever seen.

He described it as a *'crouching form like a rock with matted hair all over it and pale, flat eyes'.* I have other reports from that area of hulking man shaped shadows that are seen in a local quarry, and indeed the annals of forteana are full of such events world wide.

Whereas I have presented a case for the zoological identity of some of the more well known BHM sightings across the world, I would not suggest for one moment that these people are part of a relict population of *Homo erectus* who are waiting on the genetic sidelines before emerging into the real world to make their existence obvious, I would maintain that such apparitions fall firmly into the category of BHM sightings worldwide.

Other British, and specifically South Western British sightings of anomalous phenomena which although they are presently classed elsewhere amongst the pantheon of fortean phenomena, are the Ape and Monkey Ghosts such as 'The Man Monkey of Lincolnshire', and more appositely the ghost ape of Marwood in Devon and 'Martyn's Ape' of Athelhampton in Dorset, which although they are explicable within the terms of purely regional folklore as 'animal ghosts', exhibit in my opinion, characteristics analogous to those exhibited by the smaller BHM pehenomena of parts of the United States.

Unlike the phenomena in America, however these British phenomena each have a convenient little folk story to explain their presence in the occult infrastructure of the region. The Ghost Ape of Marwood was, when alive a pet of a local landowner who one day grabbed the landowner's young son and climbed a tree with him, refusing to come down, whereas the well known spectre of "Martyn's Ape" is supposed to have its origins in the unfortunate pet of an earlier female scion of the Martyn family who was either accidentally walled up alive during building work, or entombed (also alive) when the daughter either committed suicide in a locked, secret room or was walled up by an unforgiving parent, (depending on which account you read).

It is my supposition that rather than the apparitions being a result of these, rather far fetched stories, the stories were rather invented by local people to explain the sightings of monkey shaped apparitions, or small BHM as we should really refer to them, that had been seen in the vicinity since times immemorial.

I noted some other reports in my articles for *Fortean Times* (December 1995), *Encounters* (July 1996) and *Sightings* (May 1997. One specific incident which has disturbing parallels with the events recounted by Terry Hooper occurred in South Devon during the late summer and early autumn of 1996.

In south Devon, between the towns of Paignton and Brixham lie Churston Woods. These woods have long been of interest to me because of the sightings of mysterious small carnivores which appear to be a relict population of Beech Martens (a species thought extinct in Britain since the last Ice Age). Fifteen separate witnesses over a six week period in August/September 1996 reported seeing what they described as 'a green faced monkey' running through the woods. Although some of the descriptions were very vague most of them described a tailless animal between four and five feet tall with a flat, olive-green face. Although there are primates with 'green' faces, (for example the olive baboon and some of the west African vervet monkeys), none of these correspond in the slightest to the descriptions of a humanoid or chimp like creature which was seen both swinging through the trees and running through the woods.

I think that it is no coincidence that there were a number of UFO sightings in the area at the time and also several crop circles in the vicinity - one of which contained the mutilated corpses of several pigeons. (See Pete Glastonbury's article in *Goblin Universe* #5).

Although it is certain that bona fide primates do on occasion escape from private collections and zoos, and may well live for a while in the British countryside, I feel that the Barking Beast of Bath, like the other examples that I have presented above is a puzzling zooform phenomenon rather than an out of place animal *per se*.

HERE BE PANTHERS

by Richard Freeman

In bygone days, map makers drew monsters over unknown regions of the world and wrote 'here be dragons'. As I found out to my cost on a recent crypto-excursion, today's map makers are hardly more competant.

The whole sorry episode began with an article in the *Yorkshire Evening Post* on October the third 1996. Mr John Lisowiece and his family reported seeing '*an all black animal much larger than a dog but smaller than a calf*' running at 45 degrees across a field towards their car which was caught in traffic at the time. He described a long curling tail and a big round head.

The exotically namd family were returning to Leeds from Ilkley, and saw the cat close to Sherbourn. Shortly before an unnamed man had seen a panther close to Selby and another couple had spotted the beast on September 22nd in the same general area.

Another witness, Mr Stephen Johnson and friends picked up the beast's eye in his car headlights. Mr Johnson claimed that the animal showed no fear of his car and continued down the hedgerow. As it turned out, Mr Lisowiece resides in the same area as I do, and so I decided that an interview would be a good idea. Sadly, the Lisowieces (say that ten times when you've had a few), were ex-directory. Even so I had a crypto-mystery, figuratively, at least, on my doorstep, and it seemed that the next logical step should be an investigation in the field.

Asking for the most detailed map of the relevant area (most sightings were around Selby, just south of York), I made a photocopy of the map. Circling all the sightings I triangulated an area just North-East of Selby. According to the map there was a large wood almost as big as the town slap bang in the area.

I invited two of Leeds' finest forteans along with me. Mr Philip Thoruley, an occultist and fellow goth, and Mr Jake Kirkwood, proprietor of Ubik, the finest second-hand book store in Leeds and a veritable mine of rare and wonderful fortean books.

EDITOR'S NOTE: Hey Richard - Graham and I *are the only ones allowed to do gratuitous 'product placement'* in this mag..

I procured a hessian sack and an excessively large leg of lamb as bait. Armed with cameras we were ready. I had my packed lunch and sensible shoes. The expedition could begin.

It was at this point that things all went 'pear shaped'.

EDITOR'S NOTE: I have never actually understood that expression...

The supposedly detailed map had failed to show a labyrinth of tiny back roads in which Jake (who was travelling separately) got completely lost and was not seen again until the following day!

Philip and I drove around fruitlessly for about an hour,. During this time it dawned on me that this was by no means ideal panther country! The whole area was arable farmland with no tasty livestock and absolutely no cover. Eventually we asked a local, and showed him the point on the map where we were trying to get to. The woods, or so it turned out, were several miles away from the place they were supposed to be according to the map.

At last we found the woods. (Bishopswood as it was properly known). Looking around we found the prints of several large dogs and a muntjac deer. I placed the leg of lamb into the hessian sack after making several large gashes into the meat.

We dragged the sack through the woods from several areas back to our observation point. Then we tied the sack out of reach of fox or badger and hid in an old trailer to watch. In a scenario with which most cryptozoologists must be familiar absolutely NOTHING happened! Not a sausage!

To be honest, I had doubts about the whole area. If there were 'panthers' here I think that they were just moving through, perhaps straying in from the more wooded Yorkshire coast.

The thing that was really staggering was the innaccuracy of the maps. I find myself wondering whether, if I have this much trouble with maps in tame old Yorkshire, what will the wilds of Tasmania be like when I go there later this year in search of the elusive thylacine?

ANOTHER YORKSHIRE ABC
by Henry Moncrieff.

EDITOR's NOTE: Bearing in mind the contents of Richard's penultimate paragraph, it is ironic that very soon afterwards I received the following account of an ABC sighting taken in extracts from a letter from Yorkshire based artist Henry Moncrieff. He telephoned me one morning in late winter to tell me that only twenty minutes before he had seen what appeared to be a puma walk leisurely across an area of scubland at the bottom of his garden. I asked for some background details on the area, and a week or so later he wrote to me..

Near my house is the main cliff edge. Below is the undercliff covered in thick brush, ten foot or so high. This extends for about half a mile from my house and is about 100 yards wide until it meets the sea. Foxes and badgers live in the gorse and the hawthorn scrub.

My girlfriend and I inspected the area but we didn't find any fur or prints. It had been a very hard frost that night and naturally the ground was frozen solid. However, the previous week, my girlfriend saw large unidentified paw prints in the snow whilst out walking. Two nights before the sighting we were in bed reading at about 1.30 a.m. We heard something large move over a loose drain cover just outside our window. We are used to foxes and badgers in the garden. We leave out meat scoops for them. Whatever it was that moved over the loose drain cover was heavier than either.

On speaking to local woodsmen and poachers in the area, it seems that at least five well attested sightings of this animal have been made. (See sketch map below). To illustrate these sightings all the people that I have spoken to have described the same animal that I saw. The sightings listed are between November and January. If more information comes my way I will pass it on to you...

Picture Courtesy H. Moncrieff.

DANISH LEATHERBACK TURTLES

by Lars Thomas

EDITOR's NOTE: When we met at the 1996 Unconvention, Lars and I had a long discussion about stranded cetaceans in northern waters. In passing I mentioned my interest in European records of marine chelonia and much to my surprise he informed me that there were several records of the Leathery Turtle or Luth from Danish waters. I asked or more information, and true to his word, a few days later, I received the following letter and map...

"In December 1948, a dead leatherback turtle was found on the beach at Ylderby Vesterstrand, It disappeared at the next high tide, drifted roughly 100m abd was washed up on the shore south of Skalderviken in Sweden on the 22 January 1948. This was the first Danish record.

The second, and to date the most recent, is from the 2nd September 1965 when a fisherman caught a live specimen almost five kilometres west of Nymindegab in western Denmark. We do have a third Danish record, but this specimen, although caught by a Danish fisherman was in British territorial waters".

EDITOR'S NOTE: We would be interested to receive any other accounts of out of place chelonia from northern waters, and are presently compiling a major project on the subject.

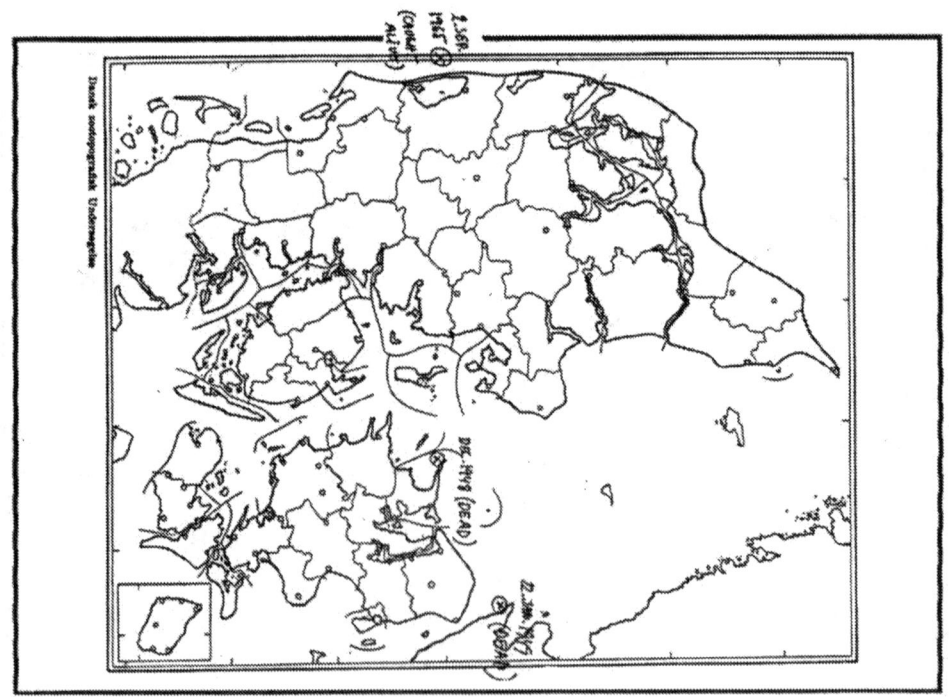

DANISH LEATHERBACK TURTLES

NORTH OF THE BORDER

For those of a vagabond disposition and an enquiring mind, coupled with a pair of sturdy thighs, a barrel chest, a rugged jaw, a grip like steel, a glint in the eye, and a suspicious bulge in the Levi's; there's only one place to be, it's...

with Tom Anderson
(Ja! Ja! I have never done it like zis before...)

Shine on Algae Bloom..

As I write the first harbingers of spring are with us. Birdsong at eventide, a carpet of crocus underfoot and yet another twist in Nessie's tail.

Following the claims of a charter boat skipper to have discovered a huge cavern at 240 metres, an Aberdeen sub-sea firm intends to survey the area using robotic cameras at an alleged cost of £10,000.

A laudable endeavour and somewhat fortuitous given the current exchange rate against the dollar and the adverse affect that this has had on this year's level of accomodation in the Inverness area. As your northern correspondent narrowly missed being a roadkill statistic last summer due to a German tour bus driver's attitude *(for you Tommy ze Summer is over)* I can hardly wait for this year's influx of Camcorder Man (similar to Cro-Magnon but less social skills)......

****

Every Christmas I am sent a very glossy, up-market fishing magazine from a 'Gentleman's Outfitters' in Pall Mall. The mechanics of impaling a fish on a hook receives considerably less attention than what to wear whilst doing it.

Almost as much print is dedicated to the fisherman's 'oneness with nature' and sundry articles extolling the writer's powers of observation.

I have yet to read one of these pieces where the author has failed to recognise some leaf, toadstool, dropping etc.

How disappointing, then for me to have to report my countrymen's shortcomings at a trout fishery near Stirling. For some time the anglers had been observing two crows, one quite large, swooping down to catch dead fish. These should not be available in a well run fishery, but I digress.

The RSPB finally identified it as a turkey vulture, wingspan getting on for two metres, ringed on left leg and doubtless an escapee.

Lack of thermals over the Atlantic precluding broad winged raptorsmaking it over. If nothing else this reinforces my scepticism regarding eyewitness accounts involving distance, ratios and volume, especially over water.

It also does nothing to dispel any doubts remaining over the veracity of fighermen's tales.

Pearl's a Singer

(as recorded by Whelkie Brooks?)
by Dr Karl P.N. Shuker.

EDITORIAL NOTE: The entire editorial team would like to disassociate themselves from the above pun which was the sole responsibility of Dr Shuker.

After our esteemed editor's recent tour de force in *Fortean Studies Volume Three*, dealing with those musical murids of mellifluous melody (or singing mice, to thee and me), I felt compelled to share with you the wondrous saga of Molly, the singing oyster.

T'was a dark and stormy night - possibly - in 1840 when Mr Pearkes was first captivated by Molly's bewitching siren song, resonating from the dank depths of a cask of shellfish newly-delivered to his store in Vinegar Yard, Drury Lane. In reality, it sounded more like a high-pitched whistle, but if you should ever find yourself being serenaded by an oyster, you will no doubt appreciate why Pearkes did not worry unduly about such phonic technicalities. Here was a star in the making, a veritable diva from the sea depths!

And so it was that Molly was proudly displayed by Pearkes in Vinegar Yard for all to see, and hear. During the next few weeks, regally ensconced within a tub of oatmeal and brine, this shrill soprano of the shellfish world delighted an ever-present audience with her reedy repertoire - enchanting the acting company from the nearby Theatre Royal, and featuring in several contemporary newspaper stories. Indeed, as befitting this molluscan megastar, she even inspired a song, penned by a popular music-hall artiste named Sam Cowell; and her lamellibranchian likeness was faithfully committed to paper and ink within the pages of Punch who referred to her as 'a phenomenal bivalve'.

Not surprisingly, Pearkes received several sizeable offers of money from circus owners and theatre managers to purchase Molly, but he declined to relinquish his protege, thereby ensuring that this nacreous nightingale sang on in Vinegar Yard.

Tragically, however, in the unfathomable workings of the Universe, dragons and shellfish-sellers may well live forever (or seem to), but not so little boys or singing oysters. And so it was that came the sad day when Molly whistled no more, having departed this mortal maelstrom for a bright, celestial sea transcending her humble oatmeal abode.

But why - and how - did she acquire her unique vocal talent? Popular opinion was sharply divided, between the satirical and the scientific. On one hand, in the words of Douglas Jerrold, perhaps our fair Molly *'...had been crossed in love and whistled only to keep up appearances'*. On the other hand, her musical attributes probably owed their origin to a simple fluke in the shell's architecture - water passing across her gills creating a whistling sound through a small hole that had somehow formed in her shell.

Whatever the answer, however, Molly's place in the annals of contemporary music must surely be assured. As to her favourite arias: whether or not they included such treasures as "Shell be coming round the mountain when she comes" or "Thank heaven for little pearls" is not recorded. (And neither, as far as I am aware, are they. Such is life.)

A World Beating Musical *Mus musculus* -

The International Singing Mouse Contest of 1937.

By Nick Smith

EDITOR'S NOTE: I have been interested in the phenomenon of singing mice for ages, and am the author of what I hope is the definitive work on the subject (no-one else would be silly enough to write about them) which was published in Steve Moore's excellent *Fortean Studies Volume Three* late last year. During my research I discovered that there was a BBC archive recording of these wee cowering but not very timorous beasties still in existence. I was unable to visit the sound archives in person so I sent along our intrepid Home Counties representative to investigate...

In the National Sound Archives of the British Museum is a two minute long recording (reference MP9810) of an extract from the BBC's and the CBC (Columbia)'s 1937 joint broadcast of the results of their search for the greatest international (seemingly only North American and British) singing mouse. It is a bizarre, amusing, but ultimately insubstantial and uninformative fragment, demonstrating only that singing mice must be very elusive creatures.

The English announcer kept the regulation BBC plum in his mouth throughout, but his jocular, giggly tone, wasn't what I expected of 1930's radio. He gave no details of the contest, nor of the age, sex or type (I assume that they were all the house mouse *M.musculus*), of the performers, their keepers and home lives, or indeed, how their special talents had been nurtured. We were merely told the competitor's names and the places that they represented.

First, was a lackadaisickal duet by the British contingent, Mickey of London and Chrissie of Wales. They made no more than usual unmelodious mouse noises: a hubbub of chirping, chattering and cheeping, along with scratchings and rustlings as they skittered around.

This unimpressive effort was obviously not going to take a lot of beating and, sure enough, the announcer, now in a cacchinatory fit, then introduced the winner of the contest, Mikey of New York, accompanied by Johnny of Toronto and Minnie of Minnesota. Their recording ended with thir performance, two of them making the familiar mousey squeaks and scrabblings, while another, presumably Mikey manically piped a reedy tremelo, his tempo and volume at least ten times faster and louder than his colleagues.

But, despite the strident Mikey's zeal, the BBC and the CBC were over generous in describing any of this cat- (mouse?) erwauling as 'singing'. Even Mikey was not musical, melodious or harmonious as these terms are commonly understood compared, either with other animals such as birds or cetaceans, or more anthropomorphically.

He didn't follow a human singer or instrumentalist, unlike Murri the German tom-cat made famous through radio broadcasts of his word-perfect renditions of nursery rhymes or even speak a human language (albeit 'parrot-fashion') in a sing song voice like Pepe the Chihuahua. (Both cases cited by Michell and Rickard, 1982).

However, even more worrying than this apparent laxness of the judging criteria of such a prestigious international context, was the extreme speed and fervour of the champion's performance. Was this rodent raving mad, or just raving? Had Mikey been taking performance enhancing amphetamines? Should there have been random drug testing of the competitors?

I've never heard anything like this outrageously frantic exhibition before, and, judging by his reaction, neither had the announcer.

Whatever it was, it certainly wasn't singing, and after listening to him, the idea of a rodent proto-Lou Reed producing a tuneless speed-freak racket in New York during 1937 sounds as plausible as anything else that I can come up with.

Thanks to Tony at the National Sound Archives.

Reference.

Michell and Rickard (1982). *Living Wonders* (Thames and Hudson, London).

EDITOR'S NOTE: The full background to this remarkable competition can be found in my paper for *Fortean Studies*. In it I discuss the folklore and the factual accounts of singing mice and come to the conclusion that whereas most of these creatures are in fact suffering from a debilitating upper respiratory tract infection, there may in fact be a genuine fortean element to some of these reports. It has always saddened me that I never heard the BBC recording for myself, and Nick Smith's ridiculous critique of it makes me more determined to head for the National Sound Archives at the earliest possibility.

The Zebro - Prehistoric Survivor or Feral Donkey?
by Angel Morant Fores

For centuries, scholars have tried to find out the origin of the word 'zebra', the name given by Portuguese sailors to the famous striped equid of Africa. It was not until the beginning of this century that E.Merea, a Portuguese philologist, conclusively demonstrated that the word 'zebro' was also used by the Spaniards and the Portuguese to designate a local wild ass inhabiting the Iberian peninsula.

The Zebro, together together with other wild animals such as bear, wolf, deer, etc is mentioned in several Spanish treatises on hunting, from the Middle Ages and the Rennaisance. In one of these works it is described as a *'mare-like, grey-coloured animal, with a black strip running along its back, and a dark muzzle'*. However, the taxonomic identity of the mysterious equid from which the african zebra borrowed its name has remained a matter of continued dispute.

Now, Spanish archaeologists Carlos Nores and Corina Liesau claim they have found a solution to this long-standing zoological riddle. In a recent paper published in ARCHAEOFAUNA they state that there are serious hints that the Iberian zebro might correspond to the *Equus hydruntinus* of the European Pleistocene. According to Nores and Liesau this prehistoric equid might have survived in southern Spain, and, perhaps in some remote parts of Portugal into the 16th Century.

There has been some controversy regarding the precise affinities of *Equus hydruntinus*. Whilst most authors think that it was an ass-Equus (Asinus) - a few others, on the basis of its dental remains, argue that its closest living relatives are the african zebras. Until recently *Equus hydruntinus* was believed to have died off 12,000 years ago, but

but some Neolithic and Bronze Age sites in Spain have yielded bones of a small equid that may contribute to change this long held view. Although the identification of such remains is difficult due to their fragmentary nature, two German palaeontologists who have examined them are confident that they belong to *Equus hydruntinus*.

Worth mentioning here is that the African domestic ass or 'donkey', an animal whose remains could easily be mistaken for *Equus hydruntinus* was not introduced to the Iberian peninsula until the 8th Century B.C.

Recently Nores and Liesau found a more complete sample of bones of a similar equid at a Bronze Age site in southern Spain. The remains have been forwarded to the Department of Zoology at the University of Madrid for evaluation.

SOURCES

Carlos Nores and Corina Leisau (1992) *La Zoologia historica como complemento de la argueozoologia El casa del Zebro*. ARCHAEOFAUNA, 1: 61-71.

EDITOR'S NOTE: It never ceases to surprise me that so many new discoveries are being made in the field of horses (no pun intended). One would have imagined that there was absolutely nothing left to discover as regards the Euidae, who, after all probably deserve the appelation of 'man's best friend' at least as much as do the dogs.

In recent issues of *Animals & Men* we have read about new animal discovered in Tibet and other parts of central Asia, as well as the ongoing discoveries concerning the true nature of the Quagga. If, indeed, this supposedly prehistoric horse did survive in Europe well into historical times it gives good hope for those of us researching other putative prehistoric survivors in the region.

"Here's another clue for you all the 'walrus' was ....DARREN!"

ODOBENUS UPDATE...

EDITOR'S NOTE: In A&M #11 we printed a collection of articles about mysterious, fortean and out of place walruses. Here two of the contributors to that collection; Clinton Keeling and Darren Naish, tie up a few loose ends and return to the subject of these singular pinnipeds...

THE RETURN OF THE SPACE WALRUS

by Darren Naish.

One of the aspects of contemporary walrus research that I attempted to make clear in my article in A&M11 (pp. 16-18) [1] was that the taxonomy of the only living walrus *(Odobenus rosmarus)*, is a comparatively settled issue. How ironic, therefore, that a paper published in 1996 should present data that nicely muddies it all.

Writing in *The Journal of Zoology* (240:495-499), two scientists of Oslo's Norwegian Polar Institute, O.Wilig and I Gjertz, explain how the immobilised and measured adult male walruses at Svalbard between 1989 and 1993. 41 animals were measured. Targeting especially large walruses Wilig and Gjertz obtained a sample varying from 960 kg to 1475 kg in weight, and from 258 to 380 cm in length. A weight of 1883 kg was estimated for one individual, but possible errors in weight estimation led to the suggestion that Svalbard walruses may weigh up to 2000 kg.

A two ton walrus is hard to imagine. I am tempted to recall a scene in 'Sinbad and the Eye of the

Tiger'. Turning to the application of this data, walrus literature states that Pacific walruses (*O.r.divergens*) are characterised by greater size than that obtained by Atlantic walruses (*O.r.rosmarus*). Standard lengths of Pacific walruses have been given as 250-300 cm, and weights as 800-1700 kg.

Wilig and Gjertz's new data shows conclusively that Svalbard walruses, and possibly Atlantic walruses as a whole, can obtain (if not exceed) the body size of Pacific walruses, so a distinction based on these factors is invalid. The sizes of individuals measured in the past may have been recorded without due consideration of factors determined by age, so the differences in size between the walrus subspecies could well be an illusion.

Do other features remain to support the distinction of the subspecies?

Pacific and Atlantic walruses still differ slightly in colour and some aspects of skull morphology, but some experts now stress that walrus taxonomy really is in need of revision. Far be it from me to hint, but the view that eventually emerges may well turn out to be more complex than taxonomic tradition.

According to an 1881 text by T. Southwell, walruses are the inspiration for one of the most consistently recognisable sea monsters of old maps and bestiaries, Gesner's *Vacca marina*. With a pig-like face and upward-pointing tusks, this 'sea orc' acts as an evil devourer of maidens in certain tales. Lest we stoop so low as to suggest that it represents yet another cryptid spied by ancient mariners, it is clearly a mythified walrus.

There are, to my knowledge, no recorded instances of walruses attacking humans, but these remarkable beasts have been known to eat seabirds and seals.

Some old male walruses actually become specialist seal killers - supposedly, they can be told from normal walruses by their heavier forequarters and blubber-stained tusks and skin. Indeed, there are reliable accounts from professional marine biologists of walruses grasping them with foreflippers and then dismembering them with their tusks. Both Ringed (*Phoca hispida*) and Bearded seal (*Erignathus barbatus*) remains have been found in walrus stomachs, as have, surprisingly, pieces of young walrus.

Normal walrus feeding is remarkable enough: they cruise along the seabed in search of large bivalves and, on finding one, violently suck the protrusive soft parts out of the shell.

Walruses do not dig with the tusks, as was long believed, but tusks are used in combat and in hauling out onto ice. Bivalve hunting is done by touch, using the sensitive bristles and upper lip, and the eyes are kept closed. Walrus suction feeding is made possible by unusual modification of the palate and throat musculature: this is clearly an efficient apparatus as captive walruses have, on occasion, demonstrated tremendous suction power.

Judith King (in *Seals of the World*, 1983) writes of one that managed to pull the 2.3 kg (6 lbs) metal plug from the bottom of its pool, and while the pool was full of water!

The mechanics of suction feeding are still poorly understood. A, however, it is essentially the only means a walrus has of getting food into the mouth, it must be the method employed during consumption of seal meat.

Narwhal (*Monodon monoceros*) remains have also been recorded from walrus stomachs. As with at least some of the seals, it is probable that the flesh was from scavenged carcasses, rather than individuals specifically killed by the walruses. There is one anecdotal account of a fight between an adult male Narwhal and a walrus, with the walrus winning and then eating the whale. It might be true, but then, it might not.

Note

Due to an editorial mishap somewhere along the way, the first line of may previous walrus article read *'As with so many other groups of extant animals, the sole living specimens of walrus, Odobenus rosmarus, represents but a fraction of the diversity that exists in its fossil relatives.'*

It should, of course, have read: *'As with so many other groups of extant animals, the sole living species of walrus, Odobenus rosmarus...'*

Also, 'their rank and subspecies' in line 10 should have read 'their rank as subspecies'.

PINNEPEDOPHOBIA
by Clinton Keeling.

On re-reading my article about walruses I realise one passage didn't quite make sense - probably because the subject filled me with such horror. What I meant to say was, as a claustrophobe, the thoughts of water passing over my head are frightening to say the least, and when this is coupled by the physical proximity of an animal that, to me, is gruesome and repulsive, the whole situation takes on a nightmarish quality - especially as the person relating this to me told with great glee and relish how these hyper-intelligent creatures indicated their wonderful sense of humour by holding one under the water until they felt they were about to pass out. Aaarrrhhhh...

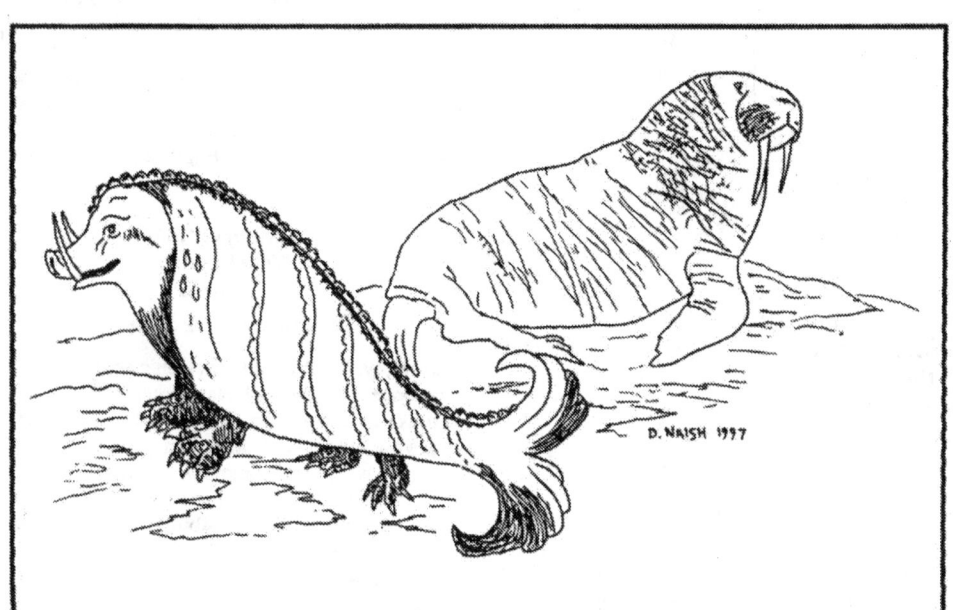

Vacca marina (after Gesner): a mythified walrus?

LETTERS TO THE EDITOR

We welcome letters on any subject of interest to readers of this magazine, although we reserve the right to edit and/or omit where appropriate. Every effort has been made not to unwittingly libel anyone or to infringe upon their copyright. Any such actions are, however, the responsibility of the individual writer and not of the Editor, the editorial team, the Centre for Fortean Zoology or the publishers of this magazine. So there!

A CALL TO ARMS

Dear Readers,

No magazine can rely solely on the efforts of a selected few, who without wages strive to spread information. Therefore, it is very important and necessary, that we help by contributing what we can. If you do not feel able to write yourself, send a clipping or make a phonecall, or send a short notice. Then, someone else will eventually do the writing.

Animals & Men now has readers in over twenty countries, and the common denominator is interest and knowledge. This is why it is so important that YOU contribute, because in your own circles you are the best and the first to know. I do not expect the editor to go to my country, Denmark, for information, as I consider it my duty to forward it with pleasure. Therefore, dear brothers in arms, please use your valuable talents, write or send clippings. The fact that a few readers have been listed as 'local correspondents' is not meant as an exclusion to all others from that area. If you feel insecure about contacting England you can write to me at:

Taarnbyvej 104,
2770 Kastrup,
Denmark.
Telephone + 45 32 50 48 78

Personally, I will try and exchange addresses and telephone numbers with Danish cryptozoologists, to make us into a more efficient network. Let us hear your wild idea ! Believe me, it is a lot more fun to be active and communicate with people of the same interests. Therefore I hope that local representatives will, if possible, list their address and telephone numbers.
Best wishes,

Erik Sorenson,
Denmark.

Animals & Men # 13 — Letters

EDITOR'S NOTE: Unfortunately it is very much the case that as investigators into the more arcane aspects of the zoological sciences we are potential targets for what can ungenerously be described as every loony on the block. I quite understand why some regional representatives are therefore unwilling to make their addresses public.

However, if any reader wishes to write to any regional correspondent, contributor or consultant of this magazine, they may do so c/o me at the Editorial Address and whilst I do not guarantee a reply, I guarantee that your letter will be passed on.

As Erik says, there are many inate problems being what is essentially a non-profit making organisation. Unfortunately, it seems that the 'powers that be' within the more mainstream branches of the zoological sciences are unwilling to give credibility to what they perceive as a bunch of dangerous anarchists, especially when it involves them parting with money.

The last time that I made such a comment it provoked a heated debate in the letters pages of this magazine, regarding our sometimes 'off the wall' approach to cryptozoological methodology.

I say now, as I said then, that I am attempting to edit a magazine which is read by people between the ages of 12 and 87, and by people who range from eminent molecular biologists to new age travellers.

It is part of the inate rationale behind this publication, and indeed behind our work on a holistic scale, that we need both ends of the spectrum in order to do what we do. Possibly this means that we shall never gain mainstream acceptance, but if this is the price we pay for being able to continue, then so be it...

MAN FROM ATLANTIS

Dear Editor,

I write with reference to the story in your latest Newsfile (A&M #11 p.9) about the family in the Philippines who claim that they have functional gill slits.

The story reminded me of another that I read recently about an eleven year old Australian boy who underwent surgery to remove 'fish-gill cartilage' from his neck [1]. Such claims of gill-like structures, supposedly vestigial evolutionary leftovers, seem to surface in the tabloid press from time to time. Indeed, many elementary textbooks perpetuate the notion that the human embryo passes through a series of evolutionary stages as it develops, one being a 'fish' stage in which the embryo has 'gill slits'.

In fact, the so-called 'gill-slits' are nothing of the sort. The human embryo DOES have pharyngeal pouches, but they have nothing to do with gills in form or function.

The pouches may superficially resemble gill slits, but they do not open into the throt (and are therefore not slits), and do not develop respiratory tissues or structures (and are therefore not gills).

In the human embryo, these pouches actually develop into various glands, the lower jaw, and structures in the inner ear. One authority states: *"Since the human embryo never has gills - branchia - the term pharyngeal arches and clefts has been adopted for this book"*. [2]

The claims of remnant fish-gills in adults and children are similarly baseless. Histological studies showed that in the Australian case mentioned above, the cartilage in the boy's neck was indistinguishable from normal human cartilage. The likely explanation is that some abnormality of embryonic development led to cartilage being incorrectly 'seeded' in the neck (i.e it is normal tissue that ended up in the wrong place). [1].

Yours sincerely,

Paul Garner,
Ely.

1. Wieland, C. (1994) 'A Fishy Story'. Creation Ex Nihilo 16(4):46-47
2. Langman, J. (1975). *Medical Embryology*. Third Edition. p.262.

EDITOR'S NOTE: I am certain that the Philippines case mentioned in A&M11 was of a similar provenance to the Australian case described by Paul Garner. However, the original newspaper report did mention that these people were able to 'breathe' underwater, and whilst it is almost certain that this is a piece of wishful thinking on the part of a gullible reporter, I would refer readers to the relevant chapter of Peter Costello's 'The Magic Zoo' where he describes historical accounts of people who did seem able to stay submerged for an extremely long period of time.

LION THROUGH OUR TEETH?

Dear Jonathan,

If the animal depicted on page 17 of A&M12 is a Barbary Lion, my name's George! Look at all that yellow in his decidedly scraggy mane, and I am sorry, but peer as I may I can see none of the characteristic extension of the mane running along the underside of the body - a dark line that's probably a shadow, yes, but no more than that! Still, as I've said for decades now, generally speaking people see what they want to see...

Contrary to popular belief, and semi-scientific dogma, the last wild Barbary Lion was shot in Algeria in 1943, and not 1922. It's full and luxuriant mane - even in wild specimens it was well endowed here - was completely black, so any present day specimens with an admixture of yellow here must automatically be suspect as far as purity goes. As Chris Moiser correctly points out, a great many present-day zoological garden Lions have Barbary blood in their make-up.

Sometimes it's very dominant - for example, the Port Lympne animals look very convincing, and I well recall the magnificent Sultan, who lived at Manchester's Belle-Vue from 1942 until 1953 when he died at an estimated age of eighteen. He was regarded as the finest lion then in the country, and was certainly mainly, perhaps even wholly, Barbary. (Incidentally, an interesting point here, but for some reason nowadays, you never hear of a certain individual lion being described as a particularly fine specimen - part of the way in which wild animal husbandry and zoological gardens are going I suppose). In the surprisingly good, indeed underrated Tunis Zoological Garden in 1973 I was shown an animal specifically stated as being a Barbary, and learned that it had come from the Frankfurt Zoological Garden.

It's the first time that I have ever seen the Barbary Lion equated in any way with the Cape Lion, which became extinct as long ago as 1865, when the last specimen was shot in Natal - and which is too often confused with the forma typica Lion still to be found in South Africa.

The Cape Lion was a very definite sub-species immediately recogniseable by being physiologically different from any other modern Lion, by means of its massive head that is perfectly rectangular in profile - as can immediately be seen in the excellent mounted specimen in the United Services Museum in London's Whitehall. Incidentally, there are only four other known mounted examples of this race - all in South Africa.

I found the article on the Mitten Crab of particular interest, as it reminded me of a tragic event, again

Animals & Men # 13

Letters

at Belle-Vue, when a Tigress named Stella, killed her keeper. This was on the 8th November 1925. Concerning this animal, I quote directly from my book 'The Life and Death of Belle Vue' (Clam Publications):

"She died after only four months in Manchester, and the autopsy revealed that she was riddled with tuberculosis which, the veterinary surgeon said, was probably caused by eating crayfish or crabs in the wild. This fascinated me for two reasons:

a. It's more than likely that the disease had affected her brain, or why else would she have been so savage to the extent of killing, without provocation, a devoted keeper?

and

b. There seems to be some kinship between TB and certain crustaceans, as between the wars, when the Mitten Crab (Eriocheir sinensis) from China contrived to establish itself in the Elbe Estuary, the German authorities rigorously forbade its use for food as it was a notorious TB carrier. Incidentally, I should like to have known the identity of the veterinary surgeon who made this observation. He was no doubt an ordinary practitioner in the Manchester area, yet I distinctly get the impression that he was far ahead of most of his contemporary colleagues when it came to knowledge of general zoology".

Interesting, I think you'll agree.

Pp. 36/7, and the 'strange' creature seen by the Lady at Borley Rectory - an excellent description of the larva of the very common Elephant Hawk Moth! Sorry to have poured cold water on a promising 'sighting'. The so called 'eyes' by the way, are dark spots just above the head to scare off predators - and people too ignorant to know what's in their gardens.

Ye Gods! I don't believe it! ... but it's there in black and white - someone talking sense about the Loch Ness Monster...

Clinton Keeling,
Guildford.

EDITOR'S NOTE: Much to our surprise, nobody got the answer to our 'phone in competition in issue eleven. The answer was 'John Lennon and Yoko Ono', who in 1969 released an appalling record called 'Life with the Lions'.

I TELL YOU WHAT I WANT WHAT I REALLY REALLY WANT ETC...

Dear Mr Downes,

I recently read your magazine for the first time, and I was very impressed. I am fifteen years old and really want to be a cryptozoologist when I leave school. How do I do it?

Best wishes,

Danny Chope,
Blackpool.

EDITOR'S NOTE: At the risk of sounding facetious, and let me assure you all that I am being nothing of the kind, the only way to do anything, is to get out and do it. I have no professional qualifications even indirectly related to this subject, but I became a fortean zoologist, mainly because I told a reporter from the Observer that I was one. In the four years since we first published Animals & Men I have managed to appear on over thirty television programmes, in over a dozen magazines, and on two CD Roms. I have published two books and written over a million words on the subject. Mind you, I've got no money and my wife walked out on me, but I never started the Centre for Fortean Zoology to make money, more because it was what I had to do. I'm sorry if that doesn't really help...

BOOK REVIEWS

Animals as Teachers and Healers - True stories of the transforming power of animals by Susan Chernak McElroy (Rider Pb 252pp £8.99)

I have usually stayed clear of 'new age' books - and this particularly horrific example is a true indication of why. What I object about is not actually what authors like this good lady have to say, but the way in which they say it. Much of this book presents 'true testimonies' from 'ordinary people' in a stomach churningly twee manner which essentially puts off any reader with pretensions to good taste before he (or she) has started.

The real problem is that Ms McElroy (I feel sure that she is a Ms) has some valid and important points to make. For example, during a reasoned discussion on predation in the wild she writes:

"*Unfortunately, some of us believe that wild animals are nothing more than competitors, vermin or garden pests. But wild animals are simply who they are, for better or worse. It is our fantasies of them as terrible predators, the fearsome varmints, the 'evil' or 'bad' animals who live in our minds that tell us about who we are*"...

However she counters this, perfectly reasonable argument, which indeed mirrors almost exactly what I, and various other writers have been saying within the hallowed pages of *'Animals & Men'* for years, with pages upon pages of dreadful poetry, and the revelation that she likes the music of John Denver!

Having finished the book (and skipped most of the poetry) I am left feeling none the wiser. It is undoubtedly true that animals have a great therapeutic value both for physical and emotional problems.

The companionship of my dog and two cats has brought me through some very bad times over the last year, and on the whole I am a supporter of the charity 'Pat-a-dog' who take tame animals to meet emotionally disturbed humans. However the style of this book is so annoying that for me at least it tended to outweigh what was often perfectly valid content.

Psychic Animals by Dennis Bardens (Capall Bann 203pp £10.95)

This book covers much of the same ground as does the one reviewed above but does so in a sober, intelligent and far more worthy manner. I was particularly interested to read some of Mr Bardens's accounts of 'Beaky' the Cornish Dolphin who both attacked and 'saved' swimmers in Cornish waters some twenty years ago. I cover the story of this remarkable cetacean in some depth in my book *'The Owlman and Others'* and it is interesting to see how Mr Bardens and I have interpreted the available facts in a complimentary, though divergent manner.

Chapter eleven, which deals with apes and monkeys (and tips a nod to Aldous Huxley at the same time) is particularly impressive.

Whereas I am still a confirmed fence sitter as regards the interstices of Extra Sensory Perception, I am quite prepared to believe that there are psychic links between humans and animals. I have experienced strong supportive evidence for this during my relationships with my own domestic pets. As the higher primates are undoubtedly our nearest relatives I am not at all surprised by the data presented here which seems to suggest that they have particularly strong links with us both on an emtional and primal level.

I am happy to reccomend this book wholeheartedly.

In Search of Frankenstein - Exploring the myths behind Mary Shelley's monster by Radu Florescu (Robson £18.95)

At Christmas Christopher Frayling presented a scholarly TV series providing an insight into the truths behind the great icons of 20th Century horror movies.

The programme on Dracula was based very heavily on this author's excellent 'In search of Dracula', and I am pleased to say that the present volume on 'the modern Prometheus' is a worthy successor. Florescu is that rare beast, a genuine academic with a flair for telling a great story. I reccomend this book to the horror movie buff and the literary historian alike.

In Search of Dracula - The enthralling story of Dracula and Vampires by Raymond McNally and Radu Florescu (299pp Robson £9.99)

No sooner had I written the above review than this timely reprint, which has been updated slightly arrived on my doormat. I cannot compare it with the original because that was one of the books that disappeared with Alison during the ritual sharing out of the books and CDs. I am overjoyed to have it back on my shelves.

If anything this is even better than the Frankenstein volume reviewed above, but perhaps I am only saying this because I have always had a fascination with things undead. McNally and Florescu present a scholarly analysis of European vampire legends, but more importantly of Bram Stoker and his classic (and much maligned) novel.

They also present a harrowing and totally dispassionate account of the life and times of Vlad Dracul (aka Vlad Tepes aka Vlad the Impaler) the Romanian Historical Hero (!!!) who is often described as being the main historical influence behind the Dracula legends. I have always been fascinated to discover that Tepes is still seen as a culture hero amongst his Wallachian people, mainly because of the sterling work he did in resisting the incoming Turks.

If you examine the historical data given and then use it as background information to explain some of the Balkan conflicts of the 20th Century then contemporary history, from the Assasination of Franz Ferdinand to the ethnic cleansing of Srebrenica suddenly starts to make more sense.

The only thing that is sadly lacking is an examination of modern 'real' vampirism. I do not mean the loony cults based around third rate Goth bands, but more the predations of such entities as those which have been reported from Highgate Cemetary. This is where the concept of the vampire meets that of the fortean zoologist.

EDITOR'S NOTE: Let me just say here, (in order to forestall a torrent of angry letters from Richard Freeman and his friends - who will probably write a number of different letters each under a variety of undead pseudonyms - I actually quite like Goth music. What I object to is this rather tacky cult of pseudo vampirism based upon the somewhat over-rated novels of Anne Rice etc.

Animals & Men # 13 — Reviews

What is even more disturbing (as shown in the news pages of the latest issue of *The Goblin Universe* is that in some people's minds fiction can blur into reality with tragic results. Vampire Love Cults are therefore not top of my agenda..

Charles Fort wrote about vampiric attacks on domestic livestock and such things have been linked with ABC and BHM phenomena as well as with more bizarre 'out of place' animals. I have written widely on the subject of such vampiric attacks, most notably in my book *The Owlman and Others* and believe that there is a definite link between them and such phenomena as cattle mutilation and the chupacabras.

Although it would have been nice to see some mention of such things in this book, it would be churlish not to give it my highest praise anyway!

V is for Vampire - The A-Z Guide to Everything Undead by David J.Skal (288pp Robson £10.99).

Without any attempt at a stupid joke (in the jugular vein) this book SUCKS! At least it does so on the first reading.

Like the 'new age' book reviewed above this book (which does indeed contain some fine and interesting material) suffers from that inate curse of modern-day publishing, in that its style once again outweighs its content, and this book's style is excrable.

The A-Z format is jokey and annoying (and quite often appears to be figuratively 'scraping the barrel' in search of cheap laughs and something to write about), and whilst there is some interesting stuff therein it is few and far between and padded out with annoying waffle.

Robson Books seem to be cornering the market in things vampiric at the moment, but they are doing themselves no favours by publishing twaddle like this. Unfortunately it will probably sell in bucketloads where the infinitely better tomes by Florescu et al will languish in undeserved obscurity!

On the track of the Sasquatch by John Green (Hancock House 64pp £8.99)
Encounters with Bigfoot by John Green (Hancock House 64pp £8.99)

Two engaging little books by the author of the classic work on North American Man-Beasts *Apes amongst Us*. They are packaged as if they were intended for children but nevertheless contain a wealth of data on North American BHM phenomena. My biggest gripe with both these books is the same as my main bone of contention with Green's classic work - namely that he insists on trying to deal with the Bigfoot/Sasquatch mystery as if it can be explained within a purely zoological frame of reference.

He is not alone in this. Grover Krantz and many other luminaries of the transatlantic cryptozoological scene have followed the same lines of reasoning. It is not, however a path which I agree with. There is such a great body of evidence to suggest that some, but by no means all of such reports should be investigated as zooform phenomena rather than as flesh and blood crypids, that I feel that it is unwise not to do so.

This criticism apart, these two books are entertainingly presented and well written and should be on the bookshelves of any BHM enthusiast.

EDITORIAL NOTE: Unfortunately because of lack of space in this issue several regular features, including Neil Nixon's "Now that's what I call Crypto" and Graham Inglis's "Out of This World" as well as the Periodicals for Review page (also compiled by Senor Inglis ex-Civil Servant of this Parish), have been held over until the next issue.

Animals & Men #13 — Sales

NEW BOOKS FROM PUBLISHERS

BARNABY David *The Elephant that walked to Manchester*. 66pp Pb.

A wonderfully bizarre little book which tells the story of the sale of Wombwell's Menagerie in 1872 the aftermath when Maharajah the Elephant walked from Scotland to his new home in Manchester. Includes many interesting vignettes on 19th Century travelling menageries. Autographed by the author. Usually £6, now £5.00

BARNABY, David & BENNETT, Clive: *The Reptiles of Belle Vue 1950-77*. 156pp A4 pb.

A wondeful insight into the workings of the reptile department at Manchester's Belle Vue Zoo, and a large body of otherwise unavailable anecdotal evidence for 'out of place' exotic reptiles in the North of England over a period of 27 years. Autographed by the authors. £7.00

BARNABY, David: *Quaggas and other zebras*

A superb book about one of the most notorious extinct animals and the desperate attempts being made to reconstitute the species. £9.00

FARRANT, D: *Beyond the Highgate Vampire*

43pp 1992 'Excellent personal history of the phenomenon by the one person really qualified to write about it. Autographed by author. £3.95

GREEN, Richard: *Wild Cat Species of the World* 163pp Pb. Illus.

The best book on the felidae since Guggisberg in the mid 70's. Includes the Onza (colour) Usually £12.50: special offer: £10

THOMAS, Dr Lars: *Ordbog Over Europiske Dyr*. Pb 180 pp.

A wnderful book which lists the common names of European mammals, birds and fish in all European languages. Essential! £7.00

SHUKER, Dr K.P.N: *In Search of Prehistoric Survivors*. 192pp hb

Arguably the most important new book on general Cryptozoology since the 1950's book 'On the track of Unknown Animals'. Shuker presents evidence for the survival of dozens of species of animals, presently known only from the fossil record. We cannot praise this book highly enough. £18.99

SHUKER, Dr K.P.N. *Dragons - A Natural History*. 120pp pb.

Gorgeously illustrated, this book must be one of the most attractive books that I have seen in many years. It is also a must for anyone with an interest in things Draconian. Shuker proves that he is not only a meticulous scientist, but a fine story teller to boot. £11.00

SHUKER Dr K.P.N. *The Lost Ark - New and Rediscovered Species of the 20th Century*. pb.

Another lavishly illustrated book which is an essential purchase for anyone interested in the advances that zoology has made over the past 95 years. NOW OUT OF PRINT £18.00

SHUKER Dr. K.P.N. *The Unexplained*

Probably the best book of general forteana to have been published for many years. I cannot recommend this book highly enough £16.99

Dr Shuker will, personally autograph his books for you at no extra cost. Telephone for details.

LEVER, Sir C: *They Dined on Eland*. 224 pp Illustrated

Excellent investigation of the 19th Century Acclimatisation Society and their founder Frank Buckland, from the author of 'Naturalised Animals of the British Isles', etc. Amusing and erudite. Lever is one of the great experts on naturalised animals and he writes with great skill and aplomb. The publishers price was £18.50; our price is £12.00

CARTER, R. *Loch Ness - the Tour*. 22pp 1996. A good and useful guidebook to Loch Ness. £1.50

STEENBURG, T.N. *Sasquatch; Bigfoot - The Continuing Mystery*. 125pp 1993 Ed.

Excellent. Reviewed in A&M 10. Well worth getting for those interested in North American BHM phenomena. £10.00

GREEN, J. *On the track of the Sasquatch*. Large format Pb. 64pp. Many illustrations. 1980. Excellent. £8.00

GREEN, J. *Sasquatch - The apes amongst us*. 492pp Lavishly illustrated.

Possibly the best book I have read on North American Man Beasts. This is a classic of cryptozoology and should be bought! £12.00

Animals & Men #13 — Sales

GREEN, J. *On the track of the Sasquatch*. Large format Pb. 64pp. Many illustrations. 1980. Excellent £ 8.00

GREEN J. *Sasquatch - The apes amongst us*. 492pp Lavishly illustrated. Possibly the best book I have read on North American Man Beasts. This is a classic of cryptozoology and should be bought! £ 12.00

FULLER, E. *The Lost Birds of Paradise*. 160pp Hb. Reviewed in A&M 10. Gorgeous illustrations. Highly reccomended. £ 30.00

BILLE, M. *Rumours of Existence*. 192pp Pb 1996. Excellent book of zoological anomalies and cryptozoology. Reviewed in A&M10. £ 12.00

GARNER, B. *Monster Monster - A survey of the North American monster scene*. 190pp Pb. Excellent book reviewed in A&M10. Highly reccomended £ 10.00

BOUSFIELD E.L and LEBLONDE P *"An account of 'Cadborosaurus willsi, New genus new species, a large aquatic reptile from the Pacific coast of North America"*.Illustrated with photographs and line drawings. Only a few copies left 32pp A4 This offer will not be repeated £3.50

BEER Trevor *The Beast of Exmoor*. The first book on the subject by the man at the centre of the investigation. Highly reccomended. 44pp with illustrations. £ 3.00

SHIELS Tony 'Doc', *Sea Headlines* The long awaited firsst fruits of the legendary Sea Heads project. Unmissable at £2.50

OUR OWN PUBLICATIONS

DOWNES, J *'Road Dreams'* A month of strange goings on. The author, his wife, and a reasonably well known rock and roll band travel across England with a bunch of engaging weirdos. pb 120pp £ 5.00

ANIMALS AND MEN BACK ISSUES

ANIMALS & MEN ISSUE 1 (out of print) Photocopy only. *Relict Pine Martens/Giant Sloths/Sumatran and Javan Rhinos/Golden Frogs/ Frog Falls*...and much more £ 2.00

ANIMALS & MEN ISSUE 2 (out of print) Photocopy only. *Mystery bears in Oxford and The Atlas Mountains/ Loch Ness/Green Lizards/Woodwose/Tatzelwurm*...and much more £ 2.00

ANIMALS & MEN ISSUE 3 (out of print)Photocopy only. *Giant Worm in Eastbourne/Lake Monsters of New Guinea/Giant Lizards in Papua/Mystery Cats/Black Dogs on Dartmoor/Scorpion Mystery*...and more £ 2.00

ANIMALS & MEN ISSUE 4 (out of print)Photocopy only *Manatees of St Helena/Lake Monster of New Britain/The search for the Thylacine/much more..news/letters etc* £ 2.00

ANIMALS & MEN ISSUE 5 (out of print)Photocopy only. *Mystery cats/Loch Ness/The Migo Video/Boars and Pumas/Hairy Hands of Dartmoor/News Reviews, obituaries, HELP etc* £ 2.00

ANIMALS & MEN ISSUE 6 *Owiman of Mawnan/Humped Elephants of Nepal/Mystery Cats/news, reviews and more* £ 2.00

ANIMALS & MEN ISSUE 7 *Mystery Whales./Strangeness in Scotland/On collecting a cryptid/ Bodmin Leopard Skull/Cryptozoological Books/News, reviews and more* £ 2.00

ANIMALS & MEN ISSUE 8. *Green Cats, Mystery Whales, Cryptozoological books, news etc* £ 2.00

ANIMALS & MEN ISSUE 9. *Hong Kong Tiger, Hoirseman in Lincolnshire, Scottish BHM, Congo Peacock, Mystery whales etc* £ 2.00

ANIMALS & MEN ISSUE 10 *Mystery moth of Madagascar, Bengal Leopard Cats, The Derry, Wild Boars in Kent, a new Irish lake monster, mystery whales and the truth about the Essex Beach Corpses etc*

ANIMALS & MEN ISSUE 11. *Mystery Walruses, Feathered Dinosaurs, Ground Sloth Survival in North America, Mystery Whales, Initial Bipedalism and much more*. £ 2.00

ANIMALS & MEN ISSUE 12. *The Barbary Lion, Feathered Dinosaurs, Chinese Crabs in the Thames, Mystery Animals of Germany, News from New Zealand, and much more* £ 2.00

STILL AVAILABLE: 'THE CASE' - 10 TRACK CD OF FORTEAN MUSIC BY JON DOWNES AND THE AMPHIBIANS FROM OUTER SPACE £10.00

THE CRYPTO SHOP APRIL 1997

POSTAGE AND PACKING

This will now be charged at cost. If you include £0.75 per book and 25p per periodical with your order then we will either refund the balance or invoice you for any extra postage due. Payment can be in cash (UK or US currency), International Money Order, Eurocheque, or a cheque drawn on a UK bank.

*Please make all cheques payable to **JONATHAN DOWNES**. Please telephone to ensure that the goods you want are still in stock. If no telephone call is received and an order is received for something that is out of stock then a credit note will be given.' If you live outside the EC please add a 10% surcharge to cover additional post and packing.*

Every effort will be taken to ensure prompt delivery within 21 days. Orders outside Europe are sent by surface mail unless aditional postage is paid.

OUR OWN PUBLICATIONS

The Owlman and Others, by Jonathan Downes.

Two decades of Owlman evidence including sightings - mostly by girls and young women - in Cornwall. This book comes about as close to the truth as anyone ever will... Many illus.   £10.00

The CFZ Yearbook 1996

The first yearbook, with nearly 200 pages of research papers and longer articles including Sky Beasts (Karl Shuker), mystery eagles (Jon Downes), Namibia's Flying Snake (Richard Muirhead), the Nnidnidification of Ness (Tony Shiels), African Man Beasts (Francois de Sarre), and much more. Many illustrations.   £12.00

The CFZ Yearbook 1997

Karl Shuker hunts anomolous aardvarks, Darren Naish figuratively shoots the Lake Dakataua Monster and Francois de Sarre asks if humans are descended from bipedal fish.

Also articles on : the pros and cons of reintroducing extinct mammal species to Scotland, a list of crypto-zoological movies, Mexican cattle mutilation and the Chupacabras - and much more. Many illustrations.
£12.00

Morgawr: The Monster of Falmouth Bay, by A. Mawnan-Peller.

The classic 1976 booklet reissued with a new introduction by Tony 'Doc' Shiels.   £1.50

The Smaller Mystery Carnivores of the Westcountry, by Jonathan Downes

Three species thought extinct; hints of several species apparently unknown to science; and a revolutionary suggestion that a species of mammal known from mainland Europe exists in England. Over 100 pages. Illustrated.   £7.50

Animals & Men #13 — Sales

THE DRAGON CHRONICLE. £1.50 each, or complete set for £7.00
#1: Death of a Godess/ Tolkein's Dragons/ Church of all worlds/Dragon Herbal/Childhood Dragons/Cryptozoology
#2: Free the Dragon/Dragon Environmental Group/Dragons and Music/A Dragons Story/The real St. George.
#3: Alchemical Dragons/The Dragon and the Pentagram/Did Dragons once Live?/What dragons do.
#4: Welsh Dragon Traditions/The Lambton Worm/ Dragons and their magic/The Sussex Dragon/Misericords.
#5: Dreams & Dragons/Dragon Circle/The Cockatrice/Dragon's Breath/Musing from a Dragon's Mage/Dragons Lost.
#6 Sky Dragons and Celestial Serpents, Celtic Dragon Myth, British Dragon Legends; History and symbolism, the Dragon Bard, Dragon wisdom

ALSO FROM DRAGON'S HEAD PRESS:

WORME WORLD-THE DRAGON TRIVIA SOURCE BOOK by A.N.Dimmick . Dragon Trivia at its worst £2.75

CATSPAWS by N.M.MacKenzie. *Books of Poetry are usually out of the scope of this book catalogue. This booklet, is, however something very different. This is an engaging collection of poems about cats, but also includes a section about 'The Golden Dawn' and some other weird stuff.* £1.75

ALL THE OTHER by N.M> MackKenzie. *Another collection of poems and other strange stuff. Described by the author as 'a collection of spells, curses and blessings with fish, flesh and good red herring for the delectation of those who have a taste for enchantment. Equally highly reccomended.* £1.75

COLEMAN Loren 'Was the First Bigfoot a Hoax? Cryptozoologys Original Sin' 27pp
£ 2.00
WILKINS Harold T: 'Mysteries and Monsters of the Great Deep' (1948) 28pp Reprint
£ 3.00

CRYPTO CHRONICLE Issues 1-5.complete set for
£ 3.00

Issue 1: Bigfoot/Roger Patterson Movie/Yeti/Sasqautch
£ 1.00
Issue 2: Mystery Cats/Pumas/Bigfoot/Orang Pendek/Moa
£ 1.00
Issue 3: Beast of Exmoor/Durham Puma/Licenced exotics/Big Cats/Sasquatch/Bigfoot/Soviet Yeti/Mystery Cats
£ 1.00
Issue 4: Loch Ness/Lake Monsters/Morgawr/Bigfoot/Out of Place Bears and Rabbits
£ 1.00
Issue 5: Wildman/Yeren/Sasquatch

£1.50

PET KEEPING BOOKLETS FROM BASSETT PUBLICATIONS DICOUNTED THROUGH THE CRYPTO SHOP

The Chipmunk and Siberian Chipmunk in Captivity
£ 3.00
Marmosets in Captivity
£ 3.00
Dwarf Hamsters
£ 1.50

Animals & Men #13 — Sales

GILROY R, 'Mysterious Australia'. pb illustrated 288pp. A fascinating collection of antipodean forteana which includes several large sections on cryptozoology and allied disciplines. The Yowie; Giant Lizards; The Thylacine Marsupial 'Panthers', Bunyips, lake and River Monsters, relict dinosaurs and much more. Highly reccomended. £11.00

10% OFF THE PUBLISHERS PRICE OF ALL FORTEAN TIMES PUBLICATIONS:

(There should be no need to introduce this re-print series which each contains invaluable and otherwise unavailable gems from across all fields of forteana)

FORTEAN TIMES 1-15 'Yesterdays News Tomorrow' 400 pp pb Includes Wolf Children, Moon Mysteries, poltergeists, Pyramids, Water Monsters, Fortean USA and lots more) £ 17.99

FORTEAN TIMES 16-25 'Diary of a Mad Planet' 416 pp pb. Includes Bleeding Pictures, Animal Attacks, Morgawr, close encounters, swarms of animals and much more. Too much to list £ 17.99

FORTEAN TIMES 26-30 'Seeing out the seventies' 320 pp pb Owlman, Mystery animals, ABC'sanimal attacks, Stigmata, Fish Falls, Fortean Phenomena in ancient pamphlets & more £ 13.50

FORTEAN TIMES 31-36 'Gateways to Mystery' 416 pp pb The touch of death, mystery blob, wildmen and hermits, mystery cats, Owlman, Doc Shiels, Morgawr, in searchg of dinosaurs, the Yeren and more £ 17.99

FORTEAN TIMES 37-41 'Heavens reprimands' 416 pp pb The Man who invented flying saucers, The Hackney Bear, Mystery Kangaroos in USA, plants growing out of peoples eyes and much more £ 17.99

FORTEAN TIMES 42-46 'If Pigs could Fly' 416 pp pb Mystery Submarines, Werewolves in Devon, horned humans, lake monsters across Europe, phantom attackers and much more £ 17.99

FORTEAN TIMES 47-51 'Fishy Yarns' 416 pp pb Australian Mystery Animals, Yeti Sightings, Lizard Man, Mystery Cats and many pages of news and much much more £ 17.99

FORTEAN STUDIES VOLUME ONE 350pp Long research papers on many subjects includes: Karl Shuker on Mystery Bats, Michel Raynal on the Giant Octopus, The Luminous Owls of Norfolk, Mike Dash on the Great Devon Mystery, Michel Meurger on Medieval French mystery cats and much more. Essential. £ 17.99

FORTEAN STUDIES VOLUME TWO 320pp Long research papers on many subjects includes: Karl Shuker on the physical evidence of British Mystery Cats, Michel Raynal and Gary Mangiacopra on Out of Place Coelecanths, Michel Meurger on Icelandic Water Monsters, Bob Rickard on Fish Falls and more £17.99

FROM THE SAME TEAM WHO PUBLISH *ANIMALS & MEN* COMES A NEW MAGAZINE - *THE GOBLIN UNIVERSE:*

ISSUE FOUR: THE BLACK DOGS OF BUCKFASTLEIGH, ST NEOT THE WEIRDEST VILLAGE IN THE WEST, THE TRUTH ABOUT FOOD ADDITIVES, THE VAMPIRE OF ST LEONARDS AND MORE THAN YOU CAN POSSIBLY IMAGINE £ 2.00

ISSUE FIVE: COMMUNICATION WITH UFOS, DECONSRUCTING A TEXT, SEA HEADS, A GREY HYPOTHESIS, ANCIENT GEOGRAPHICAL KNOWLEDGE, ANIMAL MUTILATIONS IN CROP CIRCLES AND MUCH MORE £ 2.00

animals&men

THE JOURNAL OF THE CENTRE FOR FORTEAN ZOOLOGY

Cartoon by Mark North

ISSN 1354 0637 Typeset by a Caseworker

Oh you Mambos they thrill me and kill me

ISSUE 14

JULY 1997

By the summer of 1997, I had also achieved a modicum of fame within the fortean universe, and was writing articles for up to eight publications each month. I was earning more than I ever had before, but doing my best to keep quiet about it to avoid having to pay maintenance to my ex-wife.

Richard Freeman was becoming a permanent fixture by this time, but he has always been cross that his first major dracontological article was illustrated by a drawing produced by Graham's then girlfriend, which Richard has always likened to a character from the children's TV show *The Magic Roundabout.*

animals & men

THE JOURNAL OF THE CENTRE FOR FORTEAN ZOOLOGY

Animals & Men

The Journal of The Centre for Fortean Zoology

EXCLUSIVE:
The Identity of The Beast of Le Gevaudan finally revealed

The Dragons of Yorkshire; Irish Mystery Animals; The Vampire Sheep Killer of Badminton Revisited; and much more....

Issue Fourteen

animals & men

THE JOURNAL OF THE CENTRE FOR FORTEAN ZOOLOGY

Animals & Men # 14 — Who's who and what's what

The ever changing crew of the "Animals & Men" mothership presently consists of:

Jonathan Downes: Editor
Graham Inglis: Newsfile Editor/Spin Doctor
Jan Williams: Associate Editor
Mark North: Cartoonist
Richard Muirhead: Newsagent from Nowhere
Special Agent Tina Askew: Rhine Maiden
Alyson Diffey: Editorial Assistant and Medium
Lisa Allegri: Artist

CONSULTANTS

Dr Bernard Heuvelmans (Honorary Consulting Editor)
Dr Karl P.N. Shuker (Cryptozoological Consultant)
C.H. Keeling (Zoological Consultant)
Tony 'Doc' Shiels (Surrealchemist in Residence)
Darren Naish (Palaeontology/Cetology Consultant)
Chris Moiser (Zoological Consultant)

REGIONAL REPRESENTATIVES

U.K
Scotland: Tom Anderson
Surrey: Nck Smith
Yorkshire: Richard Freeman
Somerset: Dave McNally
West Midlands: Dr Karl Shuker
Kent: Neil Arnold
Sussex: Sally Parsons
Hampshire: Darren Naish
Lancashire: Stuart Leadbetter
Norfolk: Justin Boote
Leicestershire: Alaistair Curzon
Cumbria: Brian Goodwin
S. Wales/Salop: Jon Matthias
London: Richard Askew (No relation)
Tyneside: Simon Elsdon

EUROPE
Switzerland: Sunila Sen-Gupta
Spain: Alberto Lopez Acha
Germany: Wolfgang Schmidt & Hermann Reichenbach
France: Francois de Sarre
Denmark: Lars Thomas and Eric Sorenson
Eire: The Wizard of the western world.

OUTSIDE EUROPE
Mexico: Dr R.A Lara Palmeros
Canada: Ben Roesch

DISCLAIMER

The views published in articles and letters in this magazine are not necessarily those of the publisher or editorial team, who although they have taken all lengths not to print anything defamatory or which infringes anyone's copyright take no responsibility for any such statement which is inadvertently included.

CONTENTS

3 Editorial
4 NEWSFILE
11 Obituary: Jacques Cousteau
11 In Search of Gambo by "Mungo Park"
14 The Dragons of Yorkshire - Richard Freeman
20 The Beast of Bodmin - Jon Downes
Charlie Fort and the Vampire Sheep Slayer - Terry Hooper
24 "Jackal" Update - Jon Downes
28 Big Cat Reports from Scotland - Mark Fraser
31 North of the Border - Tom Anderson
32 Three Irish Animal Stories - Richard Muirhead
34 Clinton's Cogitations - the Kelstridge Lions - Clinton Keeling
35 Now That's What I Call Crypto - Neil Nixon
36 Letters
39. Book Reviews
43. Periodicals
44. Sales
48. Cartoon by Mark North

SUBSCRIPTIONS

(Please see "Methods of Payment" on p46.)
For a Four Issue Subscription:
£8.00 UK
£9.00 EEC
£12.00 US, CANADA, OZ, NZ
(Surface Mail)
£14.00 US, CANADA, OZ, NZ
(Air Mail)
£15.00 Rest of World
(Air Mail)

'Animals & Men'

THE CENTRE FOR FORTEAN ZOOLOGY,
15 HOLNE COURT,
EXWICK, EXETER,
DEVON, EX4 2NA

Tel 01392 424811

THE GREAT DAYS OF ZOOLOGY ARE NOT DONE

Dear Friends,

Welcome to another issue of *Animals & Men*. In many ways I feel that this is the best issue that we have produced to date. There have, however, been a couple of problems. Due both to illness and equipment problems this issue is considerably later than it would have been and, because or problems with our printing equipment the production values of this issue, at least as far as the typeface are concerned are slightly lower than those which you (and we) have come to expect.

We have therefore decided that, if at all possible, as of the next issue, we will be producing the magazine on a PC rather than on our antiquated (and now rather ramshackle) AMIGAs. As long as our financial forecast goes according to plan (and let's face it so far it never has) we will be buying a second hand PC sometime before the end of August. This will give us a far greater degree of flexibility as far as production, picture and graphic manipulation, typefaces etc are concerned, and should, we hope, improve the overall quality of the magazine to a remarkable degree.

We have had a number of visitors to the CFZ this year. Erik Sorenson, Richard Freeman and Darren Naish to name but three. It is always nice to see you all, and please feel free to pop in whenever you are in the area. If (like Richard F) you end up helping me compile an article for *Fortean Studies Volume Four* or if, like Sally Parsons you end up doing my washing up, that is, I am afraid, one of the chances you have to take when you visit what is, after all a chaotic bachelor household as well as being the UK's premier cryptozoological organisation. Graham and I have recently acquired a housekeeper who is helping to civilise us to a greater extent than ever before (Hi Cara), and we also wish to welcome aboard Lisa Allegri, who has done some lovely illustrations for this issue (including the cover) and will almost certainly be a regular contributor from now on. The libellous piccie of me at the bottom of the page is by Darren Naish, who (I think) deserves credit for thinking up ruder and ruder ways to portray me within a cryptozoological context).

Our next major project, apart from the 1998 Yearbook which will, we hope at least, be available on time this year, be the *'Mystery Animals of Hong Kong'* by Richard Muirhead and Myself. We do not know what the format is likely to be, as there is almost certainly enough material to fill two volumes. We are still planning to republish 'Doc' Shiels's *"The Cantrip Codex"* before the end of the year, and we are in discussion with other authors regarding books for our longer range publication schedule.

We have decided that another long term aim of the CFZ is to move out of our somewhat squalid little abode in Exeter and into somewhere large and rural where we can have a proper visitors centre and resource library as well as holding seminars and the like. Again this is all down to finance, but there seems to be enough National Lottery money floating around for stupid schemes which neither interest nor benefit anyone, that we may well be able to get hold of some of this resource which is hitherto untapped (at least by us).

My thanks to everyone who has sent in letters, telephones and messages of goodwill regarding my forthcoming divorce. The decree nisi has now been issued and it seems that the whole unfortunate episode will be over very soon. I wish Alison all the best in her new life. She was instrumental in setting up the CFZ with me seven years ago, but she has moved on to pastures new, and so, indeed have we. Life goes on, and we at the Centre for Fortean Zoology must endeavour to move onwards and upwards...

Best wishes
Jonathan Downes

NEWSFILE

Compiled and Edited by Jonathan Downes and Graham Inglis

MYSTERY CATS

Stockport, Manchester.

Catherine Murphy, a pensioner from Stockport was 'stunned' to see a 'large cat' sleeping in her garden. "She could only see the back of the tanned-coloured creature, but knew that it was much too big to be a pet cat, and the 'wrong shape' for a dog - which could not have climbed on to the roof of the shed anyway". A photograph of an indistinct pawprint about three inches long accompanies this story but is too blurry to be conclusive either way. *Manchester Metro News 11.7.97*

Highlands, Scotland.

Farmer John MacLennan aged 69 of Dalcross reported four lambs killed, apparently by a *'large beast with brown skin'* that they saw slipping through a fence but *'were unable to identify'*. They said that the *"attacker was obviously a powerful animal with good fangs that had managed to get a good grip on the lambs"*. *Aberdeen Evening Express 10.6.97*

Mildenhall, Suffolk.

A man who wished to remain anonymous saw a big cat like animal run across the road as he was driving along the A1101 near the village of Kenny Hill. He was certain that it was neither a dog or a fox and local police have confirmed that this is only one of a number of sightings of similar animals in the area.

WHAT'S NEW PUSSYCAT?

In possibly the most exciting discovery for British Cryptozoology since that of the Kellas Cat, a cherished theorem of British mystery cat lore has been turned upon its head. Two specimens of the Jungle Cat (Felis chaus) are known from the UK. They are both road kills and it has been supposed (almost certainly) that they were escaped pets or zoo specimens. These animals have been used (very succesfully) to promote the school of thought that exotic species can and do survive in Britain. This speciesis presently found in the Middle East and North Africa, and until now there has been no suggestion that they have ever lived in the UK.

The *Daily Mail*, of Wednesday 18th June 1997, however tells a different story:

"The bones of a Jungle Cat, dating back 200,000 years have been found on the banks of the Thames at Aveley."

This, in theory at least, opens the world of British cryptozoology open to more speculation that the Shropshire and Hampshire Jungle cat specimens were in factmembers of a relict population of indigenous British wild animal. This is highly unlikely, but the new discovery has opened up a whole new metaphorical can of worms!

OUT OF PLACE

Animals & Men # 14

Newsfile

BEE IN THEIR BONNET

African Bee Eaters, beautiful birds which only ocasionally visit and breed in Britain were reported in May at Witney in Oxfordshire, attracting a horde of twitchers determined to see these rare and exciting visitors. Resident, Jayne Leonard, of Upper End, Fulbrook is quoted as saying about the obsessive birdwatchers, that *"we wondered if somebody famous had moved in and they were paparazzi"*.

PINK FLAMINGOS

A Flamingo which had escaped from a leisure park near Romsey in Hampshire turned up in Exeter in early July. The five year old Lesser Flamingo called Amy was taken to Cricket St Thomas Wildlife Park from whence she was eventually returned to her flock. *Aberdeen Press and Journal 9.7.97*

EDITOR'S NOTE (ONE): The Centre for Fortean Zoology has long been interested in these sightings of Flamingos, because although it is certain that most of the birds seen in Britain are, like this specimen, escapees from captivity, Flamingos are found in parts of southern Europe, and with the changes in climate, it is not beyond the bounds of possibility that these beautiful birds could extend their habitat into the shallow coastal estuaries of southern Britain.

EDITOR'S NOTE (TWO): There hasn't been a 'phone in' quiz for a while, but if you want to win an autographed copy of my latest book *'The Owlman and Others'* (1997), 'phone and tell me, who directed the movie 'Pink Flamingos'? Who starred in it? and why is it my favourite film? This competition is not open to Richard Freeman who presently has borrowed my own copy of this cinematic materpiece.

....AND OTHERS

It has been suggested on a number of occasions that sightings of rogue European Eagle Owls (*Bubo bubo*) may have been responsible for sightings of the notorious Owlman of Mawnan in southern Cornwall. In my aforementioned book, and at length in the pages of this magazine I have stated my reasons for believing this theory to be ill informed nonsense. It is interesting, however, to find that a bird of this species lived for several weeks on the roof of St Paul's Cathedral in London before succumbing to an untimely death! *Aberdeen Press and Journal 8.7.97; The Times 8.7.97.*

(Picture via Copyright Liberation Front)

ANIMAL ANOMALIES

STARS AND STRIPES

This doesn't really count as an 'anomaly' but it is interesting anyway. This peculiar mule-like equid is a cross between a zebra and a donkey which was bred in the United States and is now on display at a horse centre in Leominster, Herefordshire. It is interesting to note quite how much like a quagga it looks.

Picture courtesy of the Copyright Liberation Front

WHITE POWER

As noted in a recent issue of *Fortean Times* there have been a large number of albino mammals turning up this year. A pair of albino squirrels were found near Wallington in Surrey, according to the *Mail on Sunday* July 6th 1997, and the *Daily Mail*, on March 4th showed pictures of a rather stunning white (but not albino) hedgehog. The latter was probably a chinchilla mutant, but its peculiar genetic history would have been no protection against the predators who would no doubt have eaten it if it had not been rescued by a Somerset animal sanctuary. These are just two selections from our files. There have been so many odd coloured frogs turning up across the west of England that the Cornwall Wildlife Trust actually instigated a 'Golden Frog Day' to raise public awareness of these peculiar creatures.

RODENT RUCTIONS

It appears that Gerbils separated from their mates display the same symptoms of loss and grief as "depressed divorcees" according to researchers at Leeds University who are presently studying the biochemistry of depression. As someone who is in the process of undegoing a messy and traumatic divorce I could make a caustic comment here, but won't. *Aberdeen Press and Journal* 23.1.97

LAKE MONSTERS

LAKE VAN IN TRANSIT

Video footage of what is alleged to be the 'monster' living in Lake Van in Eastern Turkey was shown on several British television programmes in the early summer, and has also been posted on the Internet. Opinion seems to be divided as to whether it is any good or not, with most of the people who have described it to us (not as yet having Internet access we have been unable to see it to date) describing it in generally unfavourable terms although at least one member of the A&M posse has been quite enthusiastic about the footage. In order to spare anyone's blushes and to avoid starting arguments we will not mention names. All we can say is 'watch this space'!

LIZZY LIVES?

The creature reported in Loch Lochy, which for some reason has been nicknamed 'Lizzy' has apparently been 'seen' on sonar by an expedition involving our old friends Gary Cambell and Richard Carter. At present we have no further information save that told us in an excited late night telephone call from Mark Fraser of 'Haunted Scotland' Magazine. We will have a full report as soon as we can get hold of further details.

NESS IS MORE

A story in 'The Sun' (Saturday, July 19th 1997) claims that Stephen Spielburg has teamed up with Scottish UFO and strange phenomena researcher Malcolm Robinson to search for the elusive Loch Ness Monster. It goes into great detail about the way that they intend to capture the beast. Confusingly, however, the clipping was sent to us by Robinson himself, together with a press release which claims that

"...being 'The Sun' they have, of course, embellished the story somewhat and given it 'arms and legs'. Let me explain.

It was the Scottish Sun that phoned me and asked me what I would do if I had a million pounds in which (sic) to prove the existence of the Loch Ness Monster? I told The Sun, that funnily enough, I'd had an idea in my head for a few years on this very issue, and that I had plans of a cage in which I felt could be in the capture of 'Nessie' (sic). The Chap from the Sun asked to see my drawings, and I sent them off to the Sun AND NOT DIRECTLY TO STEPHEN SPIELBURG AS THE SUN WOULD HAVE YOU BELIEVE.

I also never said "Give me 10% of the takings of The Lost World" and I'll give you 'Nessie' on a platter". And I also never said "With Spielburg's money he could afford to drain the Loch. Typical 'added on journalism'".

I sympathise with Malcolm because I know, too my cost, that journalists and solicitors will misquote anything you say, USUALLY to your cost...

Animals & Men # 14

Newsfile

OARFISH ANOMALIES

The *Daily Mail* of May 26 1997 reported that a film crew working for *BBC Wildlife* magazine had for the first time photographed a living oarfish in its natural environment and were astounded to discover that it swims vertically stabilised by its fins, instead of horizontally as had always been supposed. This essentially deals somewhat of a death blow to people who have theorised that these remarkable fish (actually related to the herring family) could be responsible for some sightings of the 'Great Sea Serpent'. Channel Four also showed amateur video footage of a living oarfish in its death throws as part of its series 'The Deep' in July of this year. It is ironic, that like the Giant Squid, an animal which is so well known can be both so elusive and have practically nothing known about its biology or behaviour.

A rogue tiger has been attacking children in western Nepal. As of the end of January thirty five children had been killed, and therefore, although the species is rigorously protected, a Royal permit has been issued to allow the animal to be shot. Whether or not it has been we do not know. *Aberdeen Press and Journal 24.1.97*

NEW AND REDISCOVERED

EXCLUSIVE.... Karl

Shuker tells me that the long lost skin of the so-called 'Beast of La Gevaudan' which has been lost for something like three hundred years has been rediscovered in the vaults of the French Natural History Museum. It is that of a Hyena, although at present we have no idea what species it is.

Ironically the day before he told me this I had mailed my article on zooform cats to 'Enigmas' magazine which mentioned this well known case. In it I theorised that it was zooform in nature. I stick by this assertion because although there is now no doubt that a hyena did roam southern France in the 18th Century, it seems unlikely that all the predations attributed to the beast could be laid at the door of one hyena, no matter how savage!.

WHY DID THE MONKEY FALL OUT OF THE TREE? BECAUSE IT WAS DEAD!

Ross MacPhee of the American Museum of Natural History, and Donald McFarlane of Claremont McKenna College found the partial cranium of a Jamaican monkey, named *Xenothrix mcgregori*, which has the dubious distinction of being the only species of New World Primate to have become extinct since the arrival of European explorers. *Geographica April 1997*

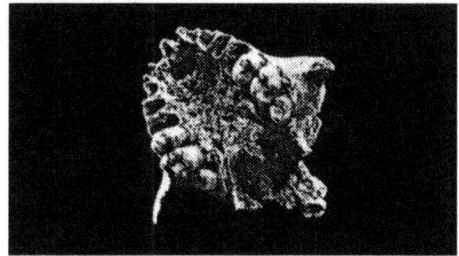

PICTURE COURTESY COPYRIGHT LIBERATION FRONT

Animals & Men # 14 — Newsfile

EMERGENT AMPHIBIANS

The Times, June 5th 1997, told the story of an intrepid amateur explorer called Martin Pickersgill who has spent a whopping £320,000 on a 10,000 mile trek across Africa from Cape Town to The Sahara, discovering four new species of reed frog on the way.

In 1983, he made a similar discovery in Natal and, in what I have always thought to be a breach of a priori scientific ethics, named it after himself. The Cryptozoology Review Summer 1997.

NEW FOX

A distinct race of the North American Mountain Fox, Vulpes vulpes macoura has been discovered in the Beartooth Plateau of Montana. It is thought that this population has remained isolated and distinct since the Ice Age. National Geographic April 1997.

REDISCOVERED TORTOISES

Despite the scepticism shown in many quarters as to whether these animals were in fact ever of a distinct species, the Nature Protection Trust of the Seychelles (NPTS) has announced that genetic tests have proved that a single pair of one Seychelles species of Giant Tortoise, and four pairs of another have been rediscovered and that a breeding prohramme is underway. For more details write to:

The Nature Protection Trust of the Seychelles
PO Box 207
Victoria
Mahe
Seychelles

Africa - Environment and Wildlife Vol 5 (#3)

NEW LEMUR POPULATION

A new population of the highly endangered Greater Bamboo Lemur (Hapalemur simus) has been discovered in rain forest in south-eastern Madagascar.

This is important as it means that the genetic diversity of specimens in captive breeding programmes can be maintained. Africa - Environment and Wildlife Vol 5 (#3)

REDISCOVERY OF THE BORNEO RIVER SHARK

The Glyphids or River Sharks are a very specialised and little known group of fish of which only two have even been given scientific names.

One further species, (presently known only as Glyphis Species B) until recently known from one specimen taken from an unknown location in Borneo a century ago was rediscovered by a IUCN investigative team.

The Cryptozoology Review has speculated that two other species, one from Australia and one from New Zealand may also be awaiting formal discovery. Known as Glyphis species B and C they are also known from two and seven specimens respectively.

Picture by Darren Naish

I'M A HOG FOR YOU BABY

In what has been a momentous period for the rediscovery of animals feared extinct, a wild pig, Sus bucculentis has been discovered in the Annamite mountains on the borders of Laos, Vietnam and Cambodia. It was first reported in 1892 from two skulls, one of which has since been lost, a partial skull was discovered recently by a team from the Australian National University headed by Dr Colin P. Groves. This proves that this

Animals & Men # 14

Newsfile

species is, contrary to expectations, alive and well. *The Cryptozoology Review Summer 1997*

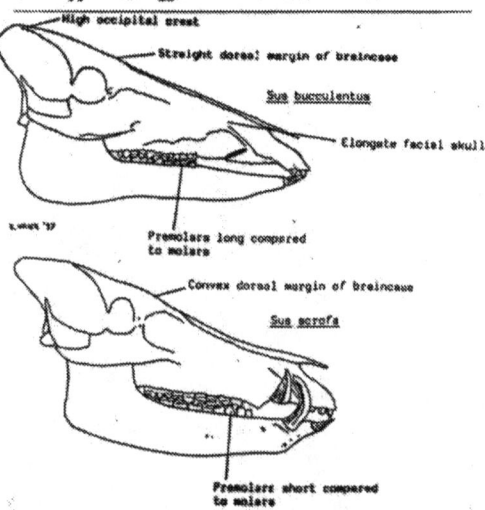

Picture by Darren Naish

OTHER STORIES

FOURTEEN MINUTES LATE
...BADGER ATE THE JUNCTION BOX AT NEW MALDEN

Privatisation seems to have made no difference to what used to be British Rail. Although, I am reliably informed that the sandwiches now taste a little better than they did, (which under the circumstances isn't saying very much), but the excuses for shoddy service still keep on coming. as *The Daily Mail* 6.3.97 said:

"*They've been delayed by leaves on the line, tracks buckled by heat, and even the wrong kind of snow. Now commuters face a new threat to punctuality*"...

This new threat, it seems comes from the humble rabbit which has increased its population so much over the past few years that in several areas throughout Railtrack's 10,000 mile system the burgeoning systems of underground rabbit warrens are causing a serious threat of subsideance.

One is reminded of the (possibly apocryphal) story about the Duke of Wellington who when visiting the then new Crystal Palace was told of the problems caused by sparrows within the vast building. He suggested "*Hawks, m'dear fellow*"...

This remedy was, or so the story goes, speedily put into action. My version of that is Polecats!

There has long been a move to introduce these delightful little carnivores into the places where a hundred years or so ago, they were relatively common. Indeed a large portion of my book "*The Smaller Mystery Carnivores of the Westcountry*" concerns this very subject.

Surely the powers that be who control Railtrack (or whatever it's called this week) would not flinch from investing a few quid into a project which would not only solve their problem but would enrich this country's depleted wildlife.

How much do you want to bet that they won't do anything of the sort?

EDITOR'S NOTE: Being in a surprisingly good mood despite a succession of disasters which have befallen me over the past few days I have decided to give away another prize in an utterly pointless 'phone in quiz. (Has anyone noticed that they are a thing that I really rather like?)

The question this time concerns the headline to the above piece. The question is Who, why, where, when, and to whom?

animals&men

THE JOURNAL OF THE CENTRE FOR FORTEAN ZOOLOGY

Animals & Men # 14 — Newsfile

Spiderman is having me for dinner tonight...

We were sent this picture (which again appears courtesy of 'The Copyright Liberation Front' by our Tyneside correspondent who saw it in a magazine about lifeboats. Unfortunately, it seems more likely that the spider landed on the lens of the camera taking the picture, rather than belonging to a hitherto unsuspected race of giant spiders like those in the classic 'Tintin and the Shooting Star' by Herge.

Herge (real name Georges Remi) was probably the best known European comic book artist of all time, but a little known fact is that for several of his books, including Tintin in Tibet and the two parter about the exploration of the moon, he employed as scientific consultant none other than Bernard Heuvelmans.

It is no wonder that Herge's depiction of the yeti in the aforementioned book was probably the most sympathetic yet within any comic book!

Newsfile Credits

The Cryptic Clipper of Mannic Productions, Simon Elsdon, Tom Anderson, Sally Parsons, COUDi, Ben Roesch, Darren Naish, Mark North, Jan Williams, Nick Smith, Chris Moiser and probably many more...

PLEASE CONTINUE SENDING US YOUR CLIPPINGS AND PHOTOCOPIES BUT PLEASE REMEMBER TO MARK THEM WITH THE DATE AND SOURCE!

OBITUARY:
JACQUES COUSTEAU
1910-1997

By Richard Freeman

Few of us fulfil but a handful of our dreams. Most lives are wasted in a mundane world that crushes souls and tramples aspirations. Jacques Cousteau was one of the few who broke his leaden chains and lived his life as life SHOULD be. He swam with whales and sharks, danced with manta rays and befriended dugongs.

His love affair with the sea began whilst he was on sick leave from The French Resistance during World War Two. He convalesced by snorkel diving on the French Riviera and thus, his life-long adventure began.

He invented the Aqualung and revolutionised undersea exploration. Shortly afterwards, he produced the film 'Silent World', together with noted Film Director Louis Malle. The film won the Golden Palm Award at the 1956 Cannes Film Festival, and Cousteau soon became a household name. He followed this with four decades of groundbreaking film and exploration from his converted minesweeper *'The Calypso'*. He did things that lesser naturalists only dream of. He saw killer whales preying on hammerhead sharks, glimpsed giant squid and not only patted saltwater crocodiles, but lived to tell the tale!

An outspoken conservationist, he warned of the dangers of pollution and whaling long before the Green Movement became popular; and more recently he attacked Jacques Chirac over his South Pacific 'Fireworks' [the French nuclear bomb tests at Maurorora Atoll].

His son called him "France's Greatest Ambassador". I would call him one of Mankind's Greatest Amb-assadors to the Natural World!

Jacques seemed like an old friend to many of us who knew him only through the medium of his spellbinding works. He reached a ripe old age and lived his exciting life to the fullest that any man could possibly do. So, it is with only a tinge of sadness that we say a fond farewell to the inimitable Jacques Cousteau.

IN SEARCH OF GAMBO
by
"Mungo Park"

EDITOR'S NOTE: As will be apparent from the contents of this article, this is a pseudonym for a member of what we, only half jokingly, refer to as the *'military wing of the Centre for Fortean Zoology'* - the people who actually DO boldly go where no other cryptofolk have gone before. So those of you expecting an article by a long dead Scottish explorer will be sadly disappointed.

I suppose that I am a closet cryptozoologist.

My day-time job requires a modest public profile, and is very susceptible to market trends. So it's jacket and tie, respect the free thinkers, and don't upset the PCP (Political Correctness Police - Not Phenecyclidine!), and keep your fingers crossed come redundancy time. Even my copy of *Animals & Men* comes in a plain brown wrapper.

A week in the Gambia with work was too good an opportunity to miss though. I had actually seen the original letter from Owen Burnham in *BBC Wildlife* back in the '80s, describing the mystery beast (Gambo) found on the sand near the Bungalow Beach Hotel. I was also aware that there had been some recent interest in the animal from the Centre for Fortean Zoology. Partly as a result of this, I had even seen a 'treasure map' that allegedly showed where the beast was buried. Although I planned to make a few late night excavations, the week dragged on; our primary work objectives on the trip were met quickly and relatively efficiently. Several time-consuming minor problems occurred though, which kept me from my late night exploring and digging.

I suppose that our delays were those that always occur in the Africa of the '90s; which when clouded with the efforts of the new government keen to eradicate corruption (not unknown to the old Government), and problems with a deteriorating exchange rate, kept us bogged down with trivia. Perhaps I was just degenerating into local speed. (G.M.T in fact stands for *'Gambia Maybe Time'*!),

Animals & Men # 14   Feature

anyway, I was at a semi-formal dinner on the last night without having had much of a chance to do anything about the conundrum buried up the road. My frustration was getting too much, in 24 hours I would be back in Europe, so I made the typical European-in-Africa excuse, (hand on tummy *"must have had too much sun..."*) and left early. Bruce (name changed to protect the guilty), another closet crypto with our group, muttered something about *"seeing if the boss is alright"*, and *"having some Immodium back at our hotel"*, and followed me out.

It had been a hot day, even on local standards, which means above 100° F. and although the evening was cooler, it was still warmer than was going to be comfortable, even for a semi-acclimatised European. We walked outside the hotel, and I acquired a stainless steel butter dish in the process. I think that Bruce made a mental note to look up my anti-malarials to see if kleptomania was a listed side effect. *"To dig with,"* I explained. He nodded. *"We'll bring it back later, OK...?"*

Thankfully the beach bums (local scroungers and professional "friends"), polite, but a persistent nuisance had dispersed after dark, (they are predominantly a rather diurnal species). I was able to get hold of a taxi fairly easily, and without attracting any attention. Although a walk down the road to the site of "Gambo's" burial was, in theory possible, it was still a bit too hot. In addition, although we would be safe from local villains (probably safer than in the UK), African roads are unlit at night, as are roadworks. My superiors at work would not have been happy with me if I'd been involved in a traffic accident, or disappeared down one of the many holes in the road.

The problem with Gambian taxis is that you normally book one from your hotel; he takes you out, waits for you, and brings you back when you are ready. The rates are cheap; the locals need the work - and, most importantly, there are often not telephones where the visitors go, and even if there were, the taxi operators usually have no telephone themselves! Clearly what we didn't need on this occasion, however, was the assistance of a taxi driver, nor did I want him as a potential witness.

Although the country was very peaceful, they had held elections only a few months before. The result was that the military government handed over to a civilian one. As this was a civilian government headed by the recently retired military leader there had been a few minor objections including an attack on a police station. I was about to dig a hole in one of the best tourist beaches in the country within thirty metres of a tourist police station, so we decided that witnesses would not be a good idea. The taxi driver, ever so keen to help, and boost his takings, offered to waive his waiting fee, so I lied through my teeth : We were on holiday, there were lots of Dutch girls staying at Bungalow Beach, hopefully we wouldn't be going back to our hotel that night. It worked, or a generous tip did, and he was gone.

A quick walk through the Bungalow Beach Hotel revealed that the back gate was closed, and the Dutch girls were either in bed or out! We re-traced our steps, left the hotel by the main entrance and circled the walled gardens to come back onto the beach. Since Owen Burnham's early monster grave digging exercise (if you'll pardon the pun), there have been a lot of additions to that beach. In particular, there have been some neat little beach bars and restaurants - and the tourist police station. Fortunately nothing appeared to have been built on the area where Owen had buried "Gambo", some were mighty close though! The police station was, thank god, only manned during daylight hours, and was now closed. The beach was almost deserted but we would be attracting attention fairly quickly. It was to be one pilot hole only, and then home if we wanted not to be caught.

I don't think that there was any offence that we were committing under Gambian law, but I was a little worried about the explanation for digging the hole and questions about our mental health that the truth would obviously raise.

From the plan I had seen, the possible area for excavation would involve a rectangle - say 5 metres wide by 10 metres long. We guessed at a place well inside this area, but in some shadow, and started digging. Stainless steel butter dishes are quite good excavation tools, they penetrate the sand well, and shift a reasonable quantity with each "dishful". It was hot digging, but we alternated, one digging and one keeping watch, and we were soon down to two feet. Further depth would be difficult: we were working at maximum reach and the hole had become too narrow to permit either of us getting into it to dig deeper.

So we stopped there. We hadn't found anything except one small pebble and a bit of palm root, but at the same time we hadn't proved that there wasn't anything there. Certainly the sand would seem to be deep enough. The possibility of a second hole somehow didn't seem right. It was still very warm and we thought that there was a couple coming up the beach so we just left the scene, back along the beach and into the Kombo Beach Hotel a little further down. A quick drink at the bar and a trip to the loo to wash off some traces of sand, and then back to our hotel by taxi.

Reconstruction of "*Gambo*" by wildlife artist Tim Harvey

When we got back to our hotel, the dinner party set had returned from the restaurant, and more by chance than anything else we met the rest of the group in the night club bar. There was a nice degree of relaxation there, a mixture of relief, the trip was over without any major disasters, the guests were gone and the local beer had been flowing for most of the evening. Bruce and I relaxed: the chances of us assisting the local constabulary with their enquiries was now virtually nil!

As ever then, just when you relax, you are found out! Two of the younger girls of our group both insisted that Bruce and I needed to dance; resistance was futile, and I suppose that the beat of the African pop was a bit hypnotic too. The problem was that we were both carrying a lot of sand in our clothes, and the faster the music got, the more we were losing onto the dance floor.

Our problem became apparent to at least one of the girls, who sidled across to me knowingly. "*I know what you've been doing*" she said. "*You snook down to the beach and built a sandcastle. I've seen you looking longingly at the beach all week*". I nodded meekly. "*Africa brings out the schoolboy in everybody.*"

Ideally the story finishes there, but in fact it doesn't. The following morning the one member of the group who didn't have a late night took a long pre-breakfast walk along the beach.

When he got near to the Bungalow Beach Hotel he noticed something of a crowd near the police station. Polite interest revealed that they were all looking at a hole about twelve to eighteen inches across and vaguely circular.

In this part of the world, circular holes in the beach tend to be made by fiddler crabs which generally fit the hole! If the crabs were starting to grow to this size then strange things must be happening!

I just hope that I don't read anything in the next year about giant Gambian Fiddler Crabs!

The Dragons of Yorkshire

by Richard Freeman

INTRODUCTION

Of all legendary monsters, the dragon is both the most widespread and the most ancient. They flap and slither through almost every culture from the Mas D'A Zil Mesolithic cultures of twelve thousand years ago to contemporary tales of winged serpents in Africa.

Several species of dragon are spoken of, so before we delve into what may be behind the legends, it may be as well to examine these differing draconian types...

1. The Heraldic or 'true' Dragon.

Also known as the *'Fire Drake'*, this was the most powerful of all dragons. A huge quadrupedal reptile, it had huge, bat-like wings. The heraldic dragon was armed with savage teeth and claws and had a mighty tail. Its most formidable weapon, however, was undoubtedly, the white-hot gusts of flame, it could spit at its victims.

Dragon illustration by Lisa Allegri

EDITORIAL NOTE: For more details on flying dragons, and historical accounts of Fire Drakes, see the article by Dr. Karl Shuker in the 1996 Yearbook.

Heraldic dragons were the most magickal of beasts. They had many powers attributed to them including shape/form changing, self healing, invisibility and mind-reading. These dragons were almost impossible to kill, being covered with scales harder than steel. They (like *'Smaug'* in *The Hobbit*, by J.R.R. Tolkein) had one tiny spot of vulnerability, but the location of this 'spot' was never the same in two different dragons.

2. The Wyvern.

A very similar beast to the creature described above, but with only one pair of legs. It is also usually depicted as being somewhat smaller than the gigantic heraldic dragons. The *Wyvern* bore a deadly barbed sting in its tail, and was believed to spread pestilence and disease in its wake.

3. The Guivre or Worm.

This is the commonest celtic dragon. The worm was a vast limbless serpent. It inhabited vast lakes, marshes and rivers. Worms killed by crushing their victims in their enveloping coils (like a constricting snake) and with their poisonous breath which they used in much the same way as the other species used their breath of fire. This poison had the ability to shrivel crops and choke both man and beast.

This type of dragon (as typified in its most famous example - *'The Lambton Worm'*) could rejoin itself together after having been hacked in two and was thus extremely difficult to kill.

4. The Lindorm or Blind Worm.

This odd creature resembled the Guivre, except for posessing a pair of hind legs. It seems not to be as much linked to water as the preceeding type and is mostly reported from Asia and Southern Europe.

5. The Amphiptere.

This was a limbless winged serpent generally reported from the Middle East and North Africa. Amphipteres are still reported today in the South East African country of Namibia.

EDITOR'S NOTE: See Richard Muirhead's article in the 1996 CFZ Yearbook, and excerpts from Karl Shuker's writings in the same volume.

The Mexican God Quetzacoatl was a giant Amphiptere, with feathers instead of scales.

6. The Eastern Dragons.

Unlike their occidental counterparts,, oriental dragons were portrayed as being beneficial in nature. They controlled the weather, the seas and the rivers. Interestingly, at different stages of their development, they seemed to resemble one or other of the standard western types of dragon.

Oriental Dragon eggs took a thousand years to hatch, and the young dragons resembled snakes. After five hundred years, they resembled giant snakes with the heads of carp. Five hundred more years and they developed a bearded reptilian head and four legs. A further five hundred years brought horns, and the final stage, after yet another five centuries, brought forth wings, with the final result looking like an ornate, but skinny analogue of the western Fire Drake.

Many scholars have argued about what lies at the root of this most universal of legends.

Fossilised dinosaur bones have been hypothesised as the remains of dragons in many areas. In China they are still known as 'Dragons Bones' and are prized in powdered form within various types of folk medicine. Dragon legends, do, however, occur in places where no fossil bones have ever been found, and moreover, some legends speak specifically in terms of live dragons and their interactions with mankind, rather than just in terms of a pile of petrified bones. It is time to cover these skeletons with some flesh!

LIVING DRAGONS

Some species of living reptile can achieve an immense size. Crocodiles are the largest and most dangerous of these.

Crocodylus porosus, the Indo-pacific crocodile is an awesome beast. The largest specimen measured by an expert was twenty-eight feet in length, but larger individuals almost certainly exist. James Montgomery, a rubber tapper in northern Borneo saw a specimen measuring over thirty three feet in length on the Sagama River during the 1950s.

The local Seluka people believed it to be 'The Father of the Devil' and threw silver coins into the river to appease it. This provides an irresistible parallel to the hordes of treasure said to be guarded by so many European dragons. Today the Ibad people of Sarawak venerate 'Bujang-Senang' - the 'King of the Crocodiles' - a twenty five foot specimen who haunts the Lumpar River and is a known man eater.

EDITOR'S NOTE: For more on this species of crocodile and its relevance to contemporary cryptozoology see Darren Naish's article in the 1997 CFZ Yearbook.

The Nile Crocodile, (*Crocodylus niloticus*), is worshipped by many tribes throughout africa. It grows to over twenty one feet in length, but reports from Central Africa's Congo rainforests suggest it may rival its Indopacific cousin in size,

Between them, these two species account for more human deaths than any other vertebrate, with the exception of man himself. They will also kill lion, buffalo, giraffe and even sharks!

Giant constricting snakes make good prototype 'worms'. *Python reticulus* - the Reticulated Python reaches thirty three feet in length and can take prey as large as a leopard.

The Anaconda (*Eunectes murinus*), is not as long, reaching a mximum recorded size of 23 feet, but is far more massive in girth and weight. It is amphibious in nature and extremely aggressive.

EDITOR'S NOTE: The supposed giant anacondas of the Amazon basin are amongst the animals discussed by Mike Grayson in his article on *'The Fortean Fauna of Percy Fawcett'* to be published in the 1998 CFZ Yearbook.

The world's largest lizard is the Komodo Dragon (*Varanus komodensis*). Discovered in 1912 on a handful of tiny Indonsesian islands it can reach twelve feet in length. This giant monitor lizard has serrated teeth containing a virulent bacteria that causes wounds inflicted by the fangs to rot and fester. If an animal survives the initial bite its wounds will seep and stink - the smell leading the giant lizard to it. Chinese pottery, hundreds of years old, has been unearthed on Komodo Island suggesting that the ancient Chinese may well have been familiar with this gigantic reptile.

However, in Australia, an even bigger monitor lizard existed until the end of the Pleistocene epoch. *Megalania prisca* grew to over thirty feet in length and was the continent's supreme terrestrial predator. Reports suggest that this terrifying creature may still stalk the wilder parts of the continent.

Reconstruction of *Megalania prisca* by Lisa Allegri after Rex Gilroy

From the 1830s onwards white men have been reporting, what the native Australians have been reporting since times immemorial. Most of these sightings emanate from New South Wales. The most important was made by a professional herpetologist called Frank Gordon in 1979. Gordon, who had been conducting field work, returned to his land-rover. On starting his engine he was astonished to see a nearby 'log' rear up and lumber away. The 'log' was a thirty foot lizard!

EDITOR'S NOTE: In his book *'Mysterious Australia'*, Rex Gilroy presents a number of pieces of anecdotal evidence for the continued survival of these giant lizards and also presents evidence that ancient Chinese navigators may have reached the island continent hundreds of years before it was 'discovered' by European explorers. This could also provide a hypothetical source for Chinese 'dragon-lore'.

Explorer John Blashford-Snell did some important work in Papua-New Guinea concerning a mythical animal called the *Artrellia*, which he hypothesised as being a giant form of Salvadori's Monitor (*V. salvadori*) which can grow to a greater length (although a smaller bulk) than the Komodo Dragon. A summary of his findings can be found in Animals & Men #3.

Victorian author, Charles Gould, postulated a similar reptile inhabiting central Asia. Gould's hypothetical 'dragon' had ribbed 'wings' like the south east Asian Flying Lizard (*Draco volens*), and a constricting tail.

The problem with all the above animals is that they live only in the tropics. Dragon legends are universal and thousands of them come from temperate or even sub arctic areas!

In 1979, Peter Dickinson offered a unique theory in his book *'The Flight of Dragons'*. Dickinson's idea was that dragons evolved from large carnivorous dinosaurs like *Tyrranosaurus rex*. They developed large, expanded stomachs filled with hydrogen gas.

The hydrogen evolved from a mixture of hydrochloric acid in the digestive juices mixed with the calcium found in the bones of their prey.

This lighter than air gas allowed them to fly.

Animals & Men # 14 — Feature

They controlled their flight by burning off excess gas as flames.

Other ideas are even more esoteric. Many people have commented on the parallels between modern 'alien abduction cases' and the folk-legends of people kidnapped by elves and taken to Fairyland. Both have elements of missing time and memory. Both feature 'implants' - high tech probes on the part of the aliens, and magic silver pins inserted by mischievous elves. They seem to be the same phenomenon, adapting to, or filtered through the collective sub-conscious fears of mankind. What were once elves and pixies are now bug-eyed aliens. Could this not be the same for dragons?

EDITOR'S NOTE: For a further exploration of these concepts see my book 'The Owlman and Others' (1997) and also my 1995 paper for 'Promises and Disappointments' #2.

There seems to be some analogue between UFOs and dragons. Both are often seen near water and both seem to be cross cultural. They bpth seem powerful and 'above' mankind. UFOs outpace places and seem to defy all attempts to capture them (Roswell shenanigans excluded!) Early dragon legends portray them as beasts of god-like power and universal consequence.

EDITOR'S NOTE: See Richard Muirhead's paper 'The Flying Snake of Namibia' in the CFZ Yearbook 1996, for more UFO/Flying Snake paralells.

It was only later that the tales of more mortal dragons and dragon slayers emerged. These can be interpreted as allegorical tales signifying Christianity's triumph over Paganism. (Can anyone really believe that a puny knight on his figurative mouse of a horse, being able to triumph over a mighty reptilian dragon?)

YORKSHIRE DRAGON LEGENDS

Theorising aside, let us now consider the visitations of a draconian nature upon the fair country of Yorkshire.

The Dragon of Wantley.

This was a 'true' winged, fire-breathing dragon. It terrorised the country surrounding Wantley, killing

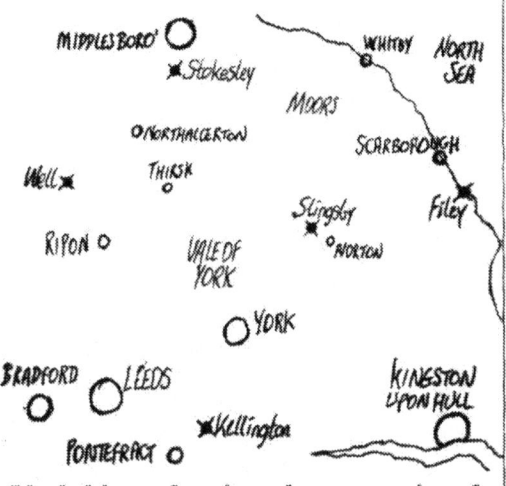

Yorkshire - showing places mentioned in text

livestock and people, and burning crops and buildings. The populace enlisted the help of a huge and grim knight known as More of More Hall.

For payment, More insisted, that prior to the battle, he be anointed by a fair-skinned, black-haired maid of sixteen. (He had impeccable taste in women!)

EDITOR'S NOTE: Hmmmm... I prefer twenty-eight year old blondes m'self!

More, had fashioned himself a suit of armour studded with spikes six inches long. He then hid in a well to ambush the dragon when it came down to drink.

The fight lasted for two days and a night, with neither opponent being able to peirce the other's armour. The dragon seized More, intent on hurling him into the air like a rag-doll, when More saw the beast's vulnerable spot and delivered a fatal kick with a spiked boot. Unromantically the one vulnerable and unarmoured place turned out to be the dragon's anus!

The legend was recorded in a lighthearted ballad dating from 1699.

Some say that the whole tale is a satire based on a

Animals & Men # 14 — Feature

Dragon Illustration by Lisa Allegri

a lawsuit over tithes in the reign of James I. The dragon being Sir Francis Wortley, who held the disputed tithes, and More being the attourney who set a lawsuit against him on the behalf of nearby gentry. The spiked armour being a document full of names and seals of men pledged to oppose Wortley.

However, several motifs in the tale, such as the spiked armour, the well, and the (almost) invulnerable dragon argue that the 1699 poem was adapted from a far more ancient legend. Wantley, it seems, may once have had a very real dragon.

The Dragon of Filey.

The hero of this tale is not a knight, a wizard or a lord, but a 'hen-pecked', meek, little taylor named Billy Biter. Whilst walkong along the cliffs one misty morning he tumbled into a ravine, that again turned out to be the lair of a 'true' heraldic-type dragon.

The dragon was about to devour him when Billy offered him the parkin (a Yorkshire delicacy somewhat akin to a cross between a treacle tart and a gingerbread man), that he had been carrying. The dragon enjoyed this piece of gooey confectionary so much that he demanded more and turned Billy loose.

Running home, he told his wife who insisted on making parkin for the dragon. As well as being domineering she was also a dreadful cook and produced the biggest and stickiest parkin in the history of Yorkshire. Billy rolled the parkin into the dragon's lair, and when the beast began to eat it, it's jaws became stuck fast. The dragon flew into the sea to wash the parkin away but was overcome by the icy waves. Its bones turned to stone and became Filey Brigg, a mile long projection of rocks that juts out to sea.

A parkin makes a nice change as a death dealing weapon from lances and swords!

This odd tale has a dramatic modern-day sequel... but more of that later.

The Worm of Loschy Mill/Slingsby/Kellington.

A confusing one this. All three tales are almost exactly the same and quite possibly have evolved fom one 'root' legend. Loschy Mill, in the parish of Stonegrave was the lair of a great worm with poisonous breath. The serpent could re-join severed segments of itself and had venomous blood.

Sir Peter Loschy, a local knight, fought the worm whilst wearing razor studded armour, and brandishing a huge sword. He was aided by his faithful dog (whose name history does not recall). The hound would grab segments of the worm whenever his master lopped one off, and ran to the neighbouring village of Nunnington. In this way the worm could not rejoin with its severed segments and was eventually killed.

However, when the knight congratulated his hound, it licked its master's hand and both master and dog died from the worm's deadly blood.

In the Slingsby version it is Sir William Wyville, and his dog who kill the worms and succumb to its blood. At Kellington, however, it is a humble shepherd and sheepdog who both perform the deed and pay for it with their lives.

The Sexhow Worm.

On a hill in this village, a worm took up residence

and demanded tribute of the milk of nine cows every day. Its venomous breath killed all those who opposed it until an anonymous knight rode into Sexhow. After a savage fight he slew the worm and went on his way asking for no reward. The villagers skinned the giant snake and displayed the hide at nearby Stokesley Church where it remained for many years. Unfortunately for cryptozoologists everywhere the skin has long since vanished.

The Handale Worm.

This beast is a bit of a hybrid. It was said to have a crested head and to breathe fire like a dragon, to have a sting like a wyvern, but to be a 'serpent' (like a worm, presumably). It haunted the woods near to Handle Priory, devouring young women. Eventually a brave peasant youth named Scaw fought the worm armed only with a sword. After a savage struggle he slew the serpent and found an Earl's daughter in its cave. Scaw married her and acquired vast estates.

The wood where the worm once lurked is now called Scaw Wood, and a stone coffin in the ruins of the priory is said to be Scaw's.

The Dragon of Well.

This 'true' dragon's reign of terror was brought to an end by a young knight named Latimer. He concealed himself in a spiked barrel, and when the dragon attempted to bite into it it succeeded only in wounding itself. Once it was sufficiently wounded, he emerged and finished the horror off. From then on a dragon appeared on the armourial bearings of the Latimer family.

The Bilsdale Dragon.

This creature is known only from the scantiest of rumous. A tumulus or barrow known as "Drake's Hill" is reputed to contain treasures guarded by a dragon (which is presumably still in residence).

Sightings of land dragons are rare today (although they persist in parts of Asia and Africa, such as the flying snakes reported from Namibia)..

EDITOR'S NOTE: See the articles by Richard Muirhead and Dr Karl Shuker in the 1996 CFZ Yearbook.

SEA DRAGONS

Dragonlike creatures are, however, still seen around the coast of Britain and some have even visited Yorkshire.

The first and most dramatic encounter took place on the 28th February 1934 on Filey Brig. (Forteans will note the "coincidence"). Fishermen had been reporting seeing a strange creature between Scarborough and Flamborough Head albeit from a distance of about three miles. On the dark, moonless night of the 28th, Coastguard Wilkinson Herbert was wandering along the Brig when in his own words:

"Suddenly I heard a growling like a dozen dogs ahead. Walking nearer I switched on my torch, and was confronted by a huge neck, six yards ahead of me, rearing up three feet high!

The head was a startling sight, - huge tortoise eyes glaring at me like saucers. The creature's mouth would be a foot wide and the creature's neck would be a yard around.

The monster was as startled as I was. Shining my torch along the ground I saw a body about thirty feet long. I thought 'this is no place for me', and from a distance I threw stones at the creature. It moved away growling fiercely, and I saw that the huge black body had two humps on it, and four short legs with huge flippers on them. I could not see any tail. It moved quickly, rolling from side to side, and went into the sea. It was the most gruesome and thrilling experience. I have seen big animals abroad but nothing like this!"

Mr Herbert's report remains one of the most spectacular on record anywhere in the world.

In 1938, Mrs Joan Borgeest saw a sea dragon off Bastington in North Yorkshire.

EDITOR'S NOTE: This is a particularly interesting lexilink, because BARGUEST, also spelt BARGHEST, BARGHAIST, BARGEST and BARN-GHAIST are names for the spectral black dogs often seen in parts of Yorkshire.

Whilst looking out to sea from the beach she suddenly saw....

"A huge creature rise; it was a green colour, with a flat head, protruding eyes, and a long, flat mouth which opened and shut as it breathed. It was a great length and moved along with a humped glide".

The beast was only ninety metres away and dived when she called to some other people. Mrs Borgeest was teased by 'freinds' and kept quiet about her story until 1961 when the BBC broadcast a radio programme about sea monsters.

In August 1945, Mr B.M.Baylis of Spilsby, and some friends of his, saw a monster.

"We were sitting on the edge of low mud cliffs at Hilston between Hornsea and Withernsea. There we saw a creature with a head and four or five rounded humps each leaving a wake.

It was moving rapidly, but quite silently along the shore, north westwards in the face of a northerly wind. Nobody at the time believed our report, but we are convinced that we saw something".

CONCLUSION.

One could write volumes on what these things could be, but space will not allow such an endeavour here. I would say, however, that in my opinion the living dragons are giant warm blooded reptiles. Dragons still have a relentless grip on our collective minds, both conscious and unconscious, and so we should not really be too surprised when they raise their scaly heads from time to time, even in this day and age.

Bibliography.

BRIGHT, M. *There are Giants in the Sea.* (1989, Robson Books Ltd, London)
DICKENSON, P. *The Flight of Dragons.* (1979, Pierrot Publications)
HEUVELMANS B. *In the Wake of the Sea Serpents* (English Ed. 1968, Rupert Hart-Davis, London).
HOLIDAY F.W. *The Dragon and the Disc.* (1973, Sidgwick and Jackson, London).
Simpson, J. *British Dragons* (1980, Batsford Books).
Whitlock, R. *Here Be Dragons* (1983, George Allen and Unwin).

THE CREDIBILITY GAP: MY SIGHTING OF THE 'BEAST OF BODMIN'
by Jonathan Downes

On the eighth of May 1997 I joined the ranks of those cryptozoologists that no-one will ever believe again because they have (more by accident than by design) seen one of the very creatures that they have dedicated their lives to searching for. It was about a quarter to eight in the evening, and I was driving along the narrow roadway between the village of Warleggan and the dual carriageway when I saw a creature cross the road about thirty feet in front of me.

I watched it for about four seconds before it disappeared and I realised that I had joined the priveliged ranks of those fortunate enough to have seen 'The Beast of Bodmin'. Although I had seen what I believe to be a mystery cat about ten years ago in the New Forest, there is the world of difference between the vague felid shaped shadow which vanished after a few seconds into a hedgerow in the middle of the night and what I saw the other evening on Bodmin Moor. For a start, I can make a positive identification of it specieswise - it was a puma!

Although I didn't see its head, the creature I saw was about two and a half feet tall with strong, gracile legs and extremely large paws. It was about four feet long with a long curved tail, about two feet long behind it. The animal was a dark, muddy, chocolate-brown with lighter coloured underparts. The end of the tail was clubbed and either dark brown or black. The ground on which I saw it was dry and it left no footprints, nor were any hair samples left on the gorse bushes through which it walked.

My camera was in its bag on the back seat of my car, but I was too shocked to use it. Like Dinsdale, Sanderson and others, I am sure that without concrete evidence no-one will ever believe me.

But it happened. So there!

CHARLIE FORT AND THE VAMPIRE SHEEP SLAYER

by Terry Hooper

Even today, ninety odd years after the event, Charles Fort's account of the *"sheep slaying mystery of Badminton"* is still quoted and regurgitated ad nauseam. When, as a local historian, I began to look into mysteries from the west of England, I knew that I had to consult my 'Fort' (*The Books of Charles Fort*, Henry Holt & Co., NY., 1941). On pages 645-646 I found the following:

"The sheep slaying mystery of Badminton; sheep were killed, and Sergeant Carter of the Gloucestershire Police said: 'I have seen two of the carcasses, myself, and can say definitely that it is impossible for it to be the work of a dog. Dogs are not vampires, and do not suck the blood of a sheep, and leave the flesh almost untouched...'

"November 25. (Bristol Mercury) It was claimed that the killer was a jackal that had escaped from a menagerie in Gloucester..."

In the Gloucester Journal, November 4th, in a long account of the depredations, there is no mention of the escape of any animal in Gloucester, nor anywhere else..."

Fort pointed out that the editor of the Gloucester Journal felt that there must be more than one beast at large. He mentions a large black dog shot at Hinton on December 16th but there had been no killings there since November 25th. Fort goes on to state that 1905-6 saw a large number of sheep killed by "unknown animals" across southern England before the killings stopped suddenly.

This case is near legendary in certain circles - the 'Vampire Killer' fascinated the curiosity of my late friend Franklyn Davin-Wilson. It was just the sort of thing that I was looking for, and so off went my letters to reference libraries in Bristol and Gloucestershire as well as to the Gloucestershire Constabulary.

By early January 1997, I had all the data that I needed. Firstly, however, I will quote the lengthy article referred to by Fort from the Gloucester Journal of Saturday, November 4th 1905:

"Gloucestershire sheep-slaying mystery. Armed farmers watch for a supposed jackal.

The inhabitants of the district around Badminton on the borders of Wiltshire and Gloucestershire may be pardoned if their nerves are somewhat shaken, says a correspondent of the 'Daily Mail'. Wild yells are heard at night, and the bloodless corpses of sheep are found in the morning, and both are attributed to the mysterious midnight marauder at present taking toll of farmer's flocks.

In the Badminton district alone there are six cases of sheep slaying attested by Sergeant Carter of the Gloucestershire Police. "I have seen two of the carcases myself", said the seargent, "and can say definitely that it is impossible for them to be the work of a dog. Dogs are not vampires and do not suck a sheep's blood and leave the flesh almost untouched". The prevailing opinion is that the animal is of the jackal type.

In the case of the ewe belonging to George Jones, a farmer, of Little Badminton, there was no blood left on the ground, though the animal's wound was a large one. It had been separated from the flock and driven to some trees.

Mr Hatherall of Oldbury, whose flock was also attacked, told the constable that he and his men had watched the flock for fifteen nights with guns. He had come to the conclusion that the same flock was never attacked twice. A light was now kept in the fields at night.

A constable named Locke said that he had heard the strange animal yelling outside his house more than once. It was not a fox's cry but more of a scream. Mrs Locke, the constable's wife, a man named Sheswell, and four or five others have actually seen the beast. Their descriptions agree remarkably well, though they are personally unknown to each other.

The animal is considerably larger than a fox, of a mouse brown colour, with a very sharp mouth and a long, bushy tail. The description ppoints very strongly to an animal of the jackal type which is assumed to have escaped from a travelling show.

On Tuesday the marauder was seen by a gentleman to enter a wood at Weston Birt, the estate of Major Holford, Equerry to the King. It has been in the district about three weeks, and the total number of its victims is ten or a dozen.

Some theories.
Satement by the Duke of Beaufort.

Writing later, the correspondent says that an extraordinary explanation has been advanced, and one which is of vital interest to fox hunters. There can be no questioning the fact that an extensive trade is done in imported foxes to be used for hunting, and also to strengthen the local breed. Herein lies the danger, for of late there has been more than one case of havoc among sheep caused by animals supposed to have been let loose as fox cubs. The most noticeable example was the Sevenoaks Jackal, which killed dozens of sheep, and was at last shot after a great drive, in which 200 persons took part. When young it takes an expert to detect the difference between a fox and a jackal.

If the Badminton jackal, if jackal it be, is to be explained in this way it has to be made clear that the Duke of Beaufort's hunt has nothing to do with its presence. The Duke said to the 'Daily Mail' correspondent on Wednesday:

"We do not do that sort of thing in this part of the country. Besides, there are plenty of foxes."

A jackal, it must be explained, does an immense amount of harm to fox hunting indirectly, for farmers set traps for the marauder, and a number of foxes are destroyed unintentionally.

"I remember", said the Duke, "when the wolf was at large in the north, very many foxes were killed by those who were on the look out for it. However, we shall be hunting in that part of the country tomorrow where the jackal is supposed to be, and we may come across it".

If the jackal was originally turned loose as a fox cub, it must have travelled many miles to get to Badminton.

Mr Sheswell probably saw the alleged jackal as closely as anyone. He was in his orchard when it came trotting through. He set his dog on it, and the jackal was knocked over. It scrambled into a bush and escaped.

With reference to young jackals imported as fox cubs, the 'Gamekeeper' in a recent issue said that young wolves also masquerade as fox cubs. "Who", it asked, "ordering foxes thought that he would receive wolves and jackals? He would as soon have thought of receiving lions and tigers. Jackals are cunning to a degree never experienced even in the case of a fox. They refuse to look at bait, and by frequently shifting quarters elude the efforts made to hunt them down". These remarks of the 'Gamekeeper' apply with special force to the Badminton pest, which never returns to a kill, and covers many miles between each of its attacks on flocks.

An Unsuccessful Hunt.

The mysterious sheep-slayer is still at large, says the Badminton correspondent of the Birmingham Gazette and Express. Farmers, Gamekeepers, Huntsmen, Clergymen, Innkeepers and Farm Labourers, all about the borders of Wiltshire and Gloucestershire, tell the same story of the raids upon sheep-folds, the strange yell, and the fighting of a strange animal a little larger than a fox, with a short tail, a mouse coloured coat, and a sharp nose.

The evidence of men in Badminton, Tetbury, Hawkesbury Upton, Didmarton, Little Badminton, Chipping Sodbury, and other places miles apart, tallies as to the description of the animal seen, and as to the nature of the night-cry, such as has never

Animals & Men # 14 — Feature

been heard in all this countryside before.

A particularly interesting point is that the huntsmen have been out to try and find foxes in Hinnegar Woods, on the Duke of Beaufort's eastate, and, although it is well known to be a famous haunt for foxes, they have not been able to find a fox there within the past few days. The theory is that this mysterious animal has scared the foxes out of their usual haunts.

The Vicar of Badminton described on thursday the cry as he had heard it during the past few days. "It seems to me", said the Vicar, "like the cry of a vixen, such as we often hear in this fox hunting district, but with a hoarseness in it, and ending up with a long howl, such as we have never heard a vixen give."

Mrs Adcock, wife of one of the Duke of Beaufort's keepers, said:

"I heard it close by Bull Wood, not far from Didmarton village, and it was different from the cry of any animal I have ever heard. It was a cry strange enough, and uncanny enough, to frighten any nervous person".

The Duke of Beaufort hunted over a portion of the stricken county on Thursday in brilliant sunshine, and a particularly fine fox was killed near Avening, but it was not large enough, or particularly rough enough to be accused of the sheep killing. The Chipping Sodbury veterinary surgeon, Mr Codrington, has been asked to make a thorough professional examination of the next sheep found dead. There were no deaths of sheep on Thursday, as far as has been discovered, but some of the farmers are still burning lights in the fields by night, and looking for a shot at the animal by day".

And that is the account that Fort is alleged to have based his references upon. The Daily Mail article could not be found, however, there was another item in the Dursley Gazette of the 18th November 1905:

"The Badminton Sheep Slayer Alleged Visits to Oxleworth and Dursley.

The 'Daily Mail' reports that the Badminton 'jackal' is at work again. On Sunday morning, one sheep was discovered at a local farm, killed in the characteristic fashion, the hind leg being gnawed through to the femoral artery, and the corpse left bloodless. Monday morning revealed three more victims, and on Thursday no fewer than eight sheep were found killed in the same mysterious manner.

On Monday night it was reported on good authority that a mysterious animal had paid an unwelcome visit to Oxleworth and caused serious damage to sheep belonging to a well known farmer.

On Thursday a Dursley gentleman told us that he had seen it that morning on Stinchcombe Hill!"

But along with the jackal theory there were others. Chief amongst these were those favourite old English scapegoats, the gypsies. The Bristol Weekly Mercury of the 25th November 1905, reported:

"A Burnham Mystery - Slaughter of Sheep.

This week a great slaughter of sheep belonging to various farmers at Berrow has taken place, the animals either being frightened to death or drowned in their attempt to escape from a dog (or dogs), and we are sorry to say that, so far, no success has rewarded the efforts to trace the owners of the animals who caused such serious mischief. How great this is may be gathered from the fact that 24 sheep were found dead, or have since died, and several more have to be added to the list before many days are over. The greatest sufferer is Mr Hill, for he has lost fourteen, and five more are in such a critical state that it is doubtful whether any of these will be saved. The loss to the owner is even greater than the figures suggest, for the whole of the victims were some very fine ewes in lamb. The other unfortunate farmers are Mr W.J.Frost, who has lost five, Mrs Hutchings, three and Mr Hawkins two. A theory that was at first suggested, that it was the work of a jackal which recently escaped from a menagerie at Gloucester, has had to be abandoned, as that animal sucks the blood of the sheep, whilst in this instance the majority of sheep had drowned themselves, or were so terribly torn and bitten about that they had to be put out of their misery. A possible explanation is that some gypsies with their dogs were encamped near the spot, but even if this were so, there is little hope of this being traced to them now".

So, Fort was guilty of a misquote and ignoring the facts that even the reporters picked up on; these killings were wholly different and there is no doubt whatsoever that dogs were involved. The Western Daily Press of the same day briefly reported:

"Berrow. Much mischief has been done to the flock of Mr H.G.Hill and other breeders by stray dogs who have mangled by night a great number of sheep. It is believed that Mr H.G.Hill alone suffered loss to the extent of £70".

The last reference to the jackal was in the Dursley Gazette of 9th December, 1905. It read:

"The Jackal Again! Considerable excitement has been prevailing in the Wotton district over the supposed appearance of the 'jackal' in the neighbourhood, and quite an organised hunt took place on Sunday afternoon, when the 'jackal' gave a dozen dogs and men a run of about ten miles, and eventually on being run down, proved to be a foxey looking collie dog which had rushed away in fright. Even then some were not satisfied, and one of the hunters remarked that he was sure that it was a 'wolf with a collar on'."

It seems very likely that, based on their descriptions, a jackal was involved at one point; whether it came from a carnival show, a menagerie, or had been shipped in as a fox cub does not matter. It was there and it killed. It may well have moved out of the area to avoid the hunts for it. It is also clear that the latter incidents did involve dogs.

Both Gloucestershire County Council, via Mr P.R.Evans, the Archivist (Searchroom), Mrs Turton and Mr Baker at Gloucester Library, and Inspector D Walker, Force Press Officer at Gloucestershire Constabulary were very helpful. However, the police records did not go that far back, and a stray dogs register only commences in 1906 (Q/Y/5/1/3).

Mrs Jefferey at the Bristol Reference Library was also very helpful in searching papers covering the days in question all of which was very useful and has provided leads to other cases.

So just what was Charles Fort up to? Did he deliberately not give full facts? The answer may be yes - but with a specific motive. Fort hated armchair experts and his humour is famous; I would like to think that this was Fort's way of making a point - it shows how people simply believed what he wrote, how journalists liked to make a good story better and much more. But you only find this out if you do the research.

The 'vampire sheep slayer' story is solved. No mystery. A pity though, it seemed such a good one!

EDITOR'S NOTE: In support of Terry's final conclusion, may I quote what Charles Fort himself, originally wrote in Lo (1931), immediately following his account of the Gloucester sheep-killings:

"I go on with my yarns. I no more believe them than I believe that twice two are four.

If there is continuity, only fictitiously can anything be picked out of the nexus of all phenomena; or if there is only oneness, we cannot, except arbitrarily find any two units with which even to start the sequence that twice two is four"...

"Jackal" Update

by Jonathan Downes.

(With help from Terry Hooper, Clinton Keeling and Dr Karl Shuker)

(As always I attempt to verify the zoological basis for articles printed in this magazine. In doing so, however, I unearthed a web of zoological intrigue and high strangeness which both corroborates and contradicts everything Terry wrote in the above article)

My first action upon reading Terry's article was to telephone Clinton Keeling who told me:

"Very few mammals do suck blood. Some

will lick the gash or wound in the prey animal causing the blood to emerge through capillary action, and causing partial exsanguination. A jackal, however, is a typical canine and an attack on a domestic animal by a jackal would be indistinguishable from that of a fox or domestic dog. Indeed it is generally thought these days that the golden jackal (Canis aureus) is, together with the wolf, one of the major ancestors of today's domestic dogs. I am, however prepared to accept that some of the jackal reports, especially those from the Sevenoaks area of Kent may be genuine, as may the Badminton ones, of 1905, although I doubt whether they have anything to do with these so-called 'vampiric' attacks".

Ironically, just after I had finished speaking to Clinton on the telephone, Terry Hooper 'phoned, with new information on the provenance of the Gloucestershire jackals, which confuse the matter further and suggest that the story is far more interesting than either Charles Fort, or indeed Terry Hooper had originally thought. Terry gained access to a number of letters held in archives at Worcester Museum, which lead us to suspect that strange canids were at large in the southern Cotswolds for about half a century before the 1905 killings:

"To the Rev. W.W.Cooper,

Pembridge 2nd February 1848. The animal that you wrote to me about has been killed at Bromsborrow, about two years ago. It is in the posession of my father who will present it to the museum. Five or six years ago, a friend of mine brought a vixen jackal with him from India. He crossed her with a fox and turned the cubs out near Newent, 12 miles from Bromsborrow. He then gave the jackal to Mr Giles at Ledbury who kept it lose for some time. Subsequently he chained it up. It broke its chain and was for some time, a year or two, lose in the wood. I am inclined to think the animal in question is bred from this vixen jackal. She must have been in the wood nearly three years when this animal was killed.

Signed

Rev R.M. Hill".

The next letter is from his father and is dated February 10th 1848 (The language is archaic and has, wherever possible, been quoted in full. We apologise for the resultant difficulty in conveying its meaning):

"My son has communicated your wishes respecting the animal which was destroyed in this parish in April 1846. I shall have it packed up tomorrow, sent on Saturday by Hart the Bromsborrow carriers (who I think puts up somewhere near the bridge), directed as you desire to Mr Reece at the museum. I likewise sent a bird that was shot last winter at Longden Marsh by the Rev W.Symonds, which he sent to me, supposing me to be a good ornithologist. It is evidently a species of goose and in many points resembles the Brent, but wants the mark on the neck that generally distinguishes that species. I am sorry that they are not better stuffed, in better repair, but such as they are, they are at your service. However if they are not thought worthy of a place in the museum (which I greatly doubt) I beg you will destroy or dispose of them in any way you choose. The animal has lost its flap eyes, tho' I believe my son told you its history. I may as well recapitulate it now to you. In April 1846 it walked into a cottage in this parish where were only a girl about 12 years of age and a young child. Seeing the child move in the cot, it dashed at it, but its attention was distracted by a kitten springing up from the hearth, which it seized and ran out of the house with whither it devoured. The older girl in the meantime locked up the door at which it long scratched and for (...illegible...) the cottage was close to the road and two men, passing by, it attacked them and one of them killed it with a pitchfork which he luckily had in his hand. The creature had been seen in the neighbouring woods and a farmyard for more than a week previous to its death. I examined the jackals in the Regent's Park last spring and it seems very much to resemble them only it is larger. A jackal had escaped from a gentleman's stable yard at Ledbury a few years ago and was for some time in the woods but was at last destroyed.

Many people think that this creature might be the produce of a dog or a fox, or of a dog and a fox. I shall be obliged to you to let me know that the box has arrived safely and what is your opinion of the contents,

Signed.

Charles Hill"

Mr Hill's letter also included an extract from the *Worcester Chronicle* (29.4.1846) which read:

"Tewkesbury. Extraordinary and dangerous adven-ture with a wolf. Last week a woman living in the parish of Ridmarley in the county of Worcester, on going into an outhouse adjoining the cottage observed what she supposed to be a large dog lying in one corner of the building, and thinking it belonged to some drover took no further notice of it. Shortly afterwards she had occasion to leave the house and the monster which was in reality a wolf, taking advantage of her absence went in and laid itself under a table. The children, three or four in number, likewise thought it was a dog but the youngest - about two months old, which was lying on a low bedstead in the corner of the room, looking up and beginning to cry - the savage animal rushed towards it and a cat belonging to the family courageously attacked the intruder. Poor puss was quickly torn limb from limb and the wolf carried her remains to the outside of the house, proceeded to devour them when the eldest child, a cripple about eight or nine years of age, had the presence of mind to shut the door. Having eaten the whole of the cat except the hind legs, the brute strove to re-enter the house. While this was going on, two men on their way to Ledbury Fair, wishing to leave their smock frocks at the cottage went towards the door but finding themselves oppoosed by the wolf. They procured a pike and a pitchfork and killed it. It was in a very poor condition and it is reported to have been seen in the neighbouring woods for some time past, having doubtless escaped from some travell-ing menagerie".

This report, which seems to be a highly dramatised version of the one reported by Mr Hill above, was apparently paraphrased from one in the Gloucester Journal.

The 'goose' referred to in Mr Hill's letter was probably a Canada Goose, a species unfamiliar in Britain at the time, but what of this peculiar canid?

Warwick Museum once had a mounted specimen of a supposed dog/fox hybrid which may have been a jackal, but the whereabouts of the specimen is presently unknown. A photograph of this animal which Terry describes as having *"a fox type head but the body of a collie with white markings on it"*, was published in *'The Countryside'* on September 27 1907. The proprietors of Worcester Museum, who provided the documentation quoted verbatim above, are still hopeful of discovering the original mounted specimen so Terry, and eventually we can obtain a photograph and hair samples.

Unfortunately none of this brings us any nearer to the identity of the Gloucester killings of 1905. They took place a year before the inception of the wild dogs register so no official records were kept. In my opinion, these so-called 'vampiric' killings, if they were truly 'vampiric' were, as Clinton Keeling has said, nothing to do with these Jackals, or complex canine hybrids that may have roamed the woods of the southern Cotswolds for a time during the 19th Century, and may have more to with the similar killings described by Theo Brown in *'Tales from a Dartmoor Village'* (1952) which several authorities, including your humble editor have likened to the recent spate of Chupacabras attacks in the hispanic areas of the world.

Having read the above paragraph to Terry down the telephone, he says that in his opinion, the killings were all down either to dogs, foxes or these strange canine hybrids and that *"Charles Fort was talking through his arse as regards the blood sucking bit!"*

Once again, placing the editorial telephone bill into serious jeapordy I telephoned Clinton Keeling who confirmed my worst suspicions.

A transgeneric hybrid between either a jackal or a dog and a fox of any species is completely impossible because the gestation period is 63 days for a jackal/dog and only about 51 days for all fox species. It is, I suppose possible that the 'jackal vixen' originally brought back from India was a female of one of the two species of fox endemic to that sub-continent but both the sespecies are smaller in build than the eurasian species (*V.vulpes*), and this seems very unlikely. Clinton confirmed that the animals that the Rev. Hill had seen at Regent's Park in the spring of 1847 were Golden Jackals

Animals & Men # 14 — Feature

(C.aureus), because they had been kept in that collection since it opened in 1828 and whilst four other species were kept later, none others were exhibited at the time.

He noted that all Jackals are pack animals which would be unlikely to fend for themselves singly and suggested that any escapee would immediately seek company of farm dogs, with whom it could, concievably interbreed.

It seems that whatever animal the Rev. Hill's friend brought back from India in about 1842, it could not have been interbred with members of the local fox population.

Writing in *Extraordinary Animals Worldwide* (1991) Dr Karl P.N.Shuker recounts a number of occurences when wolves and on occasion jackals were released into the British countryside by a local MFH eager to bolster up a flagging population of the local red fox.

He notes a jackal shot near Sevenoaks in 1905, (a good year for these curious canids) and also recounts a record from the early part of this century of coyotes released into Epping Forest. Indeed, it appears that until fairly recently the stuffed carcass of a wolf of unknown species which was shot in Epping High Street earlier this century was on display in the town's museum/town hall.

Such events are far from being uncommon within the annals of fortean zoological literature, and there are similar records from all over the United Kingdom.

My own records include both 'Scottish Hill Foxes' and (possibly) German Wolves released onto Dartmoor at the end of the 19th Century, and whilst it seems completely possible that (as Clinton Keeling suggested) the farm dog population in parts of the country has a smidgeon of jackal genes within its ancestory, and (as Theo Brown has suggested) the same thing may also apply with both North American, European and possibly even Indian wolves, the chances of the British fox population being diluted with genetic stock from any non-fox species.

Karl Shuker confirmed this and noted that the only C.aureus hybrids on record were:

"Male Golden Jackal x Female domestic dog. Hybrids tend to resemble the jackal more than the dog in colour and conformation" (Annie P Gray *'Mammalian Hybrids'* 2nd Ed 1972).

Unfortunately Annie Gray does not state which dog breed was the dam, but it is hypothetically possible, that if a jackal was brought to England from India and mated with a dog rather than a fox, and the progeny released, they could not only survive in the wild but could possibly (depending on the breed of the sire) look enough like British foxes (C.aureus is not unlike V.vulpes in essentials) to survive unsuspected alongside the local fox population.

However, as Karl Shuker pointed out, there is no evidence whatsoever to suppose that such hybrids would be viable enough to ensure a longstanding population. Such a bloodline might well only last for one or two generations.

Hypothetically, however, it is just possible that such a hybrided population existed (and may still exist) alongside, but separate to the fox population in the southern Cotswolds and elsewhere, and if so it is just about possible that an occasional animal, looking almost identical to its jackal ancestor might surface.

This could have happened in 1905.

It doesn't explain the vampiric attacks on the local sheep population though....

Just as attacks on sheep on Exmoor in 1986 were immediately blamed on a strange animal seen in the area (when in actuality they were almost certainly, largely at least, no such thing), eighty one years before the same socio-cultural scenario was being played out just up the M5 (which of course didn't exist then) in Gloucestershire.

Surely there is nothing new under the sun, and vampiric attacks on livestock, like out of place animals, (whatever their exact provenance) *"are just part of the way things are"* even if (in most cases at least) they have nothing at all to do with each other!

BIG CAT REPORTS FROM SCOTLAND

by Mark Fraser
(Editor of 'Haunted Scotland' Magazine)

Winter 1968-1969 (approx) Westwood Avenue, Ayr

The following is written down in the witnesses own words, who, at the time, was a schoolgirl of 11 or 12 years of age.

"It was around 1968-69 and it was also winter; I think it may have been January or February. I was ready to leave for school so it must have been around 8:30 am. I opened the back door and saw a huge cat on the doorstep, ready to rummage through the dustbin which sat to the right-hand side of the door. I got such a shock that I slammed the door immediately; I did not tell anyone because I knew they would not believe me. When my brothers and I left for school about five minutes later there was no sign of the cat.

It was quite unusual in the way that it looked at me when I first opened the door.. It seemed almost human-like - Its eyes, that is. The cat must have been as big as a puma or a tiger. Its head seemed enormous and its body, had I stood necxt to it, would have come up to my waist. Remember, I would have been 11 years old at the time and was at that age quite small at 4 ft 6 ins. It was stripey like a ginger tiger cat. Its whiskers seemed really long and its paws enormous. I remember its paws because it was about to rifle through the bin. I know I saw it very briefly (because I banged the door closed) but I can remember it clearly.

It was not like a tiger in the zoo, it was like a big inflated version of a house cat. There was talk of a family of wild cats near our school in the lane, but although we played on the swings I never saw them. It's funny; the cat was not in the least bit afraid when I opened the door. It did not move while I was there; it obviously moved when the door was closed. By the way, it had been snowing but the snow was not lying very thick - it was sort of slushy. I remember looking around the garden for paw prints but there was nothing to be seen."

1968-69 (2)

"This also happened around 1968-69. I had been out playing and decided to go home. At the entrance to our house there was a wall about 2½ feet in height. Sitting on the wall was a large cat, it was really big and had a large head and a thick neck that went into sort of broad shoulders a bit like a pit-bull terrier. It was hard to say how tall it was as it was sitting, but it was much larger than a house cat. The cat was a sort of black/brown colour.

"Being frightened of cats I stopped at the path and would not pass it. It seemed to be watching me, aware that I was frightened of it. It hissed and spat at me and stretched its claw s out. The claw paw was really large like an adult's hand or fist clenched. I ran off up the road. It really had strange eyes as well; they were very dark. The whole incident was frightening because the cat seemed to be aware of my fear. This was the second time I'd seen a cat."

1987

Mid summer 1987

"George", who works at Spillers near Barrhead was one night driving to work, the time was approx 9.30 to 10 pm. As he was approaching Burnhouse on a clear night he saw heading towards him on the oposite side of the road a muscular lion type animal. *"Big strong... mountain lion type."* George states the tail was 3ft long. The animal was caught in the headlights and in one bound leaped onto the white lines in the middle of the road and with another bound it reached the grass verge and was away into the fields...

Second sighting Jan 1994 approx

This time "George" was returning from work at Spillers, the time being approx 6.30 am. He was heading towards Irving and was about one and three quarter miles from the town, near the Tom Yard inn.

He first noticed greenish yellowish eyes shining in the dark, belonging to a large cat-like animal crouching on the roadside staring at the car. The animal was dark fawn, tawney in colour. The eyes first reminded "George" of two lights. The animal last thing that "George" saw. On first approaching the animal George thought the eyes belonged to a fox, he was a hundred yards away when he first saw the eyes.

1992

September 1992 approx.

The area was a forest in the Sunderland/Caithness. The gamekeeper who was the witness does not want his name known or even the location of the forest. he was doing his rounds when he suddenly sighted a large black cat: it was, he said, the size of a Labrador dog, although the head was like that of a puma or panther. It had a long sweeping tail and moved in a cat-like manner. It actually cut across the gamekeeper's path about 10 ft in front of him and then shot off into the forest. He had never seen anything like it before.

1994

February 1994

Account from Duncan and Alex Binning, who have no objections to their names being used.

They use the Dean Castle Park estates pretty regularly for walking and exercising their dogs. Late one evening in February the two dogs they had with them began acting nervously. They walked on a few paces and then heard the sound of snapping twigs coming from the trees.

The couple now became nervous themselves as they had never seen their dogs react in such a way before, the younger one of the two now having come to walk in-between the couples' legs.

As they reached the car park Duncan took a look over his shoulder and saw in the darkness behind them two yellow cat-like-eyes that belonged to a large black animal standing much taller than the dogs, 40 ft away. The couple leashed the dogs and left the area rapidly.

A few days later while out walking the dogs, the dogs began acting strangely again. Then both Duncan and Alex heard the sound of purring, like that of a domestic cat but much much louder. This time, remembering what they had seen the other evening they left quickly.

The dogs, when in the grounds before and even after the incidents, have been fine and displayed no odd behaviour at all.

Another nearby resident living in the New Farm Loch area of Kilmarnock also talked of the strange and puzzling behaviour of his dog in the early part of 1994 but cannot remember exactly which month. For years he and his faithful hound have walked the estates without incident except for the time when the dog flatly refused to enter the grounds, each evening for about a week. No amount of coaxing would entice the quivering yelping animal into the estate. Then one evening, after sniffing the air, the animal entered without any problem at all.

Summer 1994

A big cat-like creature was sighted by a girl who did not want to be named, in the Castlehill area of Ayr - near the Castlehill stables there is a wooded area. The cat was seen moving through the trees, about the size of a Labrador and black/dark grey in colour... she saw the cat briefly which was at first crouching before it stood up and moved away.

Autumn 1994

"Michael" phoned: he would not leave personal details. His wife saw a big cat or creature at the Mount House in Kilmarnock. It was a dark morning. The cat was black and had a really long tail. Michael also had a sighting himself in the same year in Kirkcudbrightshire as he was driving his lorry.

December 1994 approx

"Lee came running into the bedroom quite excited, saying there was a strange animal in the opposite field. As we jumped up to look, whatever was there was gone. Lee then became a little reluctant to say what he'd seen. The animal was bigger than any dog, except perhaps a Great Dane. He mentioned it being cat-like, maybe like a panther, and jet black. This was seen when we lived at Windy Brae: farmland near Kilwinning, near the Dalgarven Mill".

1995

August 1995 approx

Dundonald Camp. Darkness 9.30 to 10.30 pm. Mr Dunlop was driving towards Irvine when he looked into his rear view mirror to glimpse a big cat jumping into the hedge. He states it was the size of a puma, fawny, with a fairly long tail and about the size of a labrador or slightly larger.

Animals & Men # 14 — Feature

Winter 1995

Ian McCaw spotted a large brownish cat near the Louden Golf Club, Kilmarnock. *"It was heavily built, like a Dobermann, but I could see it had pointed ears like a cat and a face like a cat. There was some snow on the ground but I didn't go looking for any prints."* The Club steward admitted that he had found large cat-like prints in the snow.

1996

10-7-96 Dundonald

"D" states he had just driven past Fraser's Garden Centre when they saw a large creature running into a field. Golden coloured, long in length similar to a Great Dane. It ran across a field of cows which scattered at the animal's approach. "D" was shocked and surprised and turned the car to go back and look again but the cows had settled and the animal had gone.

He said it had moved very quickly and gracefully and seemed very powerful. "D" mentioned that for decades now, local hearsay is that a puma lives wild on the Fenick Moors.

27-7-96 Mr Wright, Dalry

At around midnight Mr Wright was travelling the Ardrossan-Dalry back road. Near the dam by the field adjacent to the boats they found a mauled sheep still alive. Mr Wright states the lamb had four claw marks or slashes down one side about one and a half inches long and bald patches around the wounds, that he did not think were caused by barbed wire. He informed the police; and returned the next day but the lamb was gone from the field.

3-8-96 approx. In Dundonald.

M was travelling the Dundonald-Kilmarnock road at approx 9pm. As he neared Fraser's Garden Centre a large cat-like animal walked along the road in front of him. It was the approx size of a fully grown *"greyhound only fuller in body and golden/fawn in colour"*. As the animal caught sight of the car it shot off very quickly into the fields.

Towards the end of August "M" saw a similar animal on the Dundonald-Drybridge road. It was dark evening time; again, the animal was walking the side of the road towards him and ran off as the car approached. It was similar in size and shape as the first sighting although "M" thought it was darker - but this could be due to the lighting conditions at the time.

7-8-96 Kilmarnock.

A driver and passenger (males, approx mid-20's) were driving to Irving from Kilmarnock in the dark, about 23:30. On the brow of a hill near Moorfield roundabout they saw, in profile only, a large *"bigger than any dog"* cat-like animal. It jumped into the middle of the road, barely touching the ground, and then onto the verge and away into the fields. A golden colour; they thought at first it was a deer but in reality they knew it was no such thing... smooth, short haired, light brown.

6-9-96 McCormack, Darvel.

Time approx 6am. Travelling to his factory in Burns Road he saw a 'big cat' moving along the railway embankment that runs along the industrial area. The animal was black and about the size of a labrador dog "but sort of feline in appearance".

8-9-96 Dobbs, Hurlford.

Mrs Mattie Dobbs and her husband, both farmers, were out walking their dog along the River Irvine on the Hurlford side and looked across and saw an unusual animal that they couldn't identify. It looked like a cat but much larger than a domestic one. Then the animal stood and as it turned to its side Mattie was amazed at its size and length - it also had a very long tail.

The way it moved and stretched put her in mind of a jungle cat. In height it was as big as their labrador. Facially it did not resemble a household cat; the face was fawny coloured the same as its body but had brown markings around the top of its head and ears (not stripey or ginger). They watched the cat walk away.

11-9-96 Galston.

Driving along the back road from Galston to Ayr, J saw a large cat-like animal near Fiveways, at the side of the road two and a half car lengths from him. He slowed and dipped his lights. J, aged 81, said that as he passed, the animal pounced at or lunged at the driver's side window, actually smashing the wing mirror. It was roughly the size of a Golden Retriever, light tan in colour with a long tail, and had cat-like facial features. J's neighbour, also a pensioner, said that, two days earlier, her car was also attacked by a large animal on the same stretch of road.

14-9-96 Dundonald.

"H" and her husband saw a large cat at Fraser's Garden centre around 11am. They'd gone to look at the horses in the field round the back, which is where they saw the cat, completely black and around the height of an adult Labrador although longer in length. The cat seemed to be walking stealthily; the donkey in the field did not seem concerned.

23-9-96 Mrs Malone, Kilmarnock.

At approximately 7 p.m Mrs Malone was out walking her 12 year old West Highland Terrier on the grass verge near the Springhill nursing home (near Mount House). Suddenly the dog crouched low *"as if it were going to stalk something; it never made a sound but stared straight ahead."*

Then, a "very large cat-like animal appeared from behind a very large tree." The creature's eyes were marine green with orange pupils.

The cat stared at her as she stood rooted to the spot with fear. The cat moved slowly from behind the tree and Mrs Malone described it as being shiny black in colour with long legs and long tail.

It was the size of a fully grown labrador. She backed away slowly, pulling the silent dog with her - she got the impression her terrier would have stood its ground.

When describing her experience a week later she said "The cat almost seemed intelligent and it was as if it were trying to 'stare me out'."

It's....
557 miles from Exeter,
469 miles from Cowering in the Minge (Wilts.)
1983 miles from Sioux Falls, Indiana.
.75 miles from the North Sea
It's..

NORTH OF THE BORDER
by
Tom 'let's send that fat fool a skull with a horn on it as a joke' Anderson

(Amazing stunts with a Combine Harvester)

Wednesday, 12th February on the A835 between Inverness and Ullapool, a lorry driver spotted what he described as a 'large dead cat' lying on the verge of the road. He described it as being 50% larger than a German Shepherd Dog and gray in colour, but when he returned to investigate it further it had gone! A white Transit Van parked nearby had also vanished.

CUB LOCAL TV NEWS PROGRAMME...

Cut to interview with 'local expert' Di Francis, standing in a field looking windswept and rained upon, who tells us that the only cat capable of reaching that size AND having gray fur is a puma. However, she points out to Grampian TV's interviewer (in my opinion wrongly), the base colour of many lynxes is also gray. She then quoted the Cannich cat - a pretty dodgy example as it was brown and had been released fairly recently before...

EDITOR'S NOTE: ALLEGEDLY Mr Anderson!!! (Anyway all forteans are used to allegations about missing and disappearing corpses of 'grays').

She believed everythingbthat the driver (name unknown and no details) claimed. I am dubious both about the missing cadaver AND the mystery van!

Lastly, if it did exist, was it a cat? As we all know, moves are afoot to reintroduce the wolf in this very area and Wolves ARE gray. *Now. If we could only tie in a certain Oxford academic with the ownership of a white van....*

I think that we should be told!

EDITOR'S NOTE: The views of Aberdeen's Mr Entertainment do not necessarily always coincide with those of the editor and his team. When they do, we don't usually admit it for fear of legal action and attracting unwanted attacks from various vigilante groups amongst the zoological community!

Giant Squid, Mystery Boar and Pregnant Snake - Three Irish Animal Stories

by Richard Muirhead

In this piece I am presenting information on three unusual animal stories from Ireland. The first, of a possible giant squid, is from 1673, the second of a boar-like animal in 1780 or 1781, and the third, a gravid grass snake of 1832. All are fairly well documented, but in the case of the "boar" (if that is what it was) more information could come to light because the text from the source found refers to other papers in Ireland referring to this animal and genealogical records could trace the eyewitnesses who were resident in the Tipperary area.

1. Squid

The account of the squid or sea monster is taken from an old issue of the journal of the Cork Historical and Archaeological Society [1]. This in turn was taken from a pamphlet at the library of the British Museum, ie the British Library, which is where I found the account of the "boar" below. The animal was actually described as being "captured" which gives the impression that it was actually taken alive. On examination of Fig. 1 the animal seems to have squid and simultaneously jelly-like features, ie the "eyes" and tentacles in the former case and the bag-like "hood" above the head. Perhaps it was a kind of octopus. Octopi are familiar from British waters, having turned up on the coast of Sussex in recent years, so it is not inconceivable that they may have turned up off Ireland. The original document makes it clear that the depiction of the animal had to be shortened due to lack of space, so it should be longer than shown. It was:

"Bigger than an ox, yet without legs, bones, fins or scales; with two heads, and ten horns of 10 or 11 feet long, on eight of which horns there grew knobs, about the bigness of a cloak button, in shapes like crowns or coronets, to the number of 100 on each horn, which were all to open, and had rows of teeth within them; and in all other parts wonderful and unparallel'd [2].

The document goes on to describe how a James Steward was riding along the sea shore at Dingel-Ichough, in Co. Kerry, when he saw the animal near the shore.

"The length of this sea monster, horns and all, was full nineteen foot, and in bulk or bigness somewhat larger than a horse... Besides, it had a natural power to contract or draw in these horns into its head (as a snal does) and extend them again at pleasure. ... Between these two smooth longest horns, and in the middle of all the rest, grew up from the great head, the little or smaller head, at about three or four foot distance; this was much in the shape of a hawk looking upwards, and had a strange mouth, and two tongues in it, and here too, no doubt it did take in much of its nourishment." [3]

The description continues with saying its colour was flesh colour, except it had a large fleshy mantle which hung loose over the body on both sides. This was bright red on the outside and perfect white within. It was cut open and the liver weighed 30lb and the fat, when boiled, hardened. Part of the animal was taken to Dublin "and presented to several persons of honour" [4] and shown publically in various parts of Ireland including Dublin.

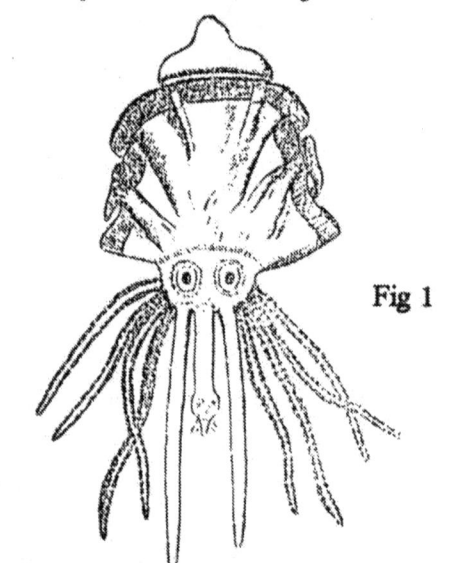

Fig 1

The opinion of Darren Naish is that this animal, when described to him in April 1997 sounded like what may be one of the first descriptions of a giant squid from this part of the world, but it had some unusual features like an extra head.

Boars?

The next account is of several anomalous animals of the same species, found in late 1780 or early 1781 in central south-west Ireland (see Fig. 2). I "found" this whilst looking through Walkers Hibernian Magazine [5] in the British Library, London. The incident involving this animal and at least five men took place near Thomastown, on the road between Tipperary to Cashel. What makes the case interesting is that the animal, although somewhat boar-like, is somewhat anomolous, to put it conservatively! The animal is shown below. The account tells how one John Carrol was travelling towards Cashel from Tipperary when he heard what sounded like the roar of a bull amongst some trees... *"After a few moments he saw the animal whose noise he had heard; its size and figure was that of an ordinary pig, but its head was armoured with spreading horns. This animal followed him at a slow pace near a mile, until coming to a place near Thomastown Pool, it entered a cave."* [6] Then the following night Mr Carrol and some other men went to the cave and found three of these animals which then took home. That is the substance of the story.

These animals sound superficially like wild boar. The trouble is, what about those horns? Wild boar have tusks, not horns. Yet according to Thompson [7], a natural historian of 19th century Ireland, the tusks of Irish wild boar were of a goodly dimensions. But a Dr Scouler [8] said that, compared with wild boar found in Scandanavian peat bogs, the Irish boar was very diminuative and was plentiful down to the 17th century, though the exact date of extinction is unknown. This information is in a book on extinct British animals [9]. If the Irish wild boar were really diminutive, would it have had prominent tusks?

Fig. 2

Snakes

Contrary to popular opinion snakes have turned up in Ireland, most probably introduced deliberately or by accident. In 1831 a Mr James Cleland introduced six grass snakes to Rath-gael in Co. Down from Covent Garden to see if they would survive. One snake was killed at Milecross about three miles distant, and it was thought at first to be a kind of eel until a naturalist identified it as a snake.

"One far-seeing clergyman preached a sermon in which he cited this unfortunate snake as token of the immediate commencement of the millenium..." [10]

Three more snakes were later killed, which left two still in the wild. The Times of Sept 8th 1832 reported, reprinting from a Belfast newspaper, that *"a gentleman called at our office last night with a female snake 3 ft 3 inches [1 m] in length, which was killed yesterday in a field at Milecross... They [grass snakes] have recently been seen in the Co. of Down."* Their presence in Ireland was attributed by some to opposition to the Reform Bill of 1832 [11].

"The futility of the popular belief that snakes cannot live in our blessed land has been most fully demonstrated - several large eggs having been found in its ovarium."

Appendix

After completing the above I came across a very unusual report from The Naturalists's Notebook for 1869 (p255). It originally appeared in Saunders's News-letter, probably sometime that year. The illustration below appears courtesy of Darren Naish, with thanks. He also agrees with me that it is no known species. This is the account in its entirety.

"Capture of a Curious Animal in a Rabbit Snare - An animal which baffles the skill of those who have seen it to define, was a few days ago found in a snare set for rabbits in the demense of the Marquis of Conyngham, at Slane, County Meath.

It is thus described by the gamekeeper:- The size of a good cat, with a tail about a foot and a quarter [18 cm] in length, covered with a strong wiry hair. The snout is sharp and pointed, something like a weazel's (sic). In the mouth there are four large tusks, two pointing upwards and two pointing downwards. A small mane of dark brown hair runs downwards. A small mane of dark brown hair runs down the whole length of the back; but the strangest thing of all is that it has twelve toes or claws on each foot, in two rows - seven on the outside row, which are exceedingly sharp, and five on the inside. In general it is more stoutly built than animals of the cat kind. Still, the body is lithe and supple, the colour throughout is dark brown, and white on the breast." - Saunders's News-letter

Could this be an as yet unidentified animal from the 9th century poem which mentions many Irish animals? This poem is featured in part in P. H. Gosse's "The Romance of Natural History", 2nd Series, 1861, pp 57-62. It concerns a prisoner offered release by his captor if he presented the latter with a pair of every Irish animal. A conversation with the current Marquis of Conyngham in July 1997 revealed that this was the first time he'd heard of the animal. Some family records had been destroyed in a fire but he had read about wild cats in Ireland "in the early 20th century."

Fig. 3

Notes and References

1. Anon. "An Account of a Sea-Monster captured at Dingleicouch in 1673" - *Cork Historical and Archaeological Society Journal* vol 3A, 1894, pp 190-192.
2. Ibid, p190.
3. Ibid, p192.
4. Ibid, p192.
5. Dea, John and Ryan, Dennis et al "Account of an extraordinary non-descript animal" in *Walker's Hibernian Magazine*, Feb 1781 (Dublin?) p73.
6. Ibid, p73.
7. Thompson, *Natural History of Ireland* vol 4 p36.
8. Dr Scouler in *Journal of the Geological Society of Dublin* vol 1 p226 and Wilde *Proceedings of the Royal Irish Academy* vol 7 p208.
9. Harting, J B. *British Extinct Animals* 1972 ed. pp93-94.
10. Chambers, R. *Book of Days* vol 1, 1883, p383.

11. Editor's note: Ireland was, at that time, ruled from London. The Reform Bill proposed reform of the anomaly-ridden voting system where, for instance, Old Sarum (an Iron-Age fort) returned two Members of Parliament whereas the emerging industrial town of Manchester had no MPs at all.

CLINTON'S COGITATIONS: THE KELSTRIDGE LIONS

by Clinton Keeling

The following event took place on 14th August 1970 when I was operating the Ashover Zoological Garden in Derbyshire. I swear to the veracity of this entire account.

At approximately 11 am I received a 'phone call from a farmer at Kelstedge a hamlet a little under a mile away. "Come quickly," he snapped in an agitated voice, "*two of your damned lions are out, and are in one of my fields.*"

How I got to the puma enclosure I don't know, as my knees felt as though they were full of warm water. We had no lions, but I deduced that Apache and Sioux, the pumas, must be out, as out of all our animals these most closely resembled lions - at least from a distance. I almost yelled out in relief when the pumas regarded me sleepily as they dozed in the sun of high summer, so I made my way to Kelstedge with something approaching interest.

First, though, a word or two about my informant. Named Muxlowe (quite a common surname in that part of the country and probably originating in the region of Kirkby Muxlowe in Leicestershire) he was an intelligent well-read man in early middle-age, and an ex-public schoolboy - although I fully appreciate that being endowed with quality grey matter and having been to a public school are by no means synonymous.

When I arrived on the scene he led me a few yards along the lane and, without a word, pointed into a field: roughly in its middle (Peakland fields are, generally, small), were two domestic dogs - golden Labradors.

Admittedly they were in a rather unusual position, as Wylie, the bitch, was lying couchant while Noble was standing upright and motionless behind her (in fact in retrospect I realise the former was on oestrus) but although I only have one eye I immediately identified them as what they were, which brings up the strangest part of the whole odd saga.

In that part of scenic Derbyshire live many townies, ex Chesterfield and Sheffield in particular, and, wanting to play the part of country people rather ex Chesterfield and Sheffield in particular, and, wanting to play the part of country people rather affectedly, make a point of keeping the variety of dog they consider to be most redolent of rural life - so consequently the golden, and to a lesser extent the black, Labrador is a common sight in the area. But not only that: Noble and Wylie were owned by Muxlowe's next door neighbour; he (M) saw them nearly every day and was on good terms with them. So work that one out if you can. After getting on for three decades, I cannot.

When I tackled Muxlowe about it subsequently he surprised me by becoming quite defensive about it. *"Well of course, it's only natural,"* more tolerant people than I have said; although in my view he hadn't a leg to stand on to defend such pure unadulterated idiocy, but on one occasion he came out with the bon mot *"anyone can have a mental abberation"*.

All the discussion and debate in the world, though, will not alter the simple fact that possibly for the hundredth time that year, Muxlowe noticed the dogs from next door - and quite genuinely saw them as two lions.

I'm reminded of the immortal words of the late Sir Arthur Keith - *"I have come to the conclusion that the existence or non-existence of the Loch Ness Monster is not a problem for zoologists but for psychologists."*

I agree.

EDITOR'S NOTE: As of this issue Clinton Keeling will be contributing a regular column to this magazine. Usually it will take the form of his 'cogitations' on the issues raised in the previous issue.

He wrote such a column for this issue, but even such a well run and delightfully organised organisation as our own sometimes makes mistakes (hint of efitorial irony there folks), and his original article disappeared into the intersteces of the CFZ Filing System, a cock up for which we can only apologise. However, the regular series, (quite possibly subtitled "Don't look Back in Anger" because even the A&M Editorial team are not immune from the odd britpop joke - and I am determined to see whether Mark North can do a cartoon of Clin as Liam Gallagher), will start in its originally planned form from issue 15.

Neil Nixon presents another extract from that highly mythical compilation album:

NOW THAT'S WHAT I CALL CRYPTO...

"The Amazing Bigfoot Diet"

Somewhere in the bargain bins of a local indie stockist or the well stocked racks of a store in imports you may encounter the work of Mojo Nixon (no relation as far as I know). Described in Rock encyclopaedia pages as *'Music's premier sex, drugs and rock 'n' roll enthusiast'*...

EDITORIAL NOTE: Such authors have obviously never met large numbers of the A&M Editorial team.

... and known primarily as a ranting oddball, who describes his own fanbase as the *'Fornication Nation'*, Mojo, like many comedians - hides a serious intent in his bizarre imagery.

Mojo's stock in trade is to combine out-to-lunch lyrics with a skeletal rock backing flexible enough to allow him to mt, rave and throw flexible timings all to hell. This opening cut of the Enigma compilation *'Unlimited Everything'* is a predictably improbable tale of life in Mojo land, in which he marries a bigfoot in the opening line and then fills in the rest of the song with headlines from US Supermarket tabloids. Superficially it's gibbering nonsense. In the context of a collection of the best of Mojo the imagery of the opening cut supports other songs which explore other folk myths, media stars and underground ideas.

Elsewhere on the album Mojo tells us that *'Elvis is Everywhere'*, Debbie Gibson is pregnants with his two headed love child, and that we should stand up against mass culture by burning down shopping malls. The infamous *"Stuffin' Martha's Muffin"* - a love song of sorts to a slow witted MTV VJ - also puts in an appearance. Looked at from this angle Mojo's relict hominoid related ditty is the perfect opener. The man celebrates the unfettered imaginitive genius behind trash culture, and the 'Bigfoot Diet' in this context is simply the endless demand for novelty and angle to triumph over reason. Food wise the diet in question never gets more specific than a command to 'Do the Bigfoot Burger' The command itself is adressed to the same audience that is swimming in tabloid story inventions.

Animals & Men # 14

If Mojo has a serious message beyond this - and he's often the last to admit it, he is telling us that Bigfoot belongs firmly in the folk-myth world of Elvis burger chefs and 'Jesus at Macdonalds'. Bigfoot is simply a talisman for the whole culture. We want Bigfoot because we want cultural icons which present us with instant wonder and amazement. Mojo's Bigfoot is what we've made him, which has very little to do with Gigantopithecus, land bridges and evolutionary possibilities. Bigfoot is ever present, and the Bigfoot agenda is about as precise as that of the tabloid papers that have printed the storis and the supermarket chains that sell the tabloids. Anything more specific would destroy Bogfoot's main use as a flexible image onto which novelty and invention can continually be hung. As the outro says:

"Bigfoot here, Bigfoot there, Goddamn Bigfoot Everywhere".

Incidentally - forget the serious bits for a moment. Mojo Nixon records always sound as though the man himself is having a seriously good time. Occasionally weird and warped, but definitely for those with a sense of humour.

EDITOR'S NOTE: I make no apologies for having included a far longer *'Now that's What I call Crypto'* than usual. It seemed important to ptint Neil's article in full, because rather than just presenting a slice of novelty pop music, and taking a wry look at one of the sillier backwaters of contemporary culture, he has (or rather his non-relative Mojo has), presented a very similar thesis as to the nature of many North American Man Beasts as did Loren Coleman in his article for Anomalist #2, and his lecture at the 1995 Unconvention.

Bigfoot, like so many other cryptozoological icons can no longer be seen purely in terms of a zoological, or even a zooform model. The terms of reference need to be broadened in order for us to accept 'him' as a true piece of socio-cultural iconography! At the risk of annoying many seasoned but somewhat 'fundamentalist' cryptozoologists, the same could be said about British Big Cats, Big Birds and even the ubiquitous 'Nessie'.

That, however, gentle readers is a whole new can of worms (or should that be 'can of media-fuelled quasi-plesiosaurs'?)

LETTERS TO THE EDITOR

The Editor welcomes letters and other communications on any subject of interestto readers of this magazine. He reserves the right to edit, omit, (and in the case of Mr Thomas Anderson) both say rude things incessantly and shamelessly malign his good name within the community, and would also like to stress that all opinions stated, both here and elsewhere in the pages of this august periodical, are the opinions of the individual writer, and not necesarily those of the editor, the editorial team, or the Centre for Fortean Zoology. Every attempt has been made not to infringe the copyright or libel laws. Any such infringements are the responsibility of the individual writer rather than the publishers or editorial team. So there!

Animals & Men # 14 — Letters

THE UNIDENTIFIED WALLABY SLASHER OF NEWQUAY - THE BACKLASH

Dear Jon,

Nice to see you at the FT UnConvention last weekend. I hope that it proved a successful event as far as the CFZ was concerned. Thinking later about your talk on the Newquay Zoo mutilations, one possible explanation did occur to me: one which I'm sure you must have considered, but were perhaps loth tpp speculate on publically. However, much in this case would be explained if the culprit were a member of the Newquay Zoo staff.

The finding of the knife in association with one of the dead waterfowl, suggests that we're looking for a human attacker, rather than a paranormal entity. Its also logical to assume that whoever was behind the waterfowl attacks was the same person involved with the later wallaby killings. If we suppose that 5this person is/was a staff member, then such an individual would have inside knowledge of the zoo's secutiry system. Also, it may explain why the other wallabies in the enclosure where the attack took place didn't seem to be peturbed: they were familiar with the sound/scent/sight of the human entering their paddock. Why was there no sign of footprints in thesand? Simple: the person had access to zoo keys and equipment, and simply raked over his prints as he made his way back to the enclosure gate, which he then locked behind him.

There's still the question of motive, but speculation on this is perhaps not productive. Sadistic attacks on animals are, alas, not all that rare. A person who wshed to commit such acts might well seek employment at a small local zoo, where the typical long hours and low pay mean that the employer can't always be over picky about who they add to the pay-roll.

As for some of the other weird stuff, such as sightings of UFO's in the same general vicinity, I suspect that if you dug deep enough at ANY given time or place, you'd turn up some reports of odd lights in the sky etc. Maybe 'Doc' Shiels doesn'r believe in coincidence, but when you think about it, 'pure' coincidence is just as likely as anything else in this goblin universe of ours. Generally I think that we should beware of jumping to paranormal conclusions, unless and until everything else has been ruled out.

Turning now to the latest issue of A&M (#13), I fear that two of the short news items on page 11 are non-stories.

(a) 'A Fishy Story', concerning the 'toppen' found in the Gambia River. I'm sure that you are correct in identiying the toppen as a misnomer for tarpon, but there is nothing unusual in the reported size of this specimen or the fact that it was found in fresh water. The tarpon (Megalops atlanticus) is well known for its ability to thrive in both salt and fresh waters. It can also survive in oxygen-poor water by coming to the surface and gulping air into its lung-like air bladder. So it's a very interesting fish but not the stuff of cryptozoological mystery.

(b). 'Everybody's got something to hide...' This item made me wonder just how many times an animal can be rediscovered. The Hairy Eared Dwarf Lemur, was for many years known only from a handful of museum specimens, but was then found alive and well in Madagascar in 1989. The quoted article from 'Die Welt' seems to be about seven years out of date!

Meanwhile, I'm gradually reading through the new CFZ Yearbook and finding it hugely enjoyable: a real credit to all concerned, and well worth the long wait!

(It sure looks as if the poor old migo has been blown out of the water by the worthy Mr Naish!)

Stay Well,

Mike Grayson,
London.

EDITOR'S REPLY: Ouch! Mike is undoubtedly perfectly correct with regards the tarpon and the lemur articles. We should have checked our source material better.... but didn't!

As far as the Newquay Zoo material, which as well as being the subject of my talk at the 1997 UnConvention, is also covered in Chapter Six of my latest book 'The Owlman and Others' (CFZ 1997), I beg to differ with him.

a. There is no evidence whatsoever that the 'kitchen knife' discovered in the zoo enclosure was used in the mutilations. The zoo staff thought so little of the discovery that for months afterwards they used it in the routine preparation of foodstuffs for the animals.

b. I feel that there is every possibility that the deaths of the waterfowl and the wallabies could have been caused by different agencies. Apart from the location, there is no real evidence linking them.

c. I agree entirely with Mike about the danger of jumping to 'paranormal conclusions'. That is why in this case I haven't done so. As stated in my book, despite the strong amount of evidence to the contrary, I believe that the whole affair MAY be a series of coincidences,

MORE FEEDBACK AND STUFF

Dear Jon,

With reference to Roderick Moore's letter concerning the Daily Mail's article reporting remains found of unidentified creatures by scuba divers in an underwater cave in the Fijian island of Matagi. (A&M12). The only other mention that I have seen of the discovery was in Fortean Times No 58 (July 1991 p.28) Summarised as follows:

Remains of four prehistoric creatures were found thirty two miles off the island in a coral cave. The remains appeared to be those of two adults, one adolescent and a juvenile. The adult skulls were approximately three feet long, with a total body length of 26-32 feet. American scientists who were shown video tapes of the remains were apparently unable to identify them. The cave was about 160 feet underwater, and the remains were between 100 and 160 feet into the cave and up a winding passage which was divided into different compartments. Apparently the isolation, depth and darkness of the site made it very different to dive.

Kevin Deacon is quoted as that the remains were either prehistoric or contemporary animals unknown to science. Frieda McHugh of Takapona, New Zealand, linked the discovery with past encounters by Fijians with a giant shark deity which they call the 'Dakuwaqua'. The article goes on to give details of a number of alleged encounters with this fearsome creature. Since then, however, I have heard no further details on the remains. Perhaps a letter to the Fortean Times would provide further details?

I would also like to mention Michael Goss's excellent item on the Canvey Island Carcass (A&M10). He mentions first learning of the event in either Weekens or Titbits magazine and was not sure of the year. I can help him here. It was an article by Mike Bennett in Titbits Magazine on the 29th 9.1981.

I hope that the above items may be of interest to fellow readers.

Yours faithfully,

Michael Playfair.
Leicester.

RICHARD FINDS SOMETHING A-FOOT (SNIGGER)

Dear Jon,

On glancing through the Bord's excellent 'Alien Animals' I noticed something I hadn't twigged onto before. On page 192 there is a photograph of Polish medium Franek Kluski apparently materialising a bird upon his shoulders. The Bord's seem to think that this looks like a bird of prey. I beg to differ. The shape of the head and the big eye make it look very much like a nightjar.

At seances, Kluski also called forth a huge 'lion-like cat' that licked participants hands with a rough tongue, and 'a creature the sitters dubbed "pithecanthropus" '. This latter beast seemed to be intermediate between man and beast and was immensely strong, lifting a man and chair up together with one hand.

So, a big cat, a man-beast and a large nightjar. Ring any bells in the zooform archetypes department?

(Does anyone know anything else about this Kluski fellow?)

Love

Richard Freeman,
Leeds and/or Nuneaton
(sometimes of King's Cross)

ER.... A POEM.

EDITOR'S NOTE: Essentially Senor Anderson (The Bard of the Highlands) dared me to print this poem, with the promise of a bottle of Talisker (a rather excellent Malt Whisky from the Isle of Skye) if I agreed. Being well known in the cryptozoological community as being someone who will do anything for a bottle of any whisky, especially the good stuff, I complied...

There's a green-eyed, scaly something,
to the west of Katmandu.
Its taxonomy an enigma, 'tis neither Vu nor Roo.
First brought to light by Ivan T,
though some doubted his sanity,
a denizen of wastelands barren,
'fishy' enough to interest Darren
(that victim of a swollen ego

Animals & Men # 14

who, having just debunked the Migo,
author of a paper brilliant,
proving it was crocodilliant,
pronouncing its dentition 'triffic,
"I'd place it in the Palaeolithic".
Published a side elevation
entitled 'Darreni cetaceean').
By now our brows were all in wrinkles,
not least that of one G.Inglis,
who, shunning academic wrangle,
approached it from an eco-angle.
Undeterred by recent failures,
blamed it all on Japanese whalers.
Then spoke the Bd, (an old curmudgeon)
"'Tis probably a mutant Gudgeon".
This 'enfant terrible' of forteans
(If he's a star, I'm a banana),
attempting to defuse this fuss,
he then did re-adjust his truss,
be-devilled by a rampant hernia
a plea was sent to far Hibernia.
Devotee of Ernst (a Kraut),
imbiber of the dusky stout.
"Sure now" said Doc on hearing rumours,
"You need witches sans their bloomers".
In Devon city, long abed,
having been 'Paignton' the town red
the editor lies wreathed in frowns
victim to life's ups and Downes.
One whose star was once ascendant,
upon the state is now dependant,
lying prone upon his sickbed.
In desperation 'phones R.Muirhead
who, normally a helpful being
ripostes "not me mate, too fortean".
By now the Bd. is on the floor,
gazing glazed at Channel 4.
His jaw drops, he lets forth a curse
it's Darren's game-show "Whales 'R' Us",
victim of a crime so heinous
(you try rhyming Odobenus).
His brief career now off the rails,
a victim of the 'Prince of Whales'.

Anon
(a.k.a Tom Anderson,
Aberdeen).

P.S. Roger McGough would have been safer, but you only get what you pay for...

BOOK REVIEWS

By Jonathan Downes except where noted.

From Flying Toads to Snakes with Wings by Dr Karl P.N.Shuker. Llewelyn Pubs $12.95US 222pp Pb.

The advent of a new Karl Shuker book is a cause for excitement amongst the cryptozoological community. For our British readers in particular the publication of this volume is no exception. The slight qualification in the last sentance is because it is essentially a collection of his cryptozoological (and fortean zoological) articles for Fate magazine - a publication which is hardly ever found in the United Kingdom. Even if, however you have a complete run of the original articles, you should still buy this book because each chapter has been meticulously updated.

The chapter on carnivorous herbivores - sheep and deer which eat young birds in order to supplement a calcium poor diet with essential minerals is particularly interesting (if gruesome), but it is hard to single out a specific chapter for praise - they are

Animals & Men # 14 — Reviews

output, this volume should (and probably eventually will be) on the shelves of every reader of this magazine.

Presently there seem to be a few problems with the British distribution of this book. We will be stocking it as soon as we can, so if you want a copy send no money now, but send us a SSAB marked clearly "SHUKER:FLYING TOADS BOOK" and we will let you know as soon as we are able to provide it.

The Cat in Ancient Egypt by Jaromir Malek (British Museum Press £9.99 Pb 144pp)

This copiously illustrated tome not only explores the position of the domestic (and wild) cat within the infrastructure, both cultural, artistic and religous in ancient Egypt.

Interestingly, as well as the domestic cat and African/Eurasian Wild Cats this book also goes into some depth in discussing the Jungle Cat (*F chaus*) and in an interesting parallel to parts of my book *'The Smaller Mystery Carnivores of the Westcountry'* (1996)

Malek goes into great deal in discussing the ancestry of the modern domestic cat. The illustrations are gorgeous and are worth the price of the book alone. Highly recommended.

Strange Northwest - Weird encounters in Alaska, British Columbia, Idaho, Oregon and Washington by Chris Bader $US11.95 Hancock House 144pp. Available in the UK from Gazelle Book Services. Telephone 91524 68765

A similar volume to 'Mysterious America' by Loren Coleman and 'Mysterious Australia' by Rex Gilroy this book covers similar areas - UFOs, lake monsters and man beasts (including some remarkable accouts of white-haired bigfoot type creatures). This stuff essentially covers much the same ground as we have read many times before. On the plus side, however, it covers some remarkable 'critters' such as a giant shrimp killed in Washington during 1948 and sightings of an Owlman type creature seen over the same area in 1947 (a weird year by anyone's standards).

Maori Art and Culture Edited by D.C.Starzecka (British Museum Publications Pb. 168 pp £17.99).

Although this book is not of direct relevance to the cryptozoologist, Maori culture as a whole is of direct interest to the cryptozoologist because of such well known and somewhat arcane cryptids as waitoreke (an otter-likecreature), the giant gcko and various cryptic birds including the Moa. This book is an essential primer intothe history of New Zealand and the related areas of Polynesia and as such is higly reccomended.

A History of the Dragon by Carl Lofmark - Edited by G.A.Wells (Gwasg Carreg Gwalch - Welsh Heritage Series No. 4) £3.50

The red dragon in the title of this excellent little book is the symbol of the Welsh nation, but the late Professor Lofmark has thoroughlly researched dragons in general to establish the beast in history. Dragons, both oriental and occidental are featured and an attempt is made to trace their common ancestry.

The question of the nature of the dragon - real or mythological - is discussed and given a firm platform from which to investigate not only dragons, but other creatures whose existence is debatable. The book is reccomended not only as a source of draconian information, but as an introduction to the study of the more mystical aspects of history.

This volume is very copiously illustrated, and contains a good bibliography of books treating of dragons and dragon-lore.

Noella MacKenzie.

Animals & Men # 14 — Reviews

THE ANOMALIST #4, Edited by Patrick Huyghe and Dennis Stacey. 144pp ISSN 1076-4208 £8.00

Available from 15 Holne Court, Exwick, Exeter, EX4 2NA

The Anomalist is a highly erudite and stylish US publication which appears in book form twice a year. This issue contains much of interest to the UFOlogist as well as other forteans. There is an article on the final years of Project Blue Book by Hector Quintinalla, which includes some fascinating insights into this much maligned organisation. There is a fascinating article exploring the possibility that The Earth has two moons, and an idiosyncratic look at the phenomenon of crop circles as well as much more non-UFOlogical writings. My favourite is a fascinating expose of the so-called De-Loys's Ape of South America which Loren Coleman (a long time hero of mine) has shown up to be a hoax with unpleasant racist connotations. Unmissable.

The Field Guide to the Extraterrestrials
by Patrick Huyghe NEL 136pp ISBN 0-340-69503-X £5.99.

There has been a need for this book for a long time, and I am pleased to see that Patrick - a writer for whom I have a lot of respect has filled this need in a succinct and professional manner. His publishers, however, have done him (and the subject) no favours by presenting, what would otherwise have been an invaluable tome, in the style of a tacky and disposable piece of fiction. The illustrations by Harry Trumbore are childish and unimpressive, and if it wasn't for the fact that the text (which describes over forty 'alien' types together with detailed case analysis, a worthy index and a good bibliography) is so good, I would be delivering the 'slag-off' review to end them all. It is a good book but it could have been sooooooo much better!!

Borderlands - The Ultimate Exploration of the Unknown
by Mike Dash
(Heinemann £16.99 501pp)

I'm always in somewhat of a dilemma when I am sent something to review which has been written or recorded by a mate. What do you do if it turns out to be dreadful? This has happened on a number of occasions, especially with CDs, and I usually end up being non-committal and evasive about the product in an attempt to get it out of the way as quickly as possible. Mike Dash is a mate of mine so I was, in theory at least, faced with the usual dilemma. I needn't have worried, however, because this book is superb. The Borderlands referred to in the title are very similar to The Outer Edge described by John Keel, and in this exhaustive work, Dash deals with a wide range of fortean and paranormal phenomena with wit, style and aplomb. The sections on UFOs, Abductions and Cryptozoology (which cites yours truly in the references), are particularly interesting, and contain much material which is new to me. This is a concise and erudite overview of fortean research over the past 50 years or more, and as such cannot be reccomended highly enough.

Lo! by Charles Fort John Brown Publishing, London 1997.
1997 revision of 1931 text. £9.99 pb 304pp including index.

Fort's motto could have been, 'nothing is sacred'. He spent 27 years at the British Museum and New York Public Library researching journals, newspapers and manuscripts to gather material on the twilight phenomena bridging science and fantasy. These mysteries, often glibly explained away by mainstream science, formed the theme for four books.

This book, his third, is written in Fort's customary abstruse style. For instance, the second chapter commences thus:

"Frogs and fishes and worms - and these are the materials for our expression upon all things. Hops and flops and squirms - and these are the motions. But we have been considering more than matter and motion to start with: we have been considering attempts by scientists to explain them. By explanation I mean organisation."

"There is more than matter and motion in our existence: there is organisation of matter and motion. Nobody takes a little clot that is central in a disease germ, as Absolute Truth, and the latest scientific discovery is only something for ideas to systematise around..."

It is often the case that the writings of a science guru are over the heads of the proletariat. If, for example, you want a quick understanding of relativity theory, then avoid reading Einstein! Similarly, Fort's musings are for the connoisseur; for the thinker; for the serious student of forteana.

This book needs to be savoured. - **Graham Inglis.**

Animals & Men # 14 — Reviews

Growing Herbs Dawn Dunn
Cassell London 1997
ISBN 0-304-34837-6 Large(ish) format, 96pp. Indexed.

This book looks at the planting and nurturing of herbs, their harvesting and preservation, and their applications - in crafts, cooking and health.

Dunn's approach is prosaic (down-to-earth?) rather than New-Age although, interestingly, you have to wait until page 31 for a definition of a herb - a soft-stemmed fragrant plant, just for the record. "Herb" is a pretty broad category, really: it includes roses, mustard and some shrubs.

Half of the book is an easy reference A-to-Z of 90 or so herbs, with typically 150 words describing the herb's appearance, ease of growth, and also its uses.

Almost every page has one or more colour photos; it seems that herbs can make a colourful addition to your garden as well as jazzing up your dinner. Even a non-gardener like me is tempted by this book to get back to nature a little bit. A good "starter" book.   - **Graham Inglis**.

Plants of Mystery and Magic
Michael Jordan   Blandford
1997   ISBN 0-7137-2645-8   £18.99
128pp indexed.

Jordan is an anthropologist and tv presenter and the first part of his latest book is an A-to-Z of 43 herbaceous plants and shrubs (including the opium poppy), with the following information for each: where and when, appearance, and traditions and associations. It's a botanical guide rather than a gardener's guide but has numerous colour photos taken by the author.

Part two covers trees with photos of their flowers and part three, also well illustrated, deals with 9 fungi including the psilocybin "magic" mushroom - and notes that these mushrooms are *"much sought after by interested parties when it first appears during the summer months."*

This is a nice enough book but seems a little over-priced at the equivalent of 10 pints of Stella lager. Anyway, I just think I'll have a little wander over Dartmoor before the end of the summer. Stretch my legs a bit after all this computer work......

Graham Inglis.

The Folklore of the Isle of Man by A W Moore Llanerch Press
ISBN 1897853 42 4 £8.95

Regular readers of my musings in these, and other pages, will be aware of my belief that contemporary sightingss of 'aliens' are part of a historical continuum of folkloric belief in faeries and elves which go back centuries. This excellent reprint of a book first published in 1891 is full of stories which will be all too familiar to the abduction/contactee scholar of the present day.

Out of the Blue   Glennyce S Eckersley   Rider, London 1997
ISBN 0-7126-7165-X   pb 163pp £6.99

Eckersley recounts many modern-day anecdotes and stories of "miracles" and coincidences from around the world and suggests that the many different things that happen to us are fundamentally interconnected and meaningful. He points out that such matters are worth investigating, as they are the nearest that most of us ever get to encountering spiritual or other-world phenomena.

For most of the book, though, Eckersley contents himself with recounting the tales, and only offers very brief discussion of "explanations" - such as the possible relationship between synchronicity and chaos theory. I wish he'd explored <u>that</u> aspect in a lot more detail, alongside all the the anecdotes.

One notable deviation from the general anecdotal style is the author's look at the coincidence of dates and places:

President John Kennedy was elected exactly 100 years after Abraham Lincoln.   Both were assassinated on a Friday in the presence of their wives. Lincoln was killed in Ford's Theatre, Kennedy in a Ford-made Lincoln convertible. Kennedy's assassin fled from a warehouse into a theatre, Lincoln's from a theatre to a warehouse. Lincoln's killer was born in 1839, Kennedy's in 1939. Both presidents were succeeded by men called Johnson. Lincoln's secretary was called Kennedy and begged him not to go to the theatre. Kennedy's secretary was called Lincoln and strongly advised him not to go to Dallas...

A pretty entertaining and thought-provoking read.

Graham Inglis.

Animals & Men # 14 — Reviews

Memories of Hell - Fortean Times 68-72 £19.99 John Brown Publications ISBN 1-870870-90-5

One of the things that never ceases to surprise me, working as a journalist within the field of paranormal research is the level of good-natured co-existence that exists between the different magazines which cover the subjects in which we are all interested. The longest running of these is the Fortean Times (founded in 1973) and this excellent volume collects together five issues from the early 1990s, including on page 54 my first ever published article (which wasn't very good). That apart - this is a valuable resource for any wannabe paranormal investigator, or indeed anyone with a taste for the bizarre!

The International Underground Directory - The Most Dangerous Book in the World ISBN 0-9 525546-6-6 Available from Megastep International. Tel. 01225 427759.

When I first received this book I was seriously disappointed. I had thought that 'the most dangerous book in the world' would look more impressive than a collection of ringbound A4 pieces of paper. However, when I read it I was stunned. The information in here; ranging from where to get a Bolivian Diplomatic Pasport to how to get your mail directed and where to buy surveillance equipment is totally stunning. In the wrong hands this stuff could be dynamite. I hope, however, that most people who buy it will just be like me and use it to spice up conversations down the boozer! (P.S a Bolivian Diplomatic Passport costs US$ 50,000). However, apart from the sheer amusement value of much of this stuff, there are a lot of pieces of equipment on sale through these pages, such as night sights, and surveillance equipment which would be invaluable to the field cryptozoologist.

Encyclopaedia of the Unexplained - Magic, Occultism and Parapsychology - the ultimate guide to the unknown, the esoteric and the unproven. Edited by Richard Cavendish (Arkana Large format Pb £16.00 304pp). ISBN 0-14-019190-9

There have been a plethora of such books in recent years, but together with Karl Shuker's The Unexplained this book is probably the best. It provides in-depth and succinct coverage of every topic that you could possibly think of, and most importantly it provides succinct thumbnail biographies of all the most important figures in the field. An invaluable work of reference this is also the sort of book which you can leave by your bath and dip into at random, sure in the knowledge that you will discover something new, challenging and wondrous.

PERIODICAL REVIEWS
BY GRAHAM INGLIS

Life beyond *Animals & Men*...

A not-entirely-regular guide to what the "opposition" (and quite a few friends too) are up to in the world of magazine publication.

Between issues 6 and 11 Jon Downes tried categorising them ("much against our fortean methodology", as he said at the time). "Periodical Reviews" then took a well-earned rest (or something) during issues 12 and 13; and now resurfaces in a manner not at all reminiscent of soggy corn-flakes in a long-neglected and overflowing washing-up bowl.

I briefly mulled over the idea of presenting the "oppo" in order of **spine colour** (from red through to purple/violet, and then on to black) - but then found that virtually all of the buggers are **white**, so that was that, really. Damn.

DEAD OF NIGHT

"Merseyside's only publication dealing with all paranormal phenomena" has quite a lot to say about some aspects of life, and it's all very entertaining.

This weighty tome (62 A4 pages for £2) covers much in the crypto world as well as ghostie bits, UFOs, cover-ups, strange human behaviour, and religious weirdities, all in a fascinating stream of small features mixed with snippets and summaries of news items...

Animals & Men # 14 — Periodicals

Dead of Night, continued...

Issue 11's crypto items include an update on Lizzie of Loch Lochy; the "monster" carcass washed up on Rhode Island; a look at the connection between the Kraken and giant squids; giant cats in Scotland; and summaries of weird and wonderful press cuttings. Excellent value.

Lee Walker, 156 Bolton Road East, New Ferry, Merseyside, L62 4RY.

PORCUPINE!

The newsletter of the Dept of Ecology and Biodiversity of Hong Kong University. Nº 16 is the first newsletter to be issued since the end of British rule of Hong Kong.

Investigative science. There's an article on types of (blue-blooded) Horseshoe Crabs, often called "living fossils" because they have changed so little over the last 450 million years. Also, more results from the South China Survey of forest areas; the application of DNA research to the understanding of biodiversity; and a tabular study of the flight periods of Hong Kong butterflies.

As well as the cartoon - the only even-slightly-political reference in this issue, there's also much more; including a look at fish introductions.

Free on application to Kadoorie Agricultural Research Centre, HKU, Lam Kam Road, Yuen Long, New Territories, Hong Kong.

We at Animals & Men extend our best wishes to the editors of Porcupine!, the people of Hong Kong and their new government (as well as its unique fauna) in its new role as a semi-autonomous Special Administrative Region of the People's Republic of China.

Belatedly, *Kung Hai Fat Choi!*

MAGONIA

"Interpreting contemporary vision and belief", the cover says, and this slender A4 magazine duly looks at the various angles of its chosen subject matter.

In issue 59 it's alien abductions (with a sceptical look at Budd Hopkin's book on the subject); satanism - is it a media myth; and plenty of book reviews, some of which are cryptozoological (or near it).

£1.25 John Rimmer, John Dee Cottage, 5 James Tce, Mortlake Churchyard, London, SW1 4 8HB

CRYPTONEWS

Issue 28 delves into the "surgeon's photo" of the Loch Ness Monster - seemingly there is much to debate on the matter. Lake monsters dominate this issue, as there's bits about "Caddy" (the somewhat prematurely-designated entity Cadborosaurus willsi), and Ogopogo, and claims of further monsters in lakes in South Africa and Austria.

For those cryptozoologists who prefer dry land, there's also a bit about the first sighting of sasquatch (bigfoot) in Quebec.

The British Columbia organisation that produce this 12pp quarterly newsletter (BCSCC) are contactable on http://www.ultranet.ca

BIPEDIA

Those intrigued by Françoise de Sarre's controversial theory of Initial Bipedalism in our 1997 Yearbook (the article that was regarded by some as the most controversial in the book) can keep abreast of things by contacting François at C.E.R.B.I., 6 Avenue George V, 06000 Nice, France.

Animals & Men # 14

Periodicals

THE NATTER JACK

The newsletter (rather a couple of leaflets, really) of the British Herpetological Society. Issue 26 looks at the global phenomenon of decline of frog and toad populations, even in protected reserves.

This issue sees the first of a series of articles on cryptozoological matters. It's on the Tatzelwurm (worm with feet) - which we at the CFZ are still hoping might be the subject of a crypto-expedition hunt one day.

Society membership details, etc: Zoological Society of London, Regents park, London, NW1 4RY.

DRAGON CHRONICLE, The Dragon Trust, PO Box 3369, London SW6 6JN. A fascinating collection of all things draconian which now appears four times a year.

COVER UP, David Colman, 39 Limefield Crescent, Bathgate, West Lothian, Scotland. BH48 1RF. The magazine of the Lothian Unexplained Phenomena Research group. UFOs, animal mutilation, ghosts, etc.

MAINLY ABOUT ANIMALS
13 Pound Place, Shalford, Guildford, Surrey GU4 8HH. Veteran zoologist Clinton Keeling edits this wonderful A5 magazine which is, as the title says, mainly about animals.

THE MILTON KEYNES HERPETOLOGICAL SOCIETY NEWSLETTER

Nice apart from the typeface (although who are we to talk, with the printer problems we've had doing this issue...)

The April issue (it's monthly, but April's the most recent we have) * reviews wildlife tv producer Nigel Marven's presentation to the society, includ-ing his following of the Quebec caribou and the Atlantic grey whales, and putting radio transmitters into rattlesnakes.

There's also pet sales ads and a follow-up on the disagreement between MKHS and Proteus Reptile Rescue, where the former now regard the latter as a "necessary evil".

* Editorial note: Normally we don't include periodicals when they haven't been sent to us for three months or so, unless:
(a) they're produced less than 4 times a year, or
(b) we really want to, or
(c) we receive £500 in a brown envelope.

CRYPTOZOOLOGIA

This mag really is only of interest to people who can read French. As my French is limited to the likes of "Un biere s'il vous plait" and "Voulez-vous coucher avec moi ce soir?" it's a bit beyond me. I gather it sometimes recasts ANIMALS & MEN items into French, though.

Available from Belge d'Etude et de Protection des Animaux Rares, Square des Latins, 49/4, B-1050 Bruxelles, Belgium.

SIGHTINGS

This mag is a glossy that is available from (or orderable from) any newsagent. It deals with sightings ranging from the UFO type to weird animals, and its contributers may sometimes sound familiar: one recent issue had about 24% of its content taken up by crypto-ish articles by Jon Downes and Karl Shuker, and a space article by me; and the book and alien tat (er, sorry, accoutrements) reviews by Jon.

The latest, vol 2 issue 3, has the lead article by Jon Downes and (from GOBLIN UNIVERSE magazine) Tina Askew on the Rendlesham UFO mystery. Also, Jon is responsible for the magazine's first-ever use of the word "bollocks" - in his reviews section, that is.

Publ: Rapide. ISSN 1363-5166 £2.95

URI GELLER'S ENCOUNTERS

A general paranormal mag who generally avoid using confusing psychedelic patterns as background for the text.

The fortean zoology content usually comes from Jon Downes who writes a monthly column called THE JD FILES. Sometimes he just goes on about the state of his car, though, or includes pictures of me poking the ground with a stick. All in rigourously paranormal context, though...

In the August issue, Jon describes his recent sighting of the "Beast of Bodmin", which is why the mag qualifies as a crypto entry here...

Publ: Paragon. ISSN 1364-1921 £2.99

ENIGMA

Another "glossy", issue 5 carries an article by some geezer called Jon Downes, writing about the Owlman of Mawnan (in Cornwall). A fair amount of the mag is psychic stuff (as indeed Owlman "him"self may be...) but there's also an article on HAARP environment-modification technology.

Publ: Newsstand. ISSN 1364-7741 £2.95

THE CRYPTOZOOLOGY REVIEW

The summer issue includes an article by Darren Naish (a major contributer to ANIMALS & MEN) on a possible second "Caddy" in carcass form.

Darren and Jon are both due to write in a forthcoming issue.

Thrice yearly. $14 sub. 166 Pinewood Avenue, Toronto ON, M6C 2V5, Canada.

Animals & Men #14 — Sales

OUR OWN PUBLICATIONS

ANIMALS AND MEN BACK ISSUES

Back issues of "Animals & Men" are available at £2 each from the editorial address. Please see "methods of payment" below.

As well as the main features detailed below, all issues of "Animals & Men" have a "Newsfile" section and letters, reviews and other shorter pieces.

ANIMALS & MEN issue 1
Relict Pine Martens, Giant Sloths, Sumatran and Javan Rhinos, Golden Frogs, Frog Falls.

ANIMALS & MEN issue 2
Mystery bears in Oxford and The Atlas Mountains, Loch Ness reports, Green Lizards, Woodwose, The Tatzelwurm.

ANIMALS & MEN issue 3
Giant Worm in Eastbourne, Lake Monsters of New Guinea, Giant Lizards in Papua, Mystery Cats, Black Dogs on Dartmoor, Scorpion Mystery

ANIMALS & MEN issue 4
Manatees of St Helena, Migo: The Lake Monster of New Britain, The search for the Tasmania Thylacine

ANIMALS & MEN issue 5
Mystery cats, Loch Ness, More on the "Migo Video", Boars and Pumas, The Hairy Hands of Dartmoor.

ANIMALS & MEN issue 6
The Owlman Special; also the Humped Elephants of Nepal, Mystery Cats, Sabre-toothed cats, Mysterious hominids of Africa, The British Nandi Bear?, Bibliography of Cryptozoology books pt 1 (Shuker)

ANIMALS & MEN issue 7
Mystery Whales, Strangeness in Scotland, On collecting a cryptid, Bodmin Leopard Skull, Bibliography of "Crypto" Books (Shuker) pt 2.

ANIMALS & MEN issue 8
Green Cats and Dogs, Mystery Whales, Quagga Project, Bibliography of Cryptozoological books (3rd & concluding part), Malayan Man Beast.

ANIMALS & MEN issue 9
Hong Kong Tiger, Horseman of Lincolnshire, Scottish BHM, Congo Peacock, Mystery whales.

ANIMALS & MEN issue 10
Mystery Moth of Madagascar, Bengal Leopard Cats, The Derry, Wild Boars in Kent, a new Irish lake monster, mystery whales and the truth about the Essex Beach Corpses.

ANIMALS & MEN issue 11
The "Walruses Special", also Feathered Dinosaurs, Ground Sloth Survival in North America, Mystery Whales, Initial Bipedalism

ANIMALS & MEN issue 12
Lions: The Barbary Lion, etc. More Feathered Dinosaurs, Chinese Crabs in the Thames, Mystery Animals of Germany, News from New Zealand.

ANIMALS & MEN issue 13
Pangolins; also Moby the Sperm Whale, Barking Beast of Bath, Yorkshire ABCs, Molly the Singing Oyster, Leatherback Turtles, Walruses.

THE GOBLIN UNIVERSE BACK ISSUES

Issues 4 to 6 of "Goblin Universe" are available at £2 each from the editorial address. Please see "methods of payment" below.

As well as the main features detailed below, all issues of "Goblin" have a "News from Nowhere" section and letters, record and book reviews and other shorter pieces.

THE GOBLIN UNIVERSE issue 4
St Neot: Weirdest village in the West?, Naked witches, hellhounds and Capel's tomb, the Vampire of St Leonards, Cattle Mutilation, and an account of psychic detective work.

THE GOBLIN UNIVERSE issue 5
Crop Circles and Animal Mutilations, Ghosts of Glamis Castle, Communication with UFOs, and The "Noosphere" and text semantics.

THE GOBLIN UNIVERSE issue 6 - out now!
Jon and Tina are shown the Rendlesham UFO crash site. Also, Mystery Planets, Cannibalism in Scotland, and D.I.Y. countries and states.

METHODS OF PAYMENT

Postage and packing is extra; pleae add 25p (30p non-UK) per magazine and 75p (80p non-UK) per book.

Payment can be made in UK or US cash, by IMO (International Money Order), Eurocheque, or by a cheque drawn on a UK bank.

Please make all cheques payable to Jonathan Downes.

Animals & Men #14 — Sales

Orders outside Europe are sent by surface mail unless aditional postage is paid.

- - -

Many of the people who produce this magazine are also involved in the Centre for Fortean Zoology, a non-profit-making organisation which promotes the study of mystery animals, new and rediscovered animal species, zooform phenomena, cattle mutilations, bizarre and aberrant animal behaviour and out-of-place animals, and much more.

We have been publishing books and magazines for over three years and the following books are still in print:

OUR OWN BOOKS

The Owlman and Others
by Jonathan Downes

Two decades of Owlman evidence including sightings - mostly by girls and young women - in Cornwall, in the vicinity of Mawnan Old Church. This book discusses the evidence - and the role of Tony 'Doc' Shiels in the case - and comes about as close to the truth as anyone ever will...

Many illustrations. £10

The CFZ Yearbook 1996

The first yearbook, with nearly 200 pages of research papers and longer articles including Sky Beasts (Karl Shuker), mystery eagles (Jon Downes), Namibia's Flying Snake (Richard Muirhead), the "Nnidnidification" of Ness (Tony Shiels), African Man Beasts (François de Sarre), The Loch Ness Monster (Neil Arnold) and much more.

Many illustrations. £12

The CFZ Yearbook 1997

Karl Shuker hunts anomolous aardvarks, Darren Naish figuratively shoots the Lake Dakataua Monster ("Migo") and François de Sarre asks if humans are descended from bipedal fish.

Also articles on : the pros and cons of reintroducing extinct mammal species to Scotland; a list of cryptozoological movies; Mexican cattle mutilation and the Chupacabras - and much more.

Many illustrations £12

Morgawr: The Monster of Falmouth Bay
by A. Mawnan-Peller.

The classic 1976 booklet reissued with a new introduction by Tony Shiels and an additional essay by Jon Downes

£1.50

The Smaller Mystery Carnivores of the Westcountry by Jonathan Downes

Over 100 pages of information on a range of small carnivores in this fascinating region of the British Isles.

Three species thought extinct; hints of several species apparently unknown to science; and a revolutionary suggestion that a species of mammal known from mainland Europe exists in England.

Illustrated. £7.50

Advertisement

CRYPTO-DOCUMENTARIES and NEWS CLIPS...

TV video, Nessie, Lizard Men, Bigfoot, etc.

A copying service from Neil Arnold.

For more information write to Neil at:

8 Gorse Avenue, Weedswood Estate, Chatham, Kent, ME5 0UQ

Cartoon By Mark North

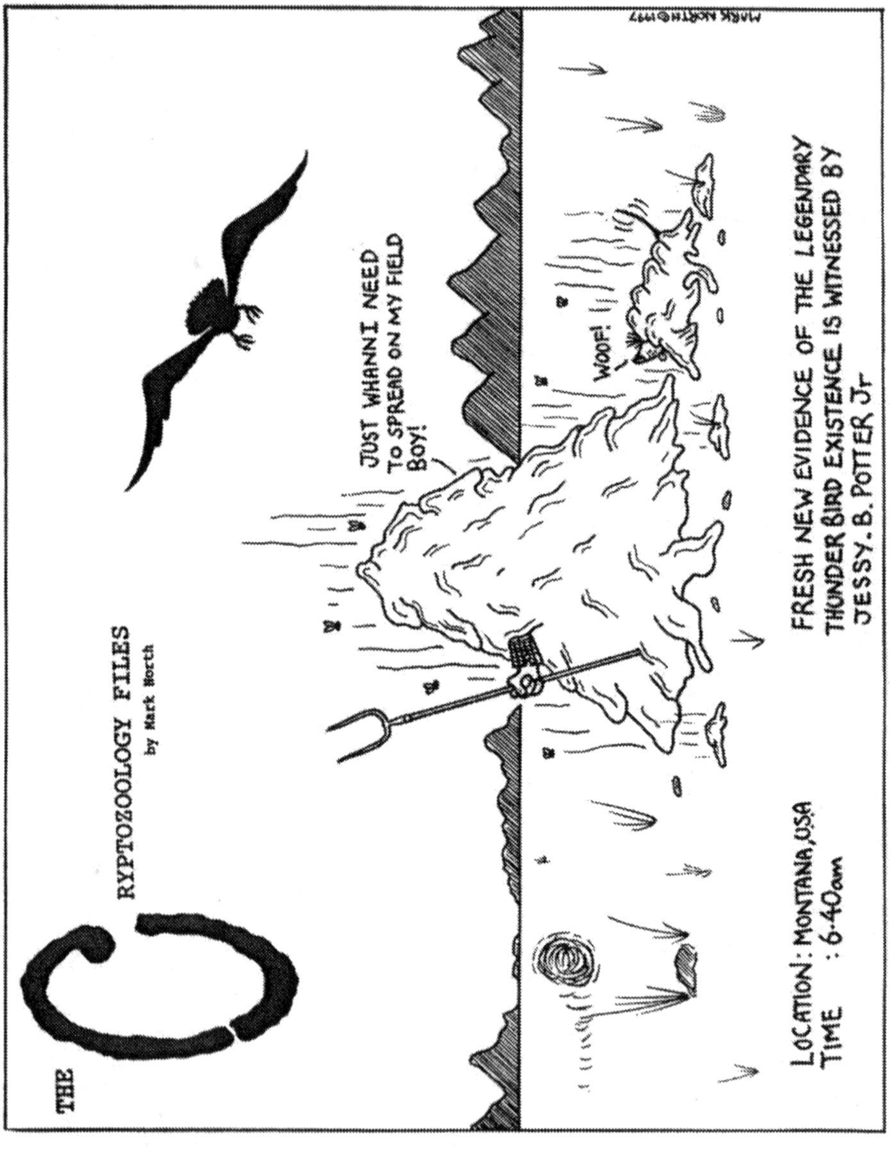

ISSN 13540637 Typeset By Ishmael²
(at the Sign of the Spouter Inn)

We smell land where there is no land...

ISSUE 15
OCTOBER 1997

THE JOURNAL OF THE CENTRE FOR FORTEAN ZOOLOGY

By this time, Richard, Graham, Mark, and I had become a fairly close-knit team, and together with stalwarts like Darren Naish, Clin Keeling and Tom Anderson, we were able to put out a pretty good magazine every three months.

It is heartening that now, a decade later, several of the contributors are still writing for us. Mark Frazer now edits our annual `Big Cat Yearbook`, and Nick Molloy has an article in the current (2008) CFZ Yearbook. However we parted company (mildly acrimoniously) with Terry Hooper in about 2002, and we haven't heard hide nor hair of Neil Nixon and Tom Anderson in years.

Looking again at it I am very proud of our achievements. Ten years later, as I write this, we have come a long way further....

animals&men

THE JOURNAL OF THE CENTRE FOR FORTEAN ZOOLOGY

Animals & Men
The Journal of the Centre for Fortean Zoology

The Bigfoot Murders; Wolverines in Wales; Visit to Loch Ness; The Migo Re-Examined; The "Waspman of Lancashire" and more..

Issue Fifteen

animals & men

THE JOURNAL OF THE CENTRE FOR FORTEAN ZOOLOGY

Animals & Men # 15

The ever-changing crew of the 'Animals & Men' mothership presently consists of:

Jonathan Downes : Editor
Graham Inglis : Newsfile Editor/Spin Doctor
Jan Williams : Associate Editor
Mark North : Cartoonist
Richard Muirhead : Newsagent from Nowhere
Special Agent Tina Askew : Rhine Maiden
Alyson Diffey : The Happy Medium
David Simons : Agent in Cyberspace

CONSULTANTS

Dr Bernard Heuvelmans
(Honorary Consulting Editor)
Dr Karl P.N.Shuker
(Cryptozoological Consultant)
C.H.Keeling
(Zoological Consultant)
Tony 'Doc' Shiels
(Surrealchemist in Residence)
Darren Naish
(Palaeontology/Cetology Consultant)
Chris Moiser
(Zoological Consultant)

REGIONAL REPRESENTATIVES

U.K

Scotland: Tom Anderson
Surrey: Nick Smith
Yorkshire: Richard Freeman
Somerset: Dave McNally
West Midlands: Dr Karl Shuker
Kent: Neil Arnold
Sussex: Sally Parsons
Hampshire: Darren Naish
Lancashire: Stuart Leadbetter
Norfolk: Justin Boote
Leicestershire: Alaistair Curzon
Cumbria: Brian Goodwin
S.Wales/Salop: Jon Matthias
Tyneside: Simon Elsdon

EUROPE

Switzerland: Sunila Sen-Gupta
Spain: Alberto Lopez Acha
Germany: Wolfgang Schmidt and Hermann Reichenbach
France: François de Sarre
Denmark: Lars Thomas and Eric Sorenson
Eire: The Wizard of the western world.

OUTSIDE EUROPE

Mexico: Dr R.A Lara Palmeros
Canada: Ben Roesch

DISCLAIMER

The views published in articles and letters in this magazine are not necessarily those of the publisher or editorial team, who although they have taken all lengths not to print anything defamatory or which infringes anyone's copyright take no responsibility for any such statement which is inadvertently included.

Who's who and what's what

CONTENTS

3 Editorial
4 Newsfile
13 The Waspman Cometh - James Lister
14 The Bigfoot Murders - Richard Freeman

BEAVER PATROL:

19 Return of the Beaver - Tom Anderson
20 Giant Beavers - Bill Petrovic
21 More British Beavers - Jonathan Downes
22 The Beast of Llangurig - Terry Hooper
25 The Tale of the Weird Warbling Whatsit of the Westcountry - Jan Scarff
28 Wherefore art thou, Nessie?- Neil Arnold
31 The Migo: Not yet Explained? - Nick Molloy
34 Clinton's Cogitations - Clinton Keeling
37 North of the Border - Tom Anderson
37 Now That's What I Call Crypto - Neil Nixon
38 Letters
31. Book Reviews
43. Periodicals
45. Sales
48. Cartoon by Mark North

SUBSCRIPTIONS

(Please see "Methods of Payment" on p 47.)

For a Four Issue Subscription:
£8.00 UK
£9.00 EEC
£12.00 US,CANADA, OZ, NZ
(Surface Mail)
£14.00 US,CANADA, OZ, NZ
(Air Mail)
£15.00 Rest of World
(Air Mail)

'Animals & Men'
THE CENTRE FOR FORTEAN ZOOLOGY,
15 HOLNE COURT, EXWICK,
EXETER, DEVON, EX4 2NA

Tel 01392 424811

WEB ADDRESS:
www.geocities.com/Area51/Lair/6501

Animals & Men # 15 — Editorial

THE GREAT DAYS OF ZOOLOGY ARE NOT DONE...

Dear Friends,

As what is euphemistically known as the 'festive season' approaches we can look back at another year at the CFZ and do what Graham and I irreverently call our *"Ring out the Old Ring in the New bit"*. It has been a very strange year - but let's face it, at the Centre for Fortean Zoology, when isn't it?

We spent the first half of the year deperately trying to catch up with the backlog inherited from 1996, but after a (I hope) succesfull appearance at the Fortean Times Unconvention in April things actually began to look up for us.

We hear with sadness that our old pal Clinton Keeling has announced that this year's Zoologica Exhibition in Sussex was his last. It was always an entertaining and informative show and we mourn its passing with sadness and regret. We very much doubt whether there will be another event quite like it!

Also on a sad note, regular visitors to us at the CFZ will be sad to hear that 'Harley' the Chinese Blue Magpie had a stroke and died in early October at the age of eight. He was a favourite with all of our visitors who used to feed him mealworms whilst discussing the wide range of peculiar subjects that visitors to our singular establishment always seem to talk about.

On a happier note (present rumoured budget cuts at Channel Four notwithstanding) Graham and I are off to Puerto Rico and Mexico in January in search of the Chupacabras - the fabled (and not so fabled) Goat sucking daemon of Hispanic America. We will be abroad for between three weeks and a month and (again all things being equal) you will be able to see the result on Channel Four's *"To the Ends of the Earth"* in the late spring of next year.

Finally, after several years of promises we now have a place on the Internet courtesy of our old mate David Symons who has set us up a web site. At the moment it is fairly rudimentary but it will get bigger and better as time goes on!

Also, our republication of Tony Shiels's classic books "The Cantrip Codex" and "13" has finally taken place, and they and the 1998 Yearbook will be available in time for Christmas. I can highly reccomend all three books (but then again I would say that wouldn't I?)

Thank you to everyone who has supported us through another year (our fourth would you believe?) and together with all the other members of the editorial team I wish you all a happy Christmas and a Peaceful New Year!

Best wishes,

Jonathan Downes.

Animals & Men # 15 — Newsfile

Newsfile

Compiled and collated by Graham Inglis with the occasional interjection by The Editor.

FOSSILS

Shake Rattle and Roll

Tyrannosaurus Rex shook its victims to death, according to research by the Russian Academy of Sciences. It used the same 'shake feeding' technique used by killer whales, sharks and crocodiles, a fossil expert told a conference - rather than skull battering. The size of their heads and powerful necks was cited as support for the idea. - *Daily Mail 12/9/97*

Only for the Birds?

The discovery of a "wishbone" in the remains of a dinosaur skeleton supports the suggestion that birds are descended from dinosaurs. Scientists from the US Museum of Natural History regard the "wishbone", or furcula, as a key characteristic of modern birds - *Daily Telegraph 2/10/97*

EDITOR'S NOTE: Darren Naish will be exploring the significance of this and other new discoveries in the worlds of dinosaurs in the next issue of *'Animals & Men'*.

NO, THEY DIDN'T...

Biologists in the US have said that a comparison of dinosaur claws with bird wings and feet contradicts the theory that birds evolved from small flesh-eating dinosaurs 150 million years ago.

Birds have retained the middle three digits during their evolution while losing the two outer ones - numbers 1 and 5. In dinosaurs, however, it was numbers 4 and 5 that were lost. Researchers at the University of North Carolina identified the respective digit positions on the basis that, in embryos, the 4th digit (the ring finger in humans) always grows directly in line with a tissue called the primary axis that later forms the bones of the arm.

Their conclusion is that dinosaurs and birds have many similar features and probably share a common ancestor - but we have yet to find its fossils - *New Scientist 1/11/97*

SNAKE 'MISSING LINK'

Researchers are setting out to prove that snakes descended from giant sea monsters. Michael Caldwell (University of Alberta) and Michael Lee (University of Sydney) say a 97 million-year-old marine fossil from Israel is a metre-long snake with tiny hind legs that evolved directly from sea lizards called mosasaurs. Caldwell said, *"We're testing a scenario that has snakes coming from the land and into the water and somewhere in the process losing their limbs, going back onto the land and specialising into a wide number of forms - and in some cases going back into the water again, as in the case of sea snakes."* - *The Toronto Star May 97*.

EDITOR'S NOTE: Darren Naish will be investigating 'snakes with legs' in the next issue of *'Animals & Men'*.

OLD CROC

The skeleton of what is believed to be the world's largest marine crocodile is to go on display at Birmingham Museum after 2000 man-hours was spent piecing its 250 bones together. The original finder gave up hope of reassembling the beast and dumped it in cardboard boxes in his garage.

The skeleton bears scars and bite marks from an underwater battle with a *lioplurodon* (a pliosaur) during the Jurassic era (165 million years ago) and has an imbedded tooth in a hind leg bone. The remains were found in Cambridgeshire six years ago - *Aberdeen Press & Journal 20/5/97*

MAN BEASTS AND BHM

ORANG PENDEK 'MISSING LINK'?

The search is on for the "missing link" - in Sumatra,

Animals & Men # 15

Newsfile

Indonesia, following recent reports of an orange coloured primate known as the *orang pendek*, that strides confidently on two legs. Reports of such a creature have occurred throughout the last 80 years. A research team has brought back hair and droppings samples, although the hair was not collected directly from the animal. These, and casts of footprints, have been passed to the Institute of Zoology in London for analysis - but the Institute needs £15,000 before they can 'afford' to examine the material.

It is currently suggested that it's a new species of orang-utan or gibbon or even a great ape, or just possibly a primitive humanoid. If the latter, it could possibly cast doubt on current evolutionary theories which suggest humans evolved from African apes.

Its habitat, in central Sumatra, is currently under threat from vast forest fires originally started by timber companies - *The Sunday Times 12/10/97*

EDITOR'S NOTE: We have been aware of these alleged Orang Pendek photographs for the best part of a year, but to date, apart from one extremely poor photocopy shown to us by a source within the BBC (who must remain anonymous) we have been unable to see the originals. The picture we saw looked like a poor quality snapshot of a siamang taken from a long distance away and is hardly conclusive evidence one way or another.

For copyright reasons (even the Copyright Liberation Front prefers to leave Mr Murdoch's News International Ltd. alone) we are unable to print the photograph of the footprint shown in the Sunday Times but we would like to point out the similarities betwen the photograph they used and the drawings included in Heuvelmans's classic *"On the Track of Unknown Animals"*, which also depict prints allegedly made by the Sumatran man-beast.

NEW AND REDISCOVERED SPECIES

ELECTRIC WARRIOR

A young girl died after being zapped by a new breed of electric eel in Brazil's Amazon estuary. The eel can deliver 650 volts into the water - *The Sun 19/9/97*

NEVER MIND THE BALAERICS

A 'lost world' has been discovered underneath the island of Majorca by workmen boring a sewage disposal sump. A partially-flooded and light-free cavern that has been undisturbed for millions of years, is home to rare worms, sea lice, shrimps, crabs and new species of crustaceans never before seen. The cavern has no direct connection with the sea - the water percolates through porous rocks.

Animals & Men # 15 — Newsfile

All the creatures are blind and find their food by homing in on chemical signals. Some have been described as living fossils. Some are similar to creatures found in other 'lost world' caverns in the Bahamas and the Galapagos Islands and it has been suggested these species evolved at the bottom of the ocean millions of years ago, when the caves were close to each other and before becoming separated by continental drift.

Prof Boxshall of the Natural History Museum in London, who has swum through the cavern, said, "Scientists have been amazed at the high levels of diversity found. Just dealing with the crustaceans, these caves are home to countless new species, numerous new genera, many new families, new orders, and even a new class."

A Blind White Crab from the Majorcan caves
(Picture courtesy the Copyright Liberation Front)

The Spanish government intends to make the cave a protected nature reserve. However, most aquatic cave habitats on Majorca are currently under threat because of over-extraction of fresh water for the tourist industry - *Daily Mail 10/9/97*

TAZZIE COME HOME!

As has been commented on widely in the fortean press it is now believed that the Tasmanian wolf (*Thylacinus cynocephalus*), believed extinct, may have survived in Irian Jaya in Indonesia. Several farmers have described how the animal attacked and killed pigs and goats. A new development to the story is, however, the local government has offered a bounty of $700 for the capture of one - *Westfalenpost 22/8/97 (from the Indonesian Observer)*

Legendary Amazon Forest Monster

600 people from Nuevo Tacna, northern Brazil, at a soccer match, reportedly saw a monster 40 metres long emerge from a forest, looking "like a giant snake, with ears about a metre in length and two aerials [antennae] like elephant trunks." Local reporter Inuma Lavi said, "While the match was going on, the earth shook, the sky got dark and the wind started blowing."

The animal reportedly destroyed a dozen trees and the area looked as if it had been run over by a bulldozer. The Navy despatched a gunship to inspect the nearby shoreline and also sent helicopters. The locals don't want the military to leave, fearing that Sachamama could return and destroy their homes. The authorities have now started an investigation.

An old legend among the indians tells of the existence of a giant animal, "*Sachamama*", that hides in the Amazon and seldom emerges, and sleeps for 10 to 30 years under water. - *De Telegraaf (Netherlands) Aug 97 and Kolner Express (Germany) 23/8/97.*

EDITOR'S NOTE: The latter paper said the event occurred in northern Peru, not Brazil. This entity would seem to be a cryptid (if indeed cryptid it be) we have never heard of! Anyone got any additional information on it?

NOT ANOTHER ONE?

Scientists have discovered a new species of deer in the remote jungles of Viet Nam. Its height is estimated at only 0.5 m (20 in) as living specimens have not yet been obtained.

The information is based on the skulls of animals hunted by locals for food - *Westfalenpost (Germany) 23/8/97*

FOUR FROGS FOR CHRISTINE

An amateur naturalist who gave up his job, possessions and savings to go frog-hunting in Africa has discovered four new species of reed frog. Martin Pickersgill set off on a 10-month trek from Cape Town to northern Africa. His girlfriend Christine stayed behind and processed the scores of

Animals & Men # 15

Newsfile

photos that he sent back. He said, "It was a choice between Christine and the frogs and I chose the frogs." The Times 5/6/97

MORE MONKEY BUSINESS

A Dutch scientist has discovered a previously unknown species of monkey in a shelter for orphaned monkeys in Brazil. It is only 10 cm tall and weighs 160 grams. The fur is greenish-grey, the face is surrounded by white hair, and the tail is black. The monkey, which belongs to the Sagui family of monkeys, has temporarily been named 'pygmy sagui' Westfalenpost (Germany) 22/8/97

Picture courtesy of the Copyright Liberation Front

OUT OF PLACE

Unidentified Hopping Objects

As British cryptozoologists have known for years red necked wallabies are fast becoming an established member of the British fauna. Wallabies were first introduced to Britain as a curiosity before the last war but many more were imported in the 1960s after the success of the TV show *Skippy*, about a bush kangaroo which helped solve crimes.

Many of the pets bounded to freedom but would usually died because of the cold and wet weather. Wallabies come from arid semi-desert areas.

Now, though, colonies numbering up to 120 are scattered throughout Britain - and even in the Isle of Man. Colonies have been reported in the New Forest and Kent, further north in the Tyneside area - and even as far north as Loch Lomond, near Glasgow in Scotland. One wildlife expert said, "Their success indicates how much the British climate has changed in the past few years." Wallaby-bashing parties, where people go out to shoot them for fun, have been reported in the Derbyshire Peak District - *The Sunday Times* 4/5/97

EDITOR'S NOTE: Whilst most of the 'hopping' animals reported across the UK, and indeed western Europe are probably Red Necked Wallabies there have been occasional reports of larger marsupials such as the Red Kangaroo reported at Totnes in Devon during the mid 1980s. This report was never verified and could just have been one of the local wallabies with a reddish-brown tinge to its pelage. Kangaroos do, however, occasionally escape, which brings us neatly onto our next item...

TOUGHER THAN TYSON

The public has been warned not to approach two kangaroos who ran off from a travelling circus in Poland. "The animals have been taught to box and can be violent if people approach," said a police spokesman. The kangaroos are called Gin and Tonic, and Gin is said to have a terrific right hook - *London Evening Standard* 25/7/97

Animals & Men # 15 — Newsfile

WHITE POWER

An albino starling in Chester has surprised experts by being accepted by the rest of his flock, instead of being shunned or driven out - *Daily Mail 31/7/97*

RARE BIRDS SANCTUARY

Devon birdwatcher Geoffrey Gush's garden is home to the richest collection of native wild birds in any British garden, says the British Trust for Ornithology. Mr Gush has recorded seeing 117 species over the last 25 years, at his home near Ottery St Mary. "The rarest birds I have seen are the toad lark, the Richard's pipit and the waxwing," he said. He puts out many sorts of food, including meat bones, peanut granules, bits of cheese, cake, and fruit. - *Express & Echo (Exeter) 29/10/97*

GOAT RUSTLERS

Dozens of goats have disappeared from a wild herd that has roamed Exmoor for more than a century. The number of goats living near Lynton (west of Minehead) has fallen from 108 in June to just 41. Although some have been shot for straying onto private land, it has been claimed that rustlers have taken a large number for sale to the Muslim 'halal' meat trade. Some halal butchers will pay around £100 for a goat. It is not a police matter because the goats are not owned by anyone. They were introduced to Exmoor in the early 19th century by farmers - *Daily Telegraph 10/10/97*

SHARK ATTACK?

Swimmer Jenny Pickles described how she was savaged by a shark - in a water reservoir in Surrey. "This huge mouth grabbed hold of my foot ... I kicked at it with my other foot and it swam off." A marine biology student who was at the lake said the teeth marks were too big for a pike, and consulted experts. They suggested it was a shark that had been bought as a pet and then dumped in the reservoir at West Molesley, south east of Heathrow Airport - *The Sun 16/8/97*

EDITOR'S NOTE: Although there are sharks which can live in fresh water, most notably the Bull Shark *(Carcharhinus leucas)* and the various species of river sharks discussed in the last issue of *Animals & Men*, I would be extremely surprised if any of these species were available as pets, or indeed if they were, would survive in a reservoir in the home counties. The Bull shark grows up to 3.4m in length and has a reputation for ferocity. Although there are a few species of shark that can occasionally be found for sale in aquarists shops, they are much smaller and placid creatures. Until more information comes to light this story must remain a mystery.

HERE KITTY KITTY

Pudge, a 6-month-old ginger and white kitten, was rescued from near the top of a 2593 ft Lake District peak. He was found by a climber, cold and close to death, and brought down to safety. Two hours later, he was safely back home - *Daily Mail 24/9/97*

SUPER WASP

"Super wasp" *Dolichovespula media*, a large and aggressive European wasp, has started to spread across Britain. Recent trends towards warm summers and mild winters are thought to be responsible. Experts say the "invader" wasp is migrating northwards by 50 km per year and that even more exotic species could soon follow. Already, a poisonous spider, *Steatoda noblis*, which came over from the Canary Islands (probably in a consignment of fruit) has readily adapted to conditions in Britain. Its bite is more painful than that of a wasp - *Dail Mail 17/9/97*

MYSTERY CATS

Animals & Men # 15

Newsfile

Sussex.

Police officers, some armed, closed in on what was thought to be an escaped puma in a Sussex village. Mr Young, an electrician, was working at a pub in East Wittering when he reported a large beast near some trees. As eight police officers surrounded the copse, out stalked a startled black moggy. Mr Young insisted that he'd seen a puma - *Daily Mail 10/4/97*

Scotland

A report of a big cat in Stonehaven sparked a hunt after a 53-year-old school teacher, Mr Anderson, said he saw a panther-like creature on the perimeter of his school playing fields. He was about 60 m away at the time. He said it had a large head, a shiny sleek black body and a long tail, and described how it jumped a 4 ft (1.2 m) fence - *"When it jumped, it did so in a cat-like manner: going back on its haunches before leaping."* Police searched the area but found nothing. A police spokesman said that, until the creature is sighted again, there is not much that they can do. *"If we could identify and pinpoint it, we could take some action."* - *Evening Express (Aberdeen) 17/10/97*.

EDITOR'S NOTE: The possible action that the police could take was not specified.

LAKE AND SEA MONSTERS

Loch Lochy, Scotland's third-deepest loch, is being 'swept' by sonar equipment and divers in search of underwater caves or rock overhangs which could house the loch's mysterious resident monster, known by the nickname of *Lizzie*. *Press & Journal (Aberdeen) 28/7/97*

OTHER STORIES

IGUANA RESCUE 1

Firefighters ripped up floorboards and dismantled part of a wall to save Basil the iguana, who escaped from his cage and disappeared behind his owner's bath. The Leicestershire fire brigade were called after the iguana had been away from his heat lamp for 36 hours. Sensitive sound-detecting equipment was used to locate Basil trapped in a wall - *UK Teletext 18/10/97*

IGUANA RESCUE 2

A fire crew turned out in Nuneaton, Warwickshire, to rescue a 4 ft (1.2 m) iguana which got stuck up her owner's chimney. The crew were warned that pulling on her tail could cause the tail to come off. Attempts to lure the lizard out with her favourite food - a banana - scared her and she disappeared further up. As the fire crew scaled the roof, intending to get the lizard down from the top, she made her own way down to safety - *newspaper source unidentified, Aug 97*.

EDITORIAL NOTE: I think that it was Charlie F himself who first noted clusters of anomalous activity. To have two iguana rescues within a few months is, whilst neither fortean nor cryptozoological per se, an interesting enough coincidence to be worth noting. Terry Hooper, regular contributor to A&M and the founder of the Exotic Animals Register has several records of escaped iguanas from all over the UK in recent months. This is indicative of several things. Firstly, as any amateur herpetologist will tell you iguanas and their relatives are great 'escape artists' and will often disappear from the most securely sealed vivarium, and secondly, as 'exotic' pets become more fashionable and easy to obtain, some people are, no doubt, getting bored with their pets and releasing them into the wild to fend for themselves. This is both cruel and stupid, and we must commend the work of organisations like Proteus Reptile Rescue for their sterling service in captuuring abandoned creatures of this type.

Whilst on the subject of 'clusters' of anomalous stories, we have recently received a number of tales on the subject of squirrels...

COME ON PUNK MAKE MY DAY

A vengeful squirrel's aggressive quest for food in Camden has been ended - but only after six people were left in need of medical treatment. The grey squirrel was described by a councillor as *"demanding food with menaces"* after biting several people in their gardens.

The squirrel's quest for food led him to attack a workman armed with a shovel. It scratched him

Animals & Men # 15 — Newsfile

and so he flattened it with the shovel' - *London Evening Standard 25/7/97*

KINGS OF SPEED

A squirrel was caught by a roadside camera speed trap in Fife, Scotland. Technicians said the racing rodent would have had to reach around 43 mph to have triggered the camera. A local RSPCA inspector said that this particular speedster must be the squirrel world's equivalent of Linford Christie. English Nature mammal ecologist Tony Mitchell-Jones said, *"It certainly sounds ludicrous to me. I suspect there is a fault with the camera. If I were a motorist caught on that particular stretch, I might be asking questions about its accuracy."*

Scotland has another type of super squirrel, however: a new generation of tough red squirrels that have successfully resisted the advance of the bigger and more aggressive greys. It is thought that the top speed of the 'super red' is about 20 mph in short bursts - *Daily Mail 22/9/97*

EDITOR'S NOTE: The original newspaper story concerning this speedy rodent described it as a 'Flying Squirrel' (presumably because of its remarkable burst of speed). It was (of course) no such thing, but merely a specimen of *S.carolinensis* with what an old lady of my acquaintaince describes as *"the wind in its tail"*.

GLOW WORMS

The increase in urban "light pollution" is being blamed for the decline of the British glow-worm. Researchers suggest that the male insects, which are attracted to the females by their glowing bodies, are no longer able to see potential mates, due to urban sprawl and road lighting - *The Times 1/7/97*

EDITORIAL NOTE: The 'clusters' keep on coming folks! Herewith a trio of weird elephant stories...

PENSION RIGHTS FOR ELEPHANTS

Elephants working in the forest reserves of West Bengal are to be given pension rights by the provincial government. A forest department spokesman said staff elephants over 65 years old will be given good food, shelter and health care facilities - instead of being sent back to the jungle, or destroyed. Special points such as good behaviour would also be noted - *BBC Ceefax 28/10/97*

FUNKY DUNG

More than 40 firefighters were called to a zoo after a pile of elephant dung caught fire at Howlett's Animal Park in Canterbury - *UK Teletext 4/10/97*

SEXY ELEPHANTS

Female elephants on heat can be heard by potential mates up to two miles away, using low frequency rumbles below the range of most humans' hearing. Dr Langbauer of Pittsburg Zoo, USA, said, *"When the males hear the call they immediately stop what they are doing and head towards the female. They are completely silent - which isn't surprising. The females are sexually receptive for only four days every four years, so the males don't want to advertise to rivals that they are on their way."*

Scientists have also revealed that male haddocks have a secret courtship ritual: they make thumping noises similar to the sound of a revving motorbike - *Daily Mail 13/9/97*

EDITOR'S NOTE: This is certainly an issue full of 'clusters' on quasi-anomalous stories. One of the strands in this issue is Bees and Wasps.

Newsfile

As well as the Waspman of Lancashire (see Newsfile Extra) and the 'Super Wasp' described above (Out of Place) we start our miscellany of uncategorisable stories with a few close encounters of the buzzing kind...

I'M A KING BEE

The mystery of honey bee communication may have been solved. It has been known since 1946 that bees use a complicated dance to pass on directions to sources of nectar. But in the darkness of a hive, bees are unable to see each other. Scientists now believe bees use their antennae and hairs to detect tiny air movements caused by a 'dancing' bee. Professor Michelson of Denmark tested his theory with a tiny robotic bee connected to a computer, and miniature microphones in a hive. *"When we brought it into the hive and made it dance, a number of the bees obeyed the instructions given by the dancer,"* he said. *"For example, they would fly 500 metres to the south if we told them to."*

He discovered that the bees were sending out waves of air pressure over short distances. The turbulence a fraction of an inch from the bee is equivalent to standing 60 ft (15 m) from a jumbo jet as it takes off - *Daily Telegraph/Daily Mail 13/9/97*

EDITORIAL NOTE: *"I'm a King Bee"* is an old blues song by (I think) Sonny Boy Williamson, but whose first band was called "The King Bees"? The first person to telephone me with the answer wins a free four issue subscription to this superlative publication...

SEARCH-AND-DESTROY WASPS

Hertfordshire scientists are training a posse of hunter-killer wasps to tackle aphids, a pest that attacks crops. The wasp, *Aphidius ervi*, which is a quarter of the size of the common British wasp, uses aphids as incubators for its eggs.

The wasps can detect volatile chemicals given off by plants when attacked, and are being put through a training program to enhance the association of the chemical smell with the presence of aphids. Untrained wasps are relatively sluggish in their response.

The hunter-killer wasps are intended for use in greenhouses. Outdoors, it is estimated that aphids cause damage costing £100 million to crops in Britain and a genetic engineering solution is being sought as an alternative to pesticides - *The Observer 14/9/97*

TEARS OF A CLONE

Leaders of 40 nations at a Council of Europe summit in Strasbourg have pledged to ban human cloning. The summit examined various human rights and social issues. A protocol on cloning will be submitted to the European Convention on Biomedicine in a bid to guard against misuse of medical advances - *BBC Ceefax / UK Teletext 11/10/97*

CLONE ALONE

Meanwhile, Dolly the cloned sheep is to be put through a breeding programme to see how fertile she is, and whether any effects from her unusual origin can be traced in her descendants. And Polly, another cloned sheep at the Edinburgh institute, will also be bred. She contains human genetic material and scientists want to see if these genes carry on down the generations - *Daily Telegraph 22/9/97*

MARINE TOOL USE

US Researchers believe dolphins have been seen using tools to find food and for defence. Five female bottle-nosed dolphins were observed in a Western Australian bay carrying sponges on the tips of their snouts as they searched for food on the seabed. It has been suggested the sponges were being used as protection against the spines and stings of animals like stingrays and stonefish, and also to rake up prey - *Press & Journal (Aberdeen) 26/7/97*

SEAHORSE BREEDING

Scottish marine experts have bred the world's biggest seahorses for the first time in Britain. The captive breeding programme will eventually see seahorses returned to the wild in oceans where they are under threat - millions are killed each year and exported to China and Japan, where they are ground up for 'medicinal' use - *Press & Journal (Aberdeen) 29/8/97*

HIGH-JUMP CONTEST

A high-jump competition was arranged by animal behaviour scientists in New Zealand who wanted to determine how high and wide a bird sanctuary barrier should be. Stoats performed the best. The results were:

Stoats   1.9 m
Ships' rats 1.7 m
Feral cats 1.5 m

THE JOURNAL OF THE CENTRE FOR FORTEAN ZOOLOGY

Animals & Men # 15

Newsfile

Feral cats 1.5 m
Wild Possums 1.2 m
Ferrets 0.5 m
Mice 0.33 m

Another test compared tunnelling animals - a challenge won by a Norway rat, which dug a length of 0.6 m in the time limit - *Independent Aug 97*.

SECRET LIFE OF THE DEEP-SEA SHRIMP

The natural behaviour of deep-sea shrimps that swarm around volcanic springs on the ocean floor may be hard to determine, Jon Copley of University of Southampton says.

Although described as eyeless when they were discovered a decade ago, the shrimps in fact have an unusual "eye" in the form of a heat-detecting organ on their backs.

The bright lights of deep-diving submersibles would almost certainly dazzle the shrimps - and possibly modify their behaviour while being observed.

The idea that an observer's act of observing interferes with the behaviour of the object being observed is a familiar one in quantum physics.

"I want to find out more about the lives of shrimps at hot springs," Copley says, "simply because they are there - and because, like quarks and quasars, they do not always give up their secrets easily..." - *New Scientist "Forum" 1/11/97*

OIL RIG SAFE HAVENS

Endangered fish are finding safe haven among abandoned oil rigs, fishermen have claimed. Fishing is banned within 500 yards (400 m) of an oil rig and population 'hot spots' of fish have now been reported in these areas.

Environmentalists have been criticised for ignoring the effect of oil rig 'safe havens' on fish stocks and for calling for the disposal of these decommissioned rigs - *Daily Telegraph 7/9/97*

EL NINO

The El Nino effect, a largely-unpredictable lurch in oceanic and meteorological conditions, is predicted to peak in early 1998 and leave droughts and floods in its wake. The last El Nino, in 1982-83, was regarded as the most destructive weather event in modern history.

The event appears to be becoming both stronger and more frequent. The cause of it is as yet unknown, although some scientists suspect that it is linked to global warming - *Daily Mail 19/9/97*

WHALING

This years' International Whaling Commission (IWC) meeting, held in Monaco in late October, has ended without agreement over whether commercial whale-killing should resume.

As usual, Japan and Norway want to continue to kill whales for "scientific" purposes - *New Scientist 1/11/97*

KNOCK-ON EFFECTS OF GENETIC ENGINEERING

Genetic manipulation of plants has affected ladybirds and radishes, British researchers have found. Potatoes engineered to resist attack by aphids can harm ladybirds, the pest's natural predator. And genes for herbicide-resistance, spliced into oilseed rape, can 'spill over' into adjacent wild radish plants and persist for several generations.

The rape/radish hybrids had variable numbers of chromosomes but no stable varieties were found.

Experts say that neither finding poses a major environmental threat - *New Scientist 1/11/97*

Contributors:

Tom Anderson, The Cryptic Clipper of Mannic Publications,
Lionel Beer, Ade Dimmick, Hermann Reichenbach, Wolfgang Schmidt.

Please mark your clippings with *your* name, the name of the publication in which it appeared and the date of publication.

The Waspman Cometh

by James Lister

A fascinating report which seemed almost too good to be true surfaced on the Preston local radio station on Thursday, August 21st 1997.

The DJ spoke gravely of a man-sized yellow and black flying insect which displayed prominent humanoid characteristics and was apparently abducting domestic pets, and terrorising the residents of Walton-le-dale, a suburb of the town.

No attempt at elucidation was provided as to what this creature may have been; but this investigator's appetite was thoroughly whetted!

But, as someone once said, nothing odd will remain odd for long. In reality the case was interesting, although not quite as engaging as I had at first supposed.

The 'mansized' insect was just two and a half inches long (fairly short for any human!), did not possess any human features, nor was it involved in the aforementioned abducting or terrorising.

It was captured on August 20th in a garden in Walton-le-dale by two men after it had 'swooped' down on them! One can safely assume that the DJ had been using a liberal dose of artistic license when referring to her 'waspman'.

In British creepy-crawly terms, however, such a length is uncommon but I fear that the creature in question is nothing more than a very large specimen of the British Woodwasp (or Horntail - so named because of the female's noticeable ovipositor), which is not unheard of on new housing estates.

There could be a mystery here, as the two men who captured it, having consulted a book on entomology, identified the creature as a Scandinavian Woodwasp; therefore making it, to a certain extent at least, an 'out of place' animal. Nothing appeals more to a fortean sensibility than the classic juxtapositioning of exotic zoology and middle class suburbia, but there is a nagging doubt that this one could be explainable - I am hoping that someone more versed in comparative entomology than I will be able to make a definitive identification of the creature.

Some forteans will not want to discount the DJ's original news report (I among them), perhaps she was telling the truth after all, and actually alluding to another, completely separate entity. Who knows? The Waspman of Walton-le-dale sounds interesting... Owlman and Mothman hunters take note!

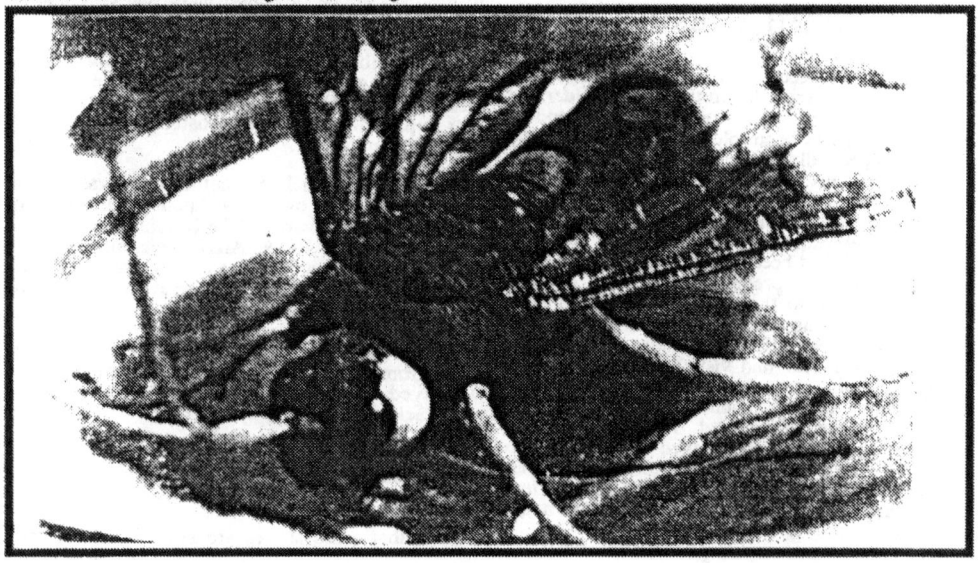

The Bigfoot Murders

by Richard Freeman

The prevailing view of Bigfoot, the Yeti and their relatives around the world is one of timid, retiring creatures, gentle despite their size and immense strength. Yet, there are those who question this assumption. These are wild animals and as such are unpredictable. The Gorilla is a peacable animal, but the Chimpanzee - another ape - is one of the most savage of all primates.

Those who question Bigfoot's temperament, pose some unnerving questions. Many people have vanished without trace in America's forested North West, and in the bleak Himalayas. Did the lost hikers merely lose their way and die? Did the missing mountaineers really fall down a crevasse? The actual accounts of Bigfoot/Yeti attacks are extremely rare, but they do exist within the literature on the subject.

In this article I will examine each in turn and see whether there is any substance to this slur on the name of one of cryptozoology's most famous icons.

NORTH AMERICA

The most famous account of a human being killed by a Bigfoot was recorded in the book *Wilderness Hunter* by former American President Theodore Roosevelt. He had been told this story in 1892 by an old trapper called Bauman. Fifty years before Bauman and his partner were trapping in the Bitterroot mountains between the Salmon and the Wisdom rivers in Idaho. This area had an unpleasant reputation, as a year before a trapper named 'Cluby' had been found dead and half eaten by some 'unknown' predator. On the first day, the pair had returned to their camp to find it in total disarray. Provisions and packs were torn apart and the ground was covered in odd tracks. The trappers assumed that a bear had visited in their absence. Later that night, some huge animal smashed into their stand-to shelter. Bauman took a shot at the stinking creature, but missed. The next day they returned once more to a wrecked camp. The men waited up that night ready to shoot the interloper. They heard its *"harsh, grating, long-drawn cries"* as the 'thing' circled their camp but remained in the shadows.

This all proved too much for the men, who decided to leave.

Bauman left to collect Beaver traps whilst his companion packed up the camp. Returning at dusk, Bauman again came upon a destroyed camp and he became understandably nervous when his friend did not answer his call. Bauman found his friend's body near a spruce log. His neck was broken and four huge fang marks lay deep in his throat. The carcass had not been eaten but had been thrown about like a toy and rolled over many times like a dog. Bauman grabbed his rifle and fled on horseback, riding through the night until he was clear of the forest. He remained deeply effected all his life and never returned to the Bitterroot woods.

There is no evidence that this story, which admittedly sounds like something out of a Hammer Horror movie had anything whatsoever to do with a Sasquatch. Bauman did not see the beats but only smelt and heard it! Even if a Bigfoot did visit the camp who is to say that it was the same creature that killed Bauman's friend? The fang marks and the fact that the body was tossed about point to a bear attack! Bears are well known, nay infamous raiders of campsites...

EDITOR'S NOTE: As anyone who has ever watched Yogi Bear cartoons will tell you...

Roosevelt had been told several tales of hairy giants in the backwoods of America. In this case I think that he put two and two together and made five!

In 1920, Albert Petka, of of Nulato, Alaska, was supposedly attacked by a 'Bushman'. Petka lived alone on his boat with only dogs for company. His pets apparently drove the monster away, but Petka died of his injuries.

Twenty-three years later another man was said to have been killed by a sasquatch. John Mire (or McQuire) also known as 'The Dutchman' staggered into the town of Ruby, Alaska, claiming to have been attacked by a 'Bushman' in his remote cabin. Mire's dogs, also chased the attacker off, but he later died of internal bleeding.

Those two cases are just too similar! Both men came from Alaska. Both men were virtual hermits who lived alone. Both had gallant dogs who chased the attacking Bigfoot away.

Both refer to the 'creature' as a 'Bushman'. Both men die shortly after telling the locals what attacked them.

I think that these stories are both fabricated tales

similar to contemporary urban myths. They surface every few years and have no basis in reality! We will see similar examples in Asia!

Finally, there is an obscure case mentioned in John Green's *"Sasquatch - The Apes Amongs Us"*. Two security guards were once found dead. Apparently they had been picked up and dashed against the ground like dolls. Huge footprints were found in the area. Obviously 'Bigfoot' was to blame. As it turned out, however, no such thing ever happened! One man had died from a stab wound and the other from a bullet. Exit Bigfoot (who is not noted for 'packing a piece')!

SOUTH AND CENTRAL AMERICA.

In the early part of this century, French naturalist Frank Blaucaneaux, the author of a book called *'Biologica Americana Centrale'*, was travelling through the Honduras with his negro servant, when they arrived at the headwaters of the Rio Mopan. The two decided to rest in a shady hollow dominated by a huge palm tree surrounded by long grass. As they prepared for their siesta the palm began to shake as if some huge creature were trying to loosen nuts growing high in the tree above them.

The Frenchman instructed his servant Miguel to look for the cause of the disturbance. Miguel became very agitated saying that it was 'a devil'. Blaucaneaux scoffed at the idea and insisted that his servant do his bidding!

Taking up a rifle, the unfortunate Miguel sallied forth. Shortly his agonised screams rent the air, followed by the sound of awful moans. Blaucaneaux ran to the aid of his servant. The man was lying beneath the palm tree with deep gashes across his face and body. His stomach was ripped open, and his entrails hung out! before he died, he told Blaucaneaux that a 'black devil' had torn him apart and then left for the forest. The Frenchman buried the body of his ill-fated servant beneath the palm tree and then attempted to follow the tracks of the beast. He followed the trail of broken twigs and bent branches for about five miles until he came upon some limestone crags.

Here, the thing had apparently entered a dried up stream bed and overturned some boulders. He followed the trail to a cave in the crags, where he found what appeared to be a handprint in the soft, white mud of the floor of the cave. It resembled a human thumb and two fingers, but all three digits bore huge claws. At this point Blaucaneaux's nerve gave out and he beat a hasty retreat.

Some time later he tried to enlist the help of the local Amerindian population in helping him to smoke the beast out of the cave, but they would have none of it and refused point blank.

Once again we do not actually have a description of anything which is identifiably a giant primate. 'Black Devil' could mean almost anything from a mad native to a melanistic jaguar. According to orthodox zoological belief there are not supposed to be any higher primates other than man in Central and South America (nor North America for that matter). A particularly curious thing about this report is the 'hand print'. All monkeys and apes have five digits, rather than three, and they have fingernails rather than the 'huge claws' described by Blaucaneaux. Surely, if a giant ape wanted to kill someone, it would bludgeon them to death with its fists, or even with a branch or a rock. The slashes on Miguel which are presumably claw markssound like something very different.

Some authorities believe that the reports of giant, hairy brutes in central and south America do not refer to primates at all, and are therefore nothing at all to do with Bigfoot, Sasquatch and their relatives in the BHM community. It has been suggested that these animals are a surviving form of Mylodont, a medium sized giant ground sloth. These were about the size of a brown bear and had skin studded with armour like bony nodules which provided protection to the whole of its body apart from the belly. Even today there are South American tribes who speak of such an animal, invulnerable apart from the belly. Mylodons possessed large, scythe-like claws for ripping down plants and for self-defence. The creature that attacked Miguel (if, indeed this is not just another traveller's yarn), sounds that it could indeed have been a Mylodon. Although we know that these animals were herbivorous we have no inkling of their temperament!

Italian archaeologist Pino Turolla was told a strange and bloody tale by an Indian guide in Venezuela. Antonio, the man in question, had gone with his two sons to the Pacaraima range. As they approached the savannah, they encountered what Antonio describes as huge lumbering beasts with smallish heads and very long arms. Three of these monsters set upon the men with clubs killing Antonio's younger son!

At first this seems like a convincing tale. Why would the old man make up such a gruesome story about the death of his own son?

However, there are a number of sequels to this story, which cause me to seriosly doubt Turolla's word!

Six months after hearing the story Turolla returned to Venezuala and persuaded the guide and some fellow Indians to take him to where the attack took place, shrill, lion-like roars terrified the Indians who would go no further. The Italian pressed on and glimpsed what he described as two blurred shadows six to eight feet tall pressed against some rocks. He said that they were erect and ape-like but that the 'creatures' vanished into the dusk.

Turolla was carrying the photograph of an 'ape' which had been taken by Francois de Loys in 1920. Turolla said that the creatures strongly resembled those in De Loys's photograph. In fact all that this famous photograph shows is a large female red faced spider monkey. The forward facing nostrils, vestigial thumbs and large exterior clitoris all identify this species plainly. This infamous photograph has been shown up as one of the most feeble hoaxes in cryptozoological history. Turolla's mention of it does not bode well for his credibility.

Turolla returned in 1970 on an expedition to the Guacamayo range near the Chancis river in Ecuador. Turolla said that he, and his South American companion had been given the location of a certain cave in the mountains by an Indian shaman. In this cave he was assured that he would find evidence to support his theories on the beginnings of South American culture. The cave, when they reached it, appeared to be man-made. After wandering by torchlight about a hundred and fifty feet into the cave, the men heard a terrifying roar that Turolla recognised as being the same that he had heard years earlier in Venezuela.

Something began to hurl huge rocks at the men who ran in terror towards the mouth of the save. On reaching the daylight they found that the Indian's heair had turned white and that Turolla himself was clutching a strange object - a tiny ancient carving of a human face on an axe head!

The whole account sounds like something out of a 'pulp' story from *The Boy's Own Paper* in about 1920! I don't believe a word of this tall tale, and I also think that the killer-ape story was a total fabrication!

EDITOR'S NOTE: I would just like to stress (having been threatened with libel actions twice during the lifetime of this magazine, that if Turolla, his heirs, publishers or anyone else reads this, that opinions expressed are those of Richard Freeman himself and now't to do with the rest of us innocent bystanders. It must be said, however, that on the face of it the story does seen a leeeeetle unlikely (to say the least). For more details on the discreditation (if that is the right word, which it probably isn't) of the Ameranthropoides loysii picture I refer you to the in depth article by Loren Coleman in issue four of The Anomalist.

ASIA

In parts of the Himalayas, the Yeti is greatly feared. In fact, one of the first thing that children are taught is that if a male Yeti chases you, one has to run uphill on account of the heavy brow which blocks the he-Yeti's view. A female Yeti finds it equally hard to run downhill due to her pendulous breasts.

In 1949, a Sherpa herdsman called Lakmpa Tenzing was reportedly torn to shreds by a Yeti in a remote pass in Nangaparbay. Details on this are, however, scant. No-one seems to know whether or not there were any witnesses. The Himalayan Black Bear, a highly aggressive species inhabits this area. Could such a bear have been the true culprit?

Of course, it is much more exciting to suppose that the luckless fellow had been killed by the mysterious man-beast of the mountains, instead of by anything as mundane as a bear!

Oddly, in other Himalayan areas the Yeti is not at all feared, and is known as an unaggressive creature, harmless unless provoked.

The 1949 case is the only reported killing that I have been able to unearth from Asia. This seems strange as it is not only the world's largest continent but in many areas it is riddled with superstition.

Perhaps, however, this is a pointer towards the TRUE nature of these animals.

I can vaguely remember having read a story of a number of soldiers having been killed and eaten by a huge Yeti somewhere in the Himalayas. The creature was subsequently shot. I cannot remember where I read this unlikely tale and I have not been able to find it again.

This story sounds remarkably like the accounts of the troll 'Grendel' in the poetic saga of '*Beowulf*', where the monster eats men one by one as they slept together in a great hall. If the Yeti was a man-eating carnivore it would surely take just one victim at a time, and would be unlikely to attack a whole group of men. If anyone can remember the source of this seemingly silly tale I would be very grateful to know.

EDITOR'S NOTE: With shades of the now infamous 'Thunderbird Photograph' I remember having read this story somewhere, but like Richard, I have no idea where. I seem to remember, however, that it was in a fairly sensationalised book about Central Asia rather than in a mainstream cryptozoological volume!

AFRICA

The darkest continent is strangely lacking in man-beast reports. After giving us the original man-beast, the once legendary Gorilla, things have been very quiet. Most African man-beasts are tiny creatures, seemingly more akin to the Sumatran Orang Pendek to the monstrous Yeti or Bigfoot. The one exception is the East African Chemosit.

This greatly feared monster is supposed to be able to claw its way through the walls of mud huts to get at its human prey. Some tribespeople wear protective headgear as The Chemosit is said to relish human brains obtained by biting through the skull. Although descriptions of this animal vary tribesmen have attributed many deaths to its predations. Some say that it is like an ape, and others that it resembles a bear, hence its common name of the Nandi Bear.

Many theories have been put forward to explain it. These include giant baboons, huge rogue hyenas, surviving chalicotheres (a peculiar clawed ungulate thought long extinct), outsized melanistic honey badgers, or even a hitherto undiscovered species of sub-Saharan bear (the only bear species known from Africa is *Ursos arctos crowtheri* - the Atlas Bear - which is now probably extinct).

EDITOR'S NOTE: For more details on the Chalicothere theory see Clinton Keeling's article in A&M 6 and for an in-depth look at the Atlas Bear see my article in A&M2.

The closest encounter between a Chemosit and a westerner took place in Kenya and involved Angus McDonald, an expatriot Briton involved in a land development scheme. He and his fellow workers were posted about a hundred miles from any sizeable settlement. They were encamped in temporary huts surrounded by scrubland at an elevation of about 9,000 feet.

One night McDonald was awoken by wild screams and a huge hairy creature lept through the window almost on top of him.

They then engaged in a frantic moonlit chase around the hut, as the evil smelling 'thing' pursued the frightened man who tried to fend it off with anything that came to hand.

After five frantic minutes, his fellow workers managed to frighten the strange assailant off by banging pots and pans. McDonald described the animal as being seven feet tall with an ape-like head and red mouth, and being greyish-brown in the moonlight.

The next morning the unnerved crew set out to track the beast down with dogs. Its tracks appeared to be roundish and ending in digits. The thing had moved bipedally before going down on all fours to make its escape into the trees around the camp.

Allowing for understandable fear induced exaggeration I believe that I know what attacked McDonald. I believe that it was a large, male Chimpanzee. Chimps are one of the most savage, unpredictable and unpleasant animals known to man...

EDITOR'S NOTE: They ARE our closest relatives...

The public image of these creatures as cute stars of the PG Tips TV commercials could not be further from the truth. The ones used on television to sell tea-bags are juvenile specimens. An adult chimp is three or four times as strong as a man and armed with savage teeth. They have been known to kill people in Africa and maim them in captivity, and they are my least favourite animal of the three hundred plus species of wild creature that I worked with during my time as a Zoo Keeper.

One of my fellow professionals had a finger bitten off in an unprovoked attack and I have heard gruesome tales of others who suffered far more serious wounds.

Chimpanzees are not usually found in Kenya. The most easterly population known is in western Tanzania. They do, however, wander. In August 1959 a surprisingly unaggressive female turned up on the western shore of Malawi's Lake Nyasa and other vagrants are also known.

EDITOR'S NOTE: This animal, which the natives named *Ufiti* (which means 'ghost') was a particularly strange animal. In many ways she seemed more akin to the chimpanzees found in western Africa rather than to other members of the East African population. She was captured in 1964 and sent to Chester Zoo where her health soon deteriorated forcing her to be destroyed. The whole question of Malawi's putative chimpanzee population has remained in abeyance ever since!!

As Bernard Heuvelmans pointed out four decades ago, the Chemosit is undoubtedly a composite animal, and personally I would suggest that large hyenas and wandering chimpanzees are the most significant components.... but no killer African Yetis!

AUSTRALIA

The only placental mammals living in Australia are rodents and bats, and all other non-marsupials were introduced by man! There have never been primates apart from H.sapiens there, and certainly no anthropoid apes. It is ironic, therefore, that in this continent where apes have never been reported there are more reports of giant man-beasts than Africa where chimpanzees and gorillas are well known!

Native Australian (Aboriginal) legend is full of tales of these strong smelling, savage horrors, many of them recounting savage battles between man beasts and humans which resulted in men being killed. These tales are not unlike those of many European troll legends, where the humans use their advanced brains to outwit their gigantic adversaries. These antipodean man-beasts are known variously as Yowies, Poolagarl, Kalkadons, Narragun and Kraitbull.

Today these creatures are reported both by the indigenous population and by white men, but they seem less aggressive.

Although there are several reports of unprovoked attacks on humans only one death has been attributed to a yowie in recent times. In 1910, two un-named men were walking in the Victoria Falls area of the Blue Mountains. A three metre (ten foot) tall, upright, gorilla-like creature charged at them from behind some bushes. It was brandishing a large rock. The Yowie hurled the rock at the men smashing the skull of one of them. The other man beat a hasty retreat, but returned the next morning with a party of armed men and dogs. All thy could find was bloodstains - the body of his unfortunate companion having been presumably carried away by the creature.

This story is totally unreferenced (like so many killer man-beast accounts) and appears in Rex Gilroy's book 'Mysterious Australia'. This is a book in which Mr Gilroy asks us to believe not only that there are no less than three species/races of Australian Man-Beast but that they share their habitat with living dinosaurs, man-eating plesiosaurs, fifty foot monitor lizards, giant marsupial lions and hidden UFO bases.

It is difficult to reconcile these views with those of mainstream cryptozoology.

The Yowie certainly exists but it is (in my view) almost certainly a zooform phenomenon rather than a bona fide flesh and blood animal.

CONCLUSION.

Most of us have a morbid fascination with murder. The stranger and more savage the killings, the greater the secret appeal that they have to us. This is why we find cannibals and serial killers like Fred West and Ed Gein so intriguing. If an animal looks like a human it also appeals to us. This is why creatures with round faces and big, forward-pointing eyes are the ones which we tend to think of as 'cute'. Chimps, monkeys, pandas, cats, pug-dogs, and bushbabies are all guaranteed to make the public go 'aaah!' This changes when the animal is bigger and stronger than us. African natives believed that Gorillas would fight Elephants, and carry off native girls to rape them. (This is a charge sometimes levelled at the Yeti today). We transpose our darkest crimes onto these hulking monsters. Thick brows, shaggy fur and glowing eyes are seen as symptoms which add up to a killer. Ugly = Evil is a strong concept for 20th Century man. In reality there is not one shred of evidence that these magnificent apes (if apes they are) have ever harmed anyone without being provoked. Some estimate that the Gorilla is twenty times as strong as a man. A large Yeti would be twice as big as a Gorilla and one an easily imagine the murderous strength that it would wield. That it has never turned this hideous strength against man is a testament to its true nature!

BIBLIOGRAPHY

BORD, J and C *The Bigfoot Case Book* (Granada, London 1987)
GILROY, R *Mysterious Australia* (Nexus, Australia 1995)
GRUMLEY, M *There are Giants in the Earth* (Panther, London 1976)
KEEL J *Strange Creatures from Time and Space* (Sphere Pb Ed, London 1975)
SHUKER K.P.N. *The Lost Ark - New and Rediscovered animals of the 20th Century* (Harper-Collins, London, 1995)
TCHERNINE O *The Yeti* (Novello/Spearman, London, 1970)
WILKINS H.T *Monsters and Mysteries* (James Pike, NYC, 1973)

Many congratulations to Richard Freeman from Connie and Raymond Marbles

BEAVER PATROL

EDITOR'S NOTE: As is our custom most issues we present a selection of short articles on different aspects of the same animal. This issue, by popular demand we approach the beaver, and to celebrate we include another in our ongoing series of telephone competitions. To win a free four issue subscription to this august magazine be the first person to telephone me, not only explaining the title of this section but singing the first few lines of the song...

RETURN OF THE BEAVER .... THE STORY SO FAR

by Tom *"don't make any Beaver Jokes"* Anderson

Under Article 22 of the E.C.Habitats Directive member states are required to consider the desirability of reintroducing species listed on Annex IV..

"...which are native to their territory (...) where this may contribute to re-establishing the species at a favourable conservation status".

Of the four species listed; the brown bear (*Ursos arctos*), the lynx (*L.lynx*), the wolf (*Canis lupus*) and the European Beaver (*Castor fiber*), only the latter was considered by Scottish National Heritage to be appropriate of further investigation in terms of reintroduction to the Scottish landscape.

This is the first attempt to reintroduce a large animal to the British countryside and is subject both to the feasibility and desirability of its implementation. Prior to its extinction in the 16th Century, the trade in beaver pelts was large enough to support a sizeable export market. Its secondary value was in the salicylic acid derived from the anal scent gland, which is, by the way, the main constituent of the drug known today as asprin.

EDITOR'S NOTE: Possibly that is why certain members of the CFZ investigative team usually need a couple of asprins after a night out looking for beaver...

Already re-established in sixteen countries, its suitability to adapt is second to the availability of sufficient habitat to support a self maintaining population.

The potential disadvantages are their impact on fish stocks, tree felling for food and building, the undermining of ditches and banks, and the flooding and blocking of waterways by damming.

The latter is less of a problem with the Scandinavian sub-species (*C.f.fiber*) which was chosen as the most suitable to be reintroduced into Poland and which only built dams on 50% of the settlement sites. This is a critical factor as there is concern that damming would impede salmon migration and that bank erosion would damage reeds with silt coverage.

The benefits of restoring elements of native biodiversity such as reducing sediment loads and slowing water flow, the preservation of wetland ecosystems and the stabilisation of groundwater levels constitute the case for the defence.

In Britain its only natural predator would be the fox. In Europe the majority of beaver deaths are attributable to man, including poaching, entanglement in fishing nets and road casualties. The only diseases that beavers carry which are transmittable to other animals are tulaemia and rabies, although Canadian animals are thought to spread the water borne disease giardia.

Current research is based on identifying habitat availability and its potential for any releases. Following this will be the more complex investigation of the effect of beaver behaviour on native fish species, in particular those of importance to anglers. This, in Scotland, is likely to be a major hurdle, both politically and economicaly sensitive.

The study has two years to run following which reccomendations will be made to government by S.N.H.

It is not hard to see the potential hazards that will follow any planned reintroduction of a species with (as yet) no cash crop potential like fur farming or hunting. The vagiaries of trout and salmon fishing and its value to small rural communities will have the riparian owners lobbying for a quid pro quo to offset any alleged erosion of their interests.

One of the prerequisites of Article 22(A) of the Habitats Directive is that reintroduction may only take place after *"proper consultation with the public concerned"*. The last word should have been ommitted. It lays the way open for vested interests to hijack the issue to their own advantage, excusing excessive water abstraction etc.

In the long term the beaver could prove of inestimable value in improving wetland conservation and land management. As the species has a high mobility and would in time colonise south of the border, its future should be seen in a national context.

SOURCES

CONROY, J.W.H and KITCHENER A c.1996 *The Eurasian Beaver (Castor fiber) in Scotland.* SNH Review no 49.
DANILOV, P.I and KAN'SHIEV, V. 1983. *The state of population and ecological characteristics of Castor fiber (Linneaus) and Castor canadensis (Kuhl) beavers in the north western USSR.* Acta Zool. Fennica 174: 95-7.
IUCN 1996 *Guidelines for reintroductions.* Species Survival Commission, Gland.
JONES, I.K. 1995 MSc thesis on reintroduction. University of Aberdeen.
LAVROV L.S. 1983. *Evolutionary development of the genus Castor and taxonomy of the contemporary beavers of Eurasia.* Acta Zool. Fennica 178: 87-90.
MACDONALD, D.W., TATTERSHALL, F.H., BROWN, E.D and BALHARRY, D. 1995 *Reintroducing the European Beaver to Britain; nostalgic meddling or restoring biodiversity.* Mammal Review 25 (4) 161-200.
PINDER, N.J. 1978. *Faunal reintroductions; A policy, a proposal format and an example; beaver in Britain.* MSc thesis, London University College 221pp.
ZUROWOSKI, W. 1992. *Building Activity of Beavers.* Acta Theriol 37(4) 403-411.

EDITOR'S NOTE: Tom Anderson also wrote about the reintroduction of this species into Scotland as part of his contribution to the 1997 CFZ Yearbook where he discussed several species including the lynx and the wolf.

GIANT BEAVERS

by Bill Petrovic

In his excellent (and unfortunately out of print) book *"Extraordinary Animals Worldwide"* Dr Karl Shuker notes that during the Pleistocene epoch which began about two million years ago, the beaver family included some truly massive species, some the size of bears. The fossil record of animals such as *Amblyrhiza* suggest they were almost totally aquatic. It is assumed that these animals became extinct long before the age of man, but Karl Shuker quotes a fascinating suggestion by rsearcher Janet Beck, who in 1972 suggested that certain animals referred to in Native American folklore could in fact have been a relict population of these enormous beasts which had survived into prehistoric times, and which had survived in the form of a folk memory amongst the Algonkian Indian tribes.

In an earlier issue of this journal, Canadian cryptozoologist Ben Roesch postulated that giant ground sloths may have existed into historical times in parts of North America, and whilst doing so he discussed quasi-folkloric accounts of 'giant squirrels'.

It is tempting, to wonder whether these accounts might not have been of ground sloths after all but instead were of giant beavers of one of the species thought to be long extinct.

Like so much in contemporary cryptozoology, however, this is probably nothing but wishful thinking!

EDITOR'S NOTE: In another of his fascinating books Dr Shuker also points out that the world's largest species of flea is a parasite found upon an obscure North American rodent called the Mountain Beaver. He also points out that this is a complete misnomer on two counts because the animal is neither a beaver or a resident of mountainous areas.

I included this fascinating vignette, not because it is of any direct importance to the rest of the subjects listed in this section but because for some reason this utterly inappropriate piece of nomenclature has always made me laugh!

MORE BRITISH BEAVERS
by Jonathan Downes.

Despite the contemporary arguments over the reintroduction of the beaver to Scotland which Tom Anderson discusses above, and despite the accidental reintroductions of the species which he described in the 1997 CFZ Yearbook, it is a little known fact that in the words of the publicity schpiel for the acclaimed Channel Four Sci Fi series 'Dark Skies' "THEY'RE ALREADY HERE!"

I have no official view as to whether the world is being populated by 'grey' aliens from Zeta Reticuli, but it is an undoubted fact that for nearly thirty years there have been beavers living and probably breeding in an out of the way part of the River Axe on the Dorset/Somerset border.

According to which account you accept, two or more animals escaped from Cricket St Thomas Wildlife Park in 1969 and set up home in the nearby countryside. I approached the wildlife park in 1991 for information but was told that no records were available, and so we are unable to verify whether the animals were of the Canadian or the European species.

There is no doubt, however, that they are there. A team from the Zoology Department at Exeter University photographed unmistakeable signs of beaver activity in the area in the early part of this decade and whilst I have not received any further reports since 1995, a newspaper report in the autumn of that year claimed that there was a small but healthy population in the area.

The Mammal Reports of the Transactions of the Devonshire Association during the mid 1950s have several cryptic accounts of what appear to be large aquatic rodents seen briefly in East Devon waterways. These may have been beavers but are more likely to have been either muskrats or most probably coypus. According to all the books on the subject of introduced and naturalised British Wildlife the coypu never colonised waterways in the westcountry. There are, however, a few accounts of what seem to be coypus from river banks in Cornwall and the museum in Exeter has, apparently, a mounted specimen of a coypu shot on the River Exe at Topsham. These may be isolated occurrences but they prove that westcountry waterways are reasonably suitable habitats in which large aquatic and semi-aquatic rodents can thrive relatively un-noticed and undisturbed.

Beavers were, as Tom Anderson has noted, quite well known in historic times and indeed in certain parts of the country their remains in a sub-fossil state are relatively common. J.E.Harting described the sub-fossil beaver skulls discovered in the Fens of Lincolnshire in his 1880 book *"British Animals Extinct within Historic Times"*. Pictures of these skulls taken from this book are reprinted below.

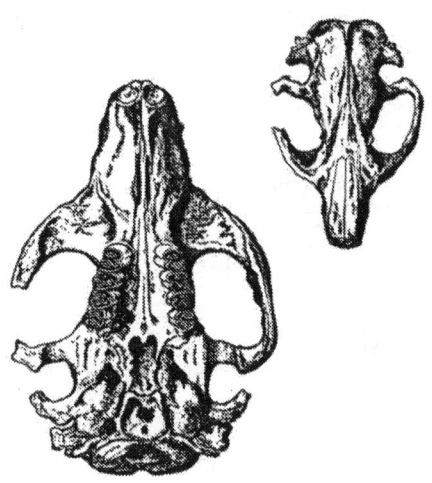

Another particularly peculiar occurence involving this singular species (or perhaps not) took place in rural Essex in (I think) 1988 when the local paper reported that two beavers had made a dam on the local river and had subsequently been shot. We contacted the newspaper concerned to be told that whilst two animals had been shot they had in fact been fox cubs and were not beavers at all. One wonders how a small reddish canid can be mistaken for a large aquatic rodent, but as Clinton Keeling is fond of saying, there's now't as queer as folk.

As always he is correct!

The Beast of Llangurig and others

by Terry Hooper

EDITOR's NOTE: Sightings of what appear to be wolverines are becoming almost commonplace in the British countryside at the moment. At the CFZ, we are currently investigating reports which MAY be of these voracious beasties from Haldon Hills, (just outside Exeter) and from a small wood near Seaton in East Devon. Terry Hooper shows, intriguingly, that the Wolverine reports may go back further than some researchers have thought...

As a member of the Society for the Investigation of the Unexplained (SITU) I often got quite a lot of unusual material through the post. Having talked to a friend in the Avon and Somerset Constabulary and indicating that I'd like to get more information on a 'mysterious beast' seen around the Llangurig area of Powys, Wales. I was told 'OK'. That usually means *"don't hold your breath"*. This time, however, within a few days I had a contact name!

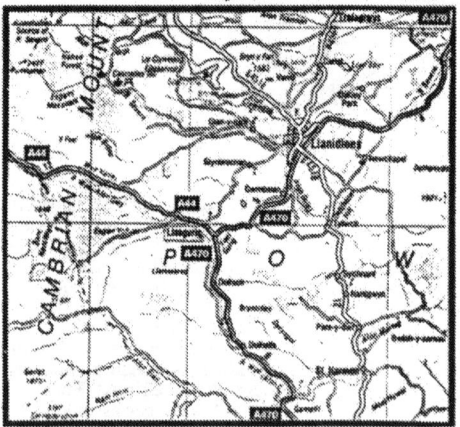

The Llangurig area of Powys

Off went my letter and I waited. In the meantime I went over what I'd learned, most of it, sadly, from the Press and TV reports.

On the morning of the 23rd October 1980, farmer Micheal Nash was working near his barn on Pant-y-Drain farm located some twoo and a half miles (4 km) from Llangurig. From amongst the straw bales of the farm came a loud snoring noise that he had heard a number of times over recent days. In front of the barn were found large pawprints in the muddy ground and Mr Nash was convinced that some strange animal was in the barn. Over a period of twenty four hours five sheep had been killed out of his three thousand strong flock; all at different times with the flesh torn away from the bones. Mr Nash did the sensible thing - he telephoned the Dyfed/Powys police.

At 14:00 hours the police arrived and summed up the situation. They then also did the sensible thing and called in armed reinforcements. The back-up force arrived and as dusk crept in Calor Gas lamps were set up - there seemed no point in keeping armed men watching a barn in the dark for a potentially dangerous animal!

Mr Nash's suggestion that the open side of the barn be closed up by using gates in case the animal tried to break out was ignored, and at 17:45 hours the police decided to drive the animal out! They *"hammered hell out of"* the asbestos side of the barn, but nothing happened. Mr Nash's viewpoint was that the beast, being disturbed and scared crept out and away unseen.

This seems quite likely. At mid-day on the 24th police moved in, armed and ready but found only wet straw where the animal had been urinating, some three or four inch (approx. 10cm) long droppings and owl pellets. Incredibly, the droppings were not saved for analysis.

By the evening of the 24th the police and accompanying media circus were gone. So, Mr Nash decided to carry on. However when he next approached the barn he again heard the snoring noise! The farmer shone his torch on the area that it appeared to come from and it all went quiet. On two or three occasions Mr Nash went back to the barn but minus his torch.

Nothing!

Whatever the animal was, it had obviously considered that its den had been compromised, and so it moved on somewhere else!

There are some interesting points to note. After the police had hammered on the barn some friends of

Wolverine

Mr Nash had seen *"peculiar animal"* cross the road about half a mile away at 23:30 hours. It was the farmer's opinion that this must have been the beast itself. A man from Shrewsbury who had seen the pawprint on TV telephoned Mr Nash to tell him that in his opinion it had been made by a Wolverine (*Gulo gulo*).

An animal food salesman who had previously worked in a zoo for two and a half years saw the print when he visited Pant-y-Drain. This man also identified the print as having been made by a Wolverine. And, true, from the photograph that I have in my possession, the print IS similar to that of a Wolverine.

So, was it indeed a Wolverine? These creatures inhabit the extensive taiga and tundra region of North America and Eurasia. Its European distribution is limited (or was) to Scandinavia, Northern Finland, and parts of the former Soviet Union. In Europe as a whole it is rare and threatened with extinction.

It has been suggested that there are at least two pairs living wild in the west of England which suggests that they are escapees from captivity, as the creature is naturally a solitary one by nature (except during the breeding season). The very fact that the Llangurig creature (if it were a wolverine) seemed to seek out a bed of straw in the barn also suggests that this is what the animal had been used to in captivity.

Some males have permanent territorial hunting grounds that can measure as much as 1000 square kilometers (about 700 square miles). They hunt both at night and during the day in these large areas and their role in the natural foodchain is quite positive as they feed mainly on weak and crippled animals - they wouldn't necesarily be a threat to sheep. In summer the Wolverine's diet consists of birds and their eggs, insects and larvae, small rodents (particularly the lemming), berries and oil rich seeds.

In winter the diet is carrion, larger mammals, ungulates and various chance tit-bits such as prey stolen from other carnivores (even wolves and bears) and its strength and apparent lack of fear are well known. It has a robust body, some two and a half to three feet (70-90 cms) long, and a tail length of six and a half to ten inches (15-25 cms) and weighs 10-20 kgs.

Animals of this species have dense brown fur with the upper facial hair and flanks paler. The Wolverine can live for fifteen-eighteen years and mates at the end of the summer of alternate years. The young (in a litter of two to five cubs) are born in the late winter after a relatively long pregnancy.

So, one can imagine, ignoring the Llangurig Wolverine (if it was a Wolverine) with no predator to threaten them and many sheep in large, rural hunting grounds as food, there is no reason to suppose that (conservatively) five or six pairs could be living in the countryside now.

I contacted the RSPCA, because one of their inspectors had seen the only professional to have studied the wounds and tracks. Their response was not unfamiliar - as was the case during my investigation of the "Barking Beast of Bath" (See A&M13), each of my letters and telephone calls were totally ignored. The Press, I found, were equally unhelpful. They had difficulty in spelling Pant-y-drain and didn't even mention the droppings found in the barn. As far as they were concerned, the creature was a lynx or a puma, and was really nothing more than a silly story to fill newspaper columns.

A response on the 11th of December from Chief Superintendent W.J.R.Edwards from the Dyfed/Powys Police HQ at Newtown was to the point and shows how little priority was given to the rival "big cat" and "wolverine" claims. Apparently, despite having personally checked the sheep, the RSPCA Inspector had not submitted a report on the incident. No veterinary surgeon had submitted a report either even though one sheep had to be put down.

The police had consulted wildlife experts who were of the opinion that the plaster casts taken from the pawprints were those of a very large dog. The police Press Releases that I was sent revealed that, on Tuesday 25th November 1980 at 08:45 hours, Mr Ernie Lloyd of Coity Farm, Cwmbelan, saw a large 'puma' type animal and reported it. Officers couldfind no animal but large paw prints were noted on the spot where Mr Lloyd said that the cat had been standing.

I heard nothing back regarding the experts' opinions on the casts made from these prints. Sheep killed, acording to Mr Lloyd, bore wounds similar to those caused by a fox.

Here I ought to point out that there is no denying that dog attacks had occurred in the Dyfed/Powys area and that at least one dog was shot whilst in the act of attacking sheep. Mr Nash had been at pains to point out that he had seen no animal at any time and definitely NOT a puma. People who say that *"dogs don't attack/wound in that way"* need to go out and see wounds from dog attacks before they speak.

But it was whilst I was corresponding with Mr Nash that he sent me a letter from a woman in Chester. It appears that the Dyfed/Powys Police also had a copy of this letter. During the summer of 1976 just outside of Conway.

Her husband and two daughters were with her, walking up the Synchant Pass when she spotted a 'good sized puma' sitting on a small mound, watching them and 'swishing' its tail about.

The woman yelled to her husband, but he couldn't see it from his position. In excitement, the woman ran with one of her daughters to get a closer look but before they reached it, the puma bounded off into the bracken and was soon lost as it went round a boulder. It was only then that the woman realised that what she was doing was dangerous and the two turned back.

They were then in for a REAL shock. Sitting calmly in the grass was a soft muted black/brown and light brown coloured cat with (possibly) spots. This cat was half the size of the puma (which was, by the way, a species with which the woman was very familiar, having spent many hours studying one at Chester Zoo).

However, she had great difficulties in identifying this second animal. She thought that it was possibly a lynx or a serval but was convined that the puma she had seen was 'protecting' it. Young pumas have spots which fade as they grow older - possibly what the woman had seen was not two different species of felid after all, but merely a female Puma and her kitten.

In recent years I have also received several reports of lynx from that part of Wales which suggests that the exotic felid population is thriving. No doubt someone COULD come up with Welsh UFO reports from that period, and they would claim that the animals described in this article were Alien Big Cats (ABC's) rather than exotic animals that some irresponsible idiot had seen fit to dump into the wild.

It should be remembered that a cat which had escaped or been released into that part of Wales would have abundant food sources for decades. When you add to that, the fact that anyone who reports a Big Cat is automatically made fun of by the newspapers there is no reason that they could not survive unmolested. I have also found over the years that losses from flocks of sheep are taken for granted when they roam free. Some drown and get carried away by rivers, some fall down old mine shafts/cliffs/gulleys and so forth and the corpses are never found. The last explanation that most farmers will come up with is that a big cat is killing their stock.

There is certainly a need for Fortean Zoologists to start logging livestock losses each year and to see whether they coincide with sightings of big cats and other out of place predators. It would be interesting to find out just how many sheep and other domestic animals are just listed as *"Missing - Cause Unknown"!*

EDITOR'S NOTE: Whereas many authorities are now beginning to give grudging credence to the idea that there are large cats of three or four different species at large in the British Countryside, the concept of wolverines, which are - after all the largest and most ferocious of the European mustelids being at large in the wilds (and not so wilds) of the United Kingdom, is a difficult concept for many people to grasp.

After a series of sightings in the Haldon Hills, we contacted Robin Khan, the Chief Forestry Officer who was very scathing about the concept of wolverines at loose anywhere in the United Kingdom. He told us that he was certain that all sightings were mididentifications and that the concept was essentially ridiculous.

A few years ago people in similar positions said the same thing to me about the existence of British Big Cats. The evidence for them is overwhelming, and indeed I have seen them on two occasions. In view of this, I am prepared to keep a cautiously open mind on the subject of British wolverines until further evidence, either for or against is produced!

The Tale of the Weird, Warbling Whatsit of the Westcountry

by Jan Scarff

On Tuesday 16th September 1997 at ten past two, BBC Radio Devon broadcast its fortnightly programme "Weird about the West", with resident host Janet Kipling and two presenters whose names will be only too familiar to readers of this magazine - Jonathan Downes and Graham Inglis. Their guest this week was Terry Hooper, who as well as being a regular contributor to 'Animals & Men' is also the founder of the national Exotic Animals Register. Their subject was strange and obscure animal sightings across the British Isles.

Listening to this broadcast were a husband and wife living in Clyst St Mary, a little village about five kilometres east of Exeter on the A 3052. For the previous three weeks they had been woken up each night at four minutes past two by a strange sound resembling the call of some exotic bird. At first they thought that it may have been revellers leaving the public house opposite, but after the fifth night they realised that this could not be the case. The lady decided to rig up her tape recorder in an attempt to capture the noise on cassette. She was not to be disappointed - at four minutes past two that morning, twenty one calls lasting approximately forty seconds were recorded, and she was convinced that they emanated from a tall poplar tree outside her bedroom window.

After hearing Terry Hooper give out his telephone number on the radio show she telephoned him to see if he was able to shed any light on the mystery. Once Terry had heard the tape he immediately telephoned Jon Downes at the Centre for Fortean Zoology. I was there with Jon at the time, and Terry told us about the strange bird calls. Jon and I discussed the matter and came up with a number of potential hypotheses:

* Peacocks from Crealy Park - a small tourist conservation centre about 1.25 km up the road from where the noise had been heard.

* Wild pheasants. This hypothesis seemed unlikely because it was the wrong time of the year for them to call.

* A cockerel. There is a farm adjacent to the property.

* An electronic intruder alarm at a utility sub-station in the locality.

Jon and I then telephoned the lady (who has asked to remain anonymous), who was very helpful and played us the tape recording down the 'phone line. My first reaction was that it was a seabird (possibly a cormorant) as they are commonly seen on the River Exe about four kms to the west of Clyst St Mary. But at four minutes past two in the morning? Jon tended to agree with my interpretation and we asked the lady for permission to investigate the events on the following Friday night. They lady was only too happy to give us her permision.

The following day we telephoned Crealy Park, explained our predicament and requested information about any peacocks, pheasants or other large birds that were in their collection. When they informed us that the only birds they kept were canaries, Jon and I were in fits of laughter at the thought of a canary the size of a turkey lurking in the wilds of Clyst St Mary. Between chuckles I thanked the lady from Crealy Park (who must have thought that we were completely bonkers) and hung up!

That evening, my son Lewis and I conducted a reconnaisance of the site. The house is by a brook and the nearest large body of moving water is the River Clyst two kilometres to the west. There are four Italian Poplar trees on the property itself with more across the road around the perimeter of the pub car park. There are two small copses of deciduous trees in the area - Cat Copse (500m to the north) and Crealy Copse (500m to the east). The lady showed us around her property and indicated the tree that she believed was the source of these mysterious bird calls. She also showed us the location of the security lighting and the passive infra-red detectors in the garden.

Next I spoke to their neighbour, who was a very pleasant ex-RAF chap about the mysterious noises that had been heard next door. He told me that he was up quite late each night working on his computer, and although he was not positive, he had a vague recollection of hearing something one night, but he wasn't able to swear to it!

He then showed me around his property and I gained permission for the members of the gained permission for the members of the investigation team to go onto his land on the following Friday night.

One of the landlords from the public house opposite then spoke to me, and in answer to my questions told me that he had never heard anything odd on any night in the previous three weeks (or, indeed at all), but added that as he always went to bed 'dog tired' after a couple of 'nightcaps', it wasn't really surprising.

Now it was time to get the investigation team together and to sort out our plan of action. I telephoned Dave Hopkins, an avid ornithologist and veteran of our Woodbury Common skywatches (see Goblin Universe #7) and Jon telephoned Alyson Diffey, a psychic and medium. The four of us arranged to meet at the pub on the friday night.

We met as arranged and after a pleasant dinner and a chat with the landlord, we stayed until closing time when the four of us took up our positions in the car park. Julian, the pub manager was so interested in what we were doing that he decided to join us. Alyson had previously 'scanned' the area and received a vision of a 'thought dragon' created by ancient druids to protect a nearby sacred site.

Time was ticking by, and so we set up our sound and video recording equipment, lights etc. At 01:30 we switched on the audio cassette recorder. At 01:50 on went the video camera and as it approached two o'clock everyone was feeling a bit apprehensive wondering what we were going to hear. The earlier jollity and sillyness had vanished as we commenced what, to us, was a very serious investigation.

By 02:05 we were all 'willing' the sound to occur, but when, by ten minutes past two nothing had happened disappointment had set in and we packed up about five minutes later, all going home for a well deserved rest!

Imagine our surprise the next morning. I telephoned the couple who had originally reported the noises and told them that we had been unable to record or video anything despite our array of (relatively) hi-tech equipment. His reply was that they had heard the sounds as usual, at 02:04 as his wife had been watching our activities through their bedroom window. The noise had been repeated twenty one times as usual, and in fact his wife had made tea for us all expecting us to join them in the house to tell them what we thought of their mysterious bird calls.

I immediately contacted Jon and was dispatched back to Clyst St Mary with two sealed, blank cassette tapes.

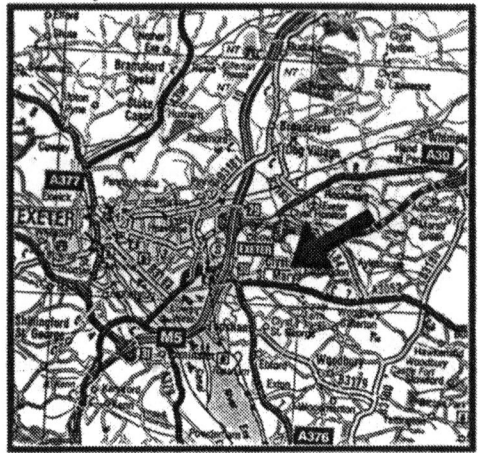

Clyst St.Mary

I spoke at length to the lady and gentleman about the apparent anomaly of the investigating team neither hearing or recording anything. They assured me again that the sounds had occurred as we were standing outside.

I left them with the two cassette tapes and requested that they record the noises for the next two nights (sunday and monday 22/3rd) starting the recording at 01:50 each night so that we were able to hear if anything else occured.

The anomalous noises occurred and were duly recorded giving Jon and me our first chance to listen to the whole thing from begining to end. When we heard the tapes, one thing was obvious - each 'call' was exactly the same length, and there was exactly the same time interval between 'calls', and, according to our informants, the calls occurred at exactly the same time each day.

As the gentleman of the house is in some not unconsiderable pain and discomfort due to receiving some unwanted radio-therapy I decided that it was up to us to get to the bottom of this intriguing mystery as soon as possible. Therefore, without further delay, I arranged that I would be present to record INSIDE the house on the night of Tuesday 23rd/Wednesday 24th.

I arrived there about half past nine that evening and

set up the video camera in the bedroom and the four track Tascam Porta-One mini-studio outside the bedroom window. This done, we retired to the lounge for coffee, biscuits and a long wait.

Earlier that day I had telephoned to ask whether they or any of their family had owned a novelty alarm clock of some sort. By this stage both Jon and I were of the opinion that the sounds appeared to have been electronically processed in some way. They assured me that to their knowledge no-one had ever brought anything of that sort into the house.

At midnight, their son Paul arrived to assist with the operation. We both listened to all three sets of recordings again and 'bashed' a few ideas around, but always coming back to the same conclusions that the sounds appeared to be 'electronically produced', and that the exact time of 02:04 HAD to be of some great significance.

01:30 came, and mindful of all the best traditions of psychic research I took the temperature in the corner of the room where the ladyassured me the sound had emanated from. I continued to do so at five minute intervals until 02:00 and it stayed at a constant 20°C.

At 01:50 both the mini-studio and the video camera were switched on. I then did a check with the window open to ensure that we had an exact point to work from on both tapes. then the battery on the video camera failed so that we had to repeat all the work that we had already done.

Finally 02:00 arrived, and you could actually taste the atmosphere in the house as our tension levels rose.

Then at 02:04 we heard the noise. Whilst Paul attempted to locate the source of the sound inside the room, I ran outside and set off all the powerful lights I had brought with me to illuminate the poplar tree.

By the time I got to the back window Paul was just opening it. I pushed it shut, listened at the glass and then opened it again. he sounds were certainly emanating from the bedroom and appeared to be coming from the dressing table.

I ran back inside to find Paul frantically pulling open every drawer from his mother's dressing table. *"The Bloody thing is moving from drawer to drawer"* he shouted excitedly. I too had the same thought and the hairs on the back of my neck began to rise. Twenty one calls and then the only sound was our breathing.

We looked at each other with Paul admitting that he'd earlier told his parents that he would only believe in this phenomenon if it was 'in his face', but that he couldn't have been any closer to it as it happened.

Just then, the voice of Paul's father came from the living room. *"Darling, are you wearing your talking watch?"* Paul's mum replied: *"No, its on my dressing table"*, whereupon her husband said, *"why not show it to the boys?"*

Paul's Dad had remembered a novelty 'talking watch' that he had received as a free gift with a pair of slippers. I, too had seen this watch and heard it 'talk', but the noises that it made were nothing like the sounds that we had heard. However, on closer examination we discovered that there was an alarm setting and it was set for 02:04AM. Coincidence? I think not!

We reset the alarm for 02:13 and waited. As Paul's mum was putting on the kettle the alarm went off and lo and behold, to much laughter, the case was solved.

I'm happy to say that the case was solved within a week of the Exeter Strange Phenomena Resarch Group/CFZ hearing of it, but the most important aspect of the whole affair is that, for the first time in over a month, two very helpful and caring people can now get a decent night's sleep!

My thanks to all involved in this investigation and I hope to get bought many a drink in the future for this story. In this case the truth wasn't 'out there' it was in the dressing table!

EDITOR'S NOTE: I have included this lengthy and amusing account, not only because it shows that we *can* do our job and because even though for a while, the case appeared paranormal in nature it began with a straightforward referral to the CFZ, but also because it sets the record straight. The day before the case was solved, Dominic Arkwright from BBC Radio 4's "Today" programme visited the CFZ to make a brief documentary about our work.

He followed us around for a day and included a brief piece about the strange bird noises of Clyst St. Mary. Now the truth (even though we didn't know it at the time) can be told.

It would be interesting to know how many other quasi fortean phenomena which have been reported in so many books and journals have equally prosaic explanations!

Wherefore art thou Nessie?

by Neil Arnold.

Maybe you're only ever a monster hunter if you've got money, suck up to the right people and have the time and money, but on the 28th and 29th of August 1996 I became a hunter, albeit for a very short time. I had realised an ambition of mine with a magickal trip that had been nothing but a dream for some fifteen years of my twenty-two year old life. For, I believe, that there is no such thing as an amateur monster hunter - let's face it, even children have seen mysterious beasts, and so armed with my trusty camcorder, my forty quid return ticket and a bag of clothes I set off.

Drumnadrochit here I come!

It was an awful twelve hour journey from my home in Kent, through London and north to the railway station at Inverness. A warning to anyone who plans a visit to Loch Ness. Drive!!!

The stench of other people's stale sweat, lack of sleep and grinding headaches may be all part of theexcitement but even Indiana Jones never had it this bad and by the time my coach arrived at Inverness I felt like a walking corpse. Nearer to Inverness the stunning scenery was reminiscent of the terrain of Canada, with its huge fir trees and gargantuan rocks. I tell you, that if a dinosaur had emerged from the forest I wouldn't have been particularly surprised because the foliage was so thick.

It seems to stare down upon you, draining you with its looming presence. The reindeer run wild and heavy mist blankets the peaks and I was still only six hours into the journey. Chewing on a Mars Bar at three in the morning was my only nourishment and there were times when I really thought that I wouldn't make it.

I think that if I'd walked all the way to Scotland then I might have appreciated the different atmospheres, but the hum of the coach became a deathly drone as the scenery flashed by my blurred vision. In a nowhere land, eight hundred miles away from home the sickness began to creep in. The vast landscapes had me in their grip, and I realised that the prize was to be at the end of the rainbow, and that I just couldn't give up.

Dark rings under my eyes welcomed the freshness of morning. The final part of the journey to Inverness was a winding route that left all other memories behind. All the other stops throughout the trip were a blur, completely unmemorable in their dreariness, but Inverness had a crispness about it and every small river that ran like a vein between the great grey rocks was like a taster for the great Loch itself. Yet, I was still a fair way from my destination and as 11:00 a.m drew near the rain lashed at the windows.

At around midday I stepped off the hell-bus carrying a heavy tent, a heavy rucksack and an even heavier head which was pounding like a bell. Yet, although the station was about as interesting as..... a station, already the tourist attractions were sparkling although I was still twenty miles from the village of Drumnadrochit. At the station, cardboard 'Nessie' signs littered the grey pavement and I boarded the waiting bus with a tingle in my hair and a paracetamol lodged half way down my throat.

The dream was becoming more and more real, but it still hadn't sunk in. But, I never would accept the fact that I was actually there until I saw the water of the great lake stretched out before me. The pull of the place lifted the haze from my head as we sped through the rural beauty of the curling lanes.

Drumnadrochit is a cheerful little place, and before I knew it, I was there and the sight that awaited me brought a lump to my throat. Through the drizzling rain a huge blanket of water stretched out before me like a velvet curtain.

As the rain poured down the water was still. Dead still. Only then did I realise that I was at one end of the enigmatic Loch and that everything else was a world away. Here was my time ... before my eyes and I'd done it all myself with the minimum of equipment. It was my obsession brought to life.

The thrill had been there throughout all the years of money worries and hope, and if it wasn't for the undoubted fact that if I'd been a 'real' monster-hunter I could have travelled anywhere that I wanted, for a few days Loch Ness and whatever was in it was MINE!

The books that I'd read and the pictures that I'd seen just hadn't prepared me for the sight or the thought that I'd finally acheived what I thought that

I had always thought that it would take too much to travel to a place that bears such an awesome legend - perhaps the greatest monster legend ever!

I set foot in the village. With my spirits high I scanned the vast horizon that consisted of unbelievable mountains, enveloping mists, green hills and truly staggering scenes that were a shock to the senses. The air throbbed with a heavy freshness and what seemed like a crackle of electricity spun an inspiring web of possibilities that were strewn before me.

The little shops that were dotted about everywhere were covered with the familiar imaged of 'Nessie' although I was soon to learn that the place lacked a serious attitude towards the 'beast' and that the place was really just a great exploitative, mickey-taking, money-making scheme.

To my left the Loch Ness Visitors Centre stood tall whilst to my right a winding road was littered with various gift shops selling everything and anything you could think of from 'Nessie' cups, models, fluffy-toys and keyrings to chocolate, pencils, rubbers, shirts and air fresheners all emblazoned with the image of cryptozoology's greatest icon.

Of course, there has to be an element of fun but the lack of serious books on the subject saddened me, and it appeared that the only real source of information was the Visitors Centre which basically told me what I'd already read in my own books. However, I was open-minded and too excited to worry.

I found my campsite, the tidy Borlum Farm which lies just a mile from Urquhart Castle and the Loch. With the weather clearing I set up camp, costing me only a couple of quid, and I munched on my rolls and biscuits before gathering my film equipment, and taking on the Loch itself.

The road which led down to the entrance of the ancient castle was long and heavy. the view was startling, across the hills and up at the steaming mountains. Scenery that didn't quite sink in, seeing as I usually live on a Council Estate in Kent! And then I caught a glimpse of something shiny beyond the fringe of trees. The cauldron of curiosity; the water that draws believer and sceptic alike.

As I made my way towards the Loch, the castle seemed like it was peering at me like some watcher of the waters standing like a guardian of that black abyss. This crumbling ruin has probably seen the mysterious beast more times than all the fortunate humans because the castle and the creature seem to exist hand in hand.

To appreciate the Loch I believe that you must become immersed for it is a dazzling sight. It is far larger than the photographs would have you believe and its grounds are just a pathway to the water's edge.

The castle casts an ominous shadow as you shuffle into its echoing confines which are illuminated wih red lighting effects.

The sound of bagpipes provide a background tune to your journey through the spindly row of trees towards the lake.

For miles and miles the darkness shimmers like a mirrored carpet and the ripples and shadows emerge from nowhere. The unseen undercurrents bubble below the surface and the blackness comes right in to the sheer banks and sloping sides of the lake which could hide many a cavern.

The sheer size of the place is frightening and a few times I found it hard to take it all in, for it engulfs the soul in a mystical essence.

Sure, I'd always believed in Nessie so coming here was always going to be stunning, but anyone who walks away not mesmerised might as well be dead because the scene is truly stunning and the setting is truly poetic.

The sides of the Loch are so steep that they are hostile terrain for any explorer, for even if the water weren't mysterious enough the rest of the place is equally inaccesible. Fishing boats were just dots on the horizon, and as I made my way towards the water's edge the mist embraced my senses. Just standing there was hypnotising.

Under the stark wing of the castle and on the edge of the chasm I wondered how anyone could just dismiss the stories of the Loch Ness Monster because, obviously, there was no way that this water could ever be truly explored. In its beauty

Loch Ness is incredibly powerful and draining. It is almost a fantasy setting but I knew that whatever lives there has to be a long way from pure fiction. I was determined, however, that there was no way that I was going to stand and watch the water and shout "Monster!" if I merely saw a duck or a floating log.

If I was to report a sighting it would have to be conclusive. If we are to count the shadows, the little bumps in the water, the shadows and the wakes as credible sightings then I saw 'Nessie' twice in under fifteen minutes but those days are long gone.

However, I maintain that 'yes, the Loch CAN play tricks on you' but Sue Blackmore must think that all the witnesses are plain stupid if they travel eight hundred miles to see a floating log and then to report it as a sighting of a real creature.

On the shore of the bay I stood and filmed. Flanked by two white rock faces and a pair of exceptionally stupid sceptics (who claimed that Nessie was a fish), I cupped the silken water in my hand, grasped a few stones and took in the aura of magic. I had never come here expecting the beast to rise before me, and I would be fooling myself if I just filmed dots in the diistance and claimed a sighting. My advice is, if you go monster-hunting don't film anything.

If you look at some of the photographs taken at the Loch they are nothing more than rubbish! Those showing long necks are still credible as are those which depict obese humps, but I feel that we should dismiss the pictures that show nothing more than waves and ripples. I filmed a large shadow that I could see through a haze of fog. Nowehere else on the Loch seemed to throw up a similar image, but I didn't become overly excited. I just laughed to myself and thought *"If that IS you, Nessie, then I need something better than this!"* The shadow seemed to stretch for about twenty feet, but as I zoomed in on it with my camera I realised that although it seemed out of character, anyone with any experience is aware of the images that this water can create.

I also filmed a classic wake that appeared to stem from nothing. There was no boat going by, no duck and no diver! Just an enormous wake .... that's all. I thought at the time that my two insignificant peices of footage would amaze some of the sad little people who seem determined to 'see' Nessie in every little ripple, but not all of us are content to 'create' our own monster images.

The familiar wake. What could have caused it, though?

People littered the water's edge. They appeared to be tourists and the sort of non-believers who basically flock to mock, but there you go. For me it was like a dream and almost a year later it still is. In all the time I was on the Lochside I never expected to see a serpent, a plesiosaur or even an otter but I knew that something was in that water and I felt honoured to be part of that place. The ghosts of aeons past still haunt those hills and each one has a story to tell people like me.

Some of the folk who sell their little oddments to the tourists are not believers and are just there to make money but there are a lot of 'normal' people at the Loch who have seen strange things but aren't interested in making money out of their experiences. Visitors to the area should take the waters with an open mind because whether you believe or not it is a beautiful place, and sometimes, although the beast does lie within the lake, you forget that you are a monster hunter and become a nature addict or a poet, even though mere words are hardly sufficient to describe the awe one feels at the breathtaking views.

The Loch simply cannot be conquered on on foot, although even if you are without transport, you can pay to sail on (or under) the waters. I, however simply found myself captivated and rooted to the spot. For a few days I was simply there to experience the things that I thought I would never see.

The shadow on the calm loch against the mountainous background. It is interesting to compare this shape with the classic McNab photograph showing the 'shape' against Urquhart Castle. McNab's 'shadow' was more protruding, though, whereas my 'shape' appeared to be certainly beneath the water.

'The Migo - Not yet explained?

by Nick Molloy

In February 1972, a Japanese newspaper report told of a strange water 'monster' alleged to inhabit Lake Dakataua on New Britain. New Britain is the largest island in the Bismark Archipelago, situated off the northeast coast of Papua New Guinea. The 'monster' is referred to locally as the migo (pronounced mee-go). Throughout this article, what is. or is alleged to be the migo will simply be referred to as M.

During January and February of 1994, a Japanese television crew succeeded in filming something upon the lake surface. So far, no portion of the film (at least to my knowledge) has been broadcast in the UK. A still from the film, was however, published in issue 102 of the *Fortean Times*.

I am a member of those select few who have been able to view and analyse the footage. What follows is an attempt to weigh up and assess some of the various interpretations of the footage to date.

The horseshoe shaped Lake Dakataua is separated from the ocean by high cliffs; as the crow flies - no more than a few hundred yards. Throughout the sixty minute plus documentary it was suggested by the film makers, that Lake Dakataua was connected to the ocean by underwater channels. Further, and leading on from this it was suggested that M gains entry to Lake Dakataua, from the oceans via these underwater channels, and is subsequently witnessed by locals.

To test out this theory exploratory dives were conducted in the ocean under the cliffs, separating Dakataua from the open sea. Further, whilst the whole of the documentary is presented in Japanese, a couple of interviews are conducted with English speakers (including Roy Mackal). One of those interviewed, informed us that a study would be undertaken to investigate whether the lake was tidal. Indeed, this man also stated that there have been sightings of M at the point where Dakataua is closest to the sea (he pointed this out on a map). The results of this investigation were either not given or presented in Japanese.

I never really want to see the mystery solved. It has been around too long for that, but sometimes I wish that some droppings or maybe an egg could be found. Mind you, a huge cavern has recently been discovered in the lake, and if it has taken this long to discover a giant cave, then how long will it take to catch one beastie?

The truth is that the scenery and the legend itself are like a brother and sister. One cannot live without the other. Who knows? Other undiscovered animals may live within the shadows, and I have to admit that it wouldn't surprise me if they did - considering the lay out of the place and the surrounding areas. If you do go to the lake you have to go with your heart open, and not with the idea of denigrating the legend.

Nessie hunter Steve Feltham has been at the lochside for many yhears now (unfortunately I never got around to finding him), but HE appreciates the place for itself and I was lucky enough to pick up one of his cute little Nessie figurines. But despite his frustration in not having gained that conclusive piece of evidence he is still there because he is following his dream, but he is not prepared to desparately try and see things that aren't there!

My biggest disappointment of the venture was the attitude of the locals and the shops nearby. Of course, they have to make their money, but they could try to treat the matter with a little more respect.

Next time I go, I am determined to have one of the boat trips as well as a tour in a bus around the shores. Of course my own sighting would be wonderful but if it ever did occur, I really don't know how I'd react.

It was just my luck that only a few days after my return a party of school-children saw a classic 'upturned boat' shape in the water AND some video was taken showing a strange frothing of the water, although as I have said already, THAT type of evidence is too fragile now.

The memory that I have of those wonderful few days is as strong as ever and I strongly reccomend the place in general to anyone who is considering a visit. If, however you are a monster-hunter then the excitement is even more intense.

I always thought that Loch Ness was too far away, but then again so are most of your dreams unless you follow them and find them!

Two main pieces of alleged M footage are shown during the documentary. One of these is undoubtedly taken at sea. This is validated by two things. The first is demonstrated by the lack of land visible in the film. Lake Dakataua is not big enough to produce this effect. The second is demonstrable by the crude map drawn on screen immediately preceeding the footage. This confirms beyond reason that the footage was not taken on Lake Dakataua, but taken instead on the neighbouring ocean.

The footage itself is less than perfect, no more than a second or two. When viewing a less than perfect copy of the film, Karl Shuker stated that:

"... it revealed what appeared to be a section of the body rapidly emerging from the water in a vertical upsurge and bearing two slender projections resembling dorsal fins or spikes before emerging again - followed immediately by the vertical emergence of what might have been a tail, with two horizontal whale-like flukes. It was clear that the object being filmed was not only animated but animate - alive".

When watching my less than perfect copy of the film, I concurred totally with Shuker's conclusions. I was convinced that the short piece of film concluded with the submergence of a large forked tail, something not seen in modern whales and more commonly associated with primitive whales or Archaeocetes. Despite the footage not being filmed on Lake Dakataua itself, it appeared to show a completely unknown animal.

EDITOR'S NOTE: The fault for both these copies must rest with me. At the time my video copying facilities were less than perfect (in fact they are not much better now - something which I intend to rectify in the short term future). Even the 'master copy' was less than clear having been transferred from the Japanese/US NTSC format to the British PAL format.

After viewing the master copy, however, it was clear that this conclusion was in error. Viewing a clearer copy of the film, what I had first interpreted as a single large animal became three silhouetted dolphins, very close together, rising, then diving in tandem with each other. Even on the master copy it was still difficult to distinguish between the animals and the first impression still led overwhelmingly to my original conclusion. It was only with carefully repeated viewings that the truth became apparent.

However, it should be stressed that diagrammatic representations of this sequence appearing with articles penned by Darren Naish in the CFZ Yearbook 1997 and The Cryptozoology Review are highly schematic. They give an example of how rising dolphins can create false impressions of larger animals. The actual footage was capable of misleading even experienced observers.

The first section of film purported to show M is of known animals. What then of the second piece of film alleged to show M?

This piece of footage was much longer and comprised a good few minutes. It was, however, shot at a considerable distance. I would estimate at least half a mile. The zoom of the camera helped to eat up this distance, but nevertheless, the resulting footage is distant and inconclusive.

My initial impression was that the footage showed a large creature with many crocodilian features. Two different portions of M were clearly visible. The front portion (head) tapered off at the front, into what looked like a rather elongated snout. A series of apparent bony ridges projected at the back of the head.

For the majority of the film it is difficult to discern the body of M. Water appeared to fill the void between the front and back portions of M. The tail of M also appeared crocodilian in feature. Numerous spikes or ridges appeared to run along the length of the back section of M.

However, if the tail appeared crocodilian in nature, it did not correspond to normal crocodilian behaviour. Throughout the footage M swims slowly across our view. I would estimate the speed of M at no more than three or four knots. Its method of propulsion certainly seems tail based. During one piece of footage it is clear that the tail propels M through the water by a side to side lateral motion of the tail (this is very well demonstrated on the master copy if you fast-forward this section). However, during other portions of this film, M appears to propel itself by vertical undulations of the tail. The back portion clearly submerges and re-appears on several occasions. This behaviour is not consistent with that of a crocodile. These vertical undulations are a strong argument for ruling out any known crocodile identities.

The length of the M shown is also a contentious issue. Mackal produced a length of 33 feet. Naish expressed that a length of 20 feet would appear more consistent. To establish a precise length would be very difficult without detailed analysis and calculations giving known constants.

Estimating distances on watery backgrounds is a

notoriously difficult procedure, given the lack of immediate background objects for comparison. During the film out only comparative items are birds that fly and land on the lake's surface and the trees on the far shore in the background.

In his book "The Loch Ness Monster - The Evidence" (Aquarian Press 1986), Steuart Cambell has suggested techniques for length and height analysis given known parameters. Constantly challenged and changing, such analyses bring to mind the phrase "Lies, damned lies, and statistics".

Moreover, and either way, M is of considerable dimensions. It is much bigger than the water birds that are close by and it would not seem unreasonable that the length is somewhere between the estimates of Naish and Mackal.

It has recently been suggested in some quarters that the creature filmed on the lake is a hoax. So the suggestion goes, the Japanese film crew have towed a very large (but dead) crocodile across our field of vision. This is supposedly supported by an alleged wake in front of the creature and the claim that a live crocodile will always react to a wake that crosses its path.

EDITOR'S NOTE: As far as I am aware the only person who actually made this claim was Richard Askew, ex of the British Earth and Aerial Mysteries Society (BEAMS) and now an independent researcher who claimed., one evening about a year ago, to Graham Inglis and myself, and later (I believe) to Darren Naish, that not only was the said crocodile dead but that it was lifted out of the water by some inflated flotation device.

I have not been overly convinced by these particular hoax assertions. After discussions the piece of film that shows the wake was pointed out to me (on a copy of the master tape).

After repeated of this film, I find myself unable to concur that the film actually shows a wake. What was pointed out to me as a possible wake, I had assumed throughout, was either heat distortion arising from the lake's surface, or possibly a discrepancy in the actual quality of the tape.

After a re-assesment of the film, my original diagnosis of heat haze/distortion still holds.

Further, if the creature is indeed dead, there remains the question of how the rear portion appears to propel the front portion undulates both vertically and horizontally during the film.

Inconclusion therefore, what have we got? The usual cryptozoological arguments have risen their proverbial heads. A hoax, versus a known animal, versus a true cryptozoological anomaly. Although it should not be ruled out, I feel that a hoax is unlikely. If a hoax has been perpetrated it is certainly one of an elaborate nature. Some sort of mechanical model would seem to be the only possible candidate which could account for M's apparent tail undulations. The film was not consistent with a dead animmal being towed. However, an interesting omission from the documentary was the disappearance of the creature. Presumably it sank back into the depths of Lake Dakataua. This was not shown and questions have to be asked why?

If the film is of an unknown animal, the resulting footage is too indistinct and fails to provide us with enough detail to conclude a definitive identity. We can only speculate on a range of possibilities. If on the other hand the film shows a known animal, it is still of considerable zoological significance.

A 1974 study of Lake Dakataua by a wildlife researcher showed that the lake was devoid of life. We are still left, therefore, with an out-of-place animal, albeit a known one. Identifying the creature, however, is speculative given the problems outlined above.

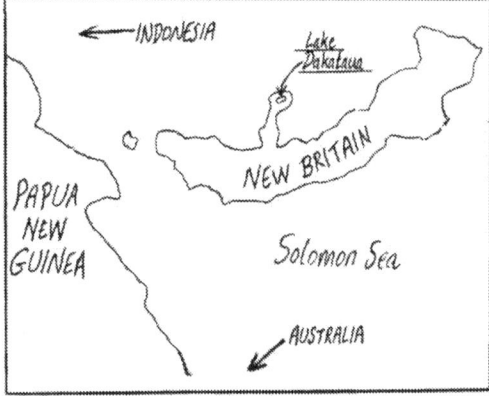

It should be further stressed that the documentary was wholly in Japanese. It may be possible that some important data was ommitted through a lack of translation. An example of this is the claim that Japanese scientists analysed the film and concluded that M was fifteen metres long. If this could be verified, the known crocodile identity could be immediately ruled out.

It has also been suggested to me that the following hoax scenario could have been perpetrated. That is whereby, a known animal, such as a crocodile, could have been deliberately transported and released into Lake Dakataua and then filmed deliberately to look ambiguous. However, a known crocodile could not produce the vertical undulations shown by the tail.

Naish has suggested that the picture enhancements led to the picture becoming 'pixelised' and this could perhaps account for the vertical tail undulations.

Yet, it should be stressed that the vertical undulations were also visible in parts of the film that had not been enhanced and were therefore not pixelised. Further, although the enhancement may pixelise the static image, it should not cause the moving image to 'jump' thus giving the impression of vertical tail undulation.

Finally, it is also alleged that the same Japanese film crew captured a smaller M on film in August 1994.

The film allegedly shows a smaller M but much closer to the camera. Given the large degree of indeterminates surrounding the first Migo film, the second film may go a long way to revealing the true nature of Dakataua's mysterious denizen(s). It may well be that the February 1994 film shows a known animal.

Thus far however, I don't think the case for an unknown animal has been satisfactorily dismissed.

EDITOR'S NOTE: I have been told, although I have not seen it in print, that certain U.S authorities on contemporary cryptozoology are now alleging that the first Migo film does not, as Darren Naish has suggested, show a specimen of the Indopacific Crocodile but in fact shows two animals of this species in the process of mating whilst a third crocodile (presumably of the same species) is following them in a display of what the smuttier members of the CFZ core team have described as a display of reptilian voyeurism.

The same sources have also claimed that the 1994 film (to which Nick Molloy refers in his article) is clearly of an Indopacific Crocodile which is why it has not been widely distributed.

I agree with Nick that the hypothesis that the film shows a dead crocodile being towed behind an unseen boat ishighly unlikely for the reasons he has given.

I also cannot bring myself to believe in the hypothesis of reptilian troilism, for the simple reason that like Darren Naish, and Nick Molloy I am convinced that the animal in the film (whatever it is) is a single living creature.

I cannot comment on the second Migo film because I have not yet seen it, so I have to agree with Nick, that in words stolen from the closing sequences of quite a few episodes of 'The X Files' the case still remains open...

Much of the present confusion surrounding the 'Migo' footage is, as already stated our fault in that the video copies made available to UK researchers were less than perfect and came from us. I must stress, however, that as regular readers of this magazine will be aware, I was an enthusiastic believer in the veracity of the film, and remained so until one evening in late 1995 when I watched it together with Darren Naish and he began to explain his theories to me.

Whereas I think that there is no doubt that the creature is crocodillan, I agree with Nick that the matter does, indeed warrant a degree of further investigation, if only because (at worst) it will identify a new population of the Indopacific Crocodile on the island of New Britain. At best, the truth may be far more exciting!

"Don't look back in anger"
sang whichever of the Gallagher
brothers it was.
"I won't"
said Clinton Keeling, as he presented us with:

CLINTON'S COGITATIONS ON ISSUE 14

Henceforth it will be my pleasure to make comments - for what they are, or are not, worth - on items in the current issue of *'Animals & Men'*. As a means of introduction, I'm a professional zoologist and ex-zoological garden curator who is primarily concerned with the education side of his work, and wild animal husbandry. For many years I was an enthusiastic, and optimistic cryptozoologist - but no longer, I'm very much afraid.

Just a couple of comments on items that appeared in issue 13. I was amused to read on page five that a *"Mountain Lion"* had been reported *"by various witnesses living wild along a railway track in West London"*. Wild witnesses indeed! On page twenty-nine Tom Anderson writes about *"the mechanics of impaling a fish on a hook..."* and so reminds me of Dr Johnson's succinct description of angling per se - *"a line with a fish on one end and a fool on the other"*.

Anyway, let's look at issue 14, which like all of *"Animals & Men"* provides much scope for a rich source/sauce of cogitation.

I thought the cartoon on page three absolutely brilliant, although I can't quite make up my mind whether it's an Ameranthropoid trying to look like our revered editor - or vice versa!

Page four - The Jungle Cat. This is an extremely adaptable species, as it's habitat ranges from riverine grassland (remember 'Reed Cat' is an alternative name), up to 6,000 feet above sea level in the Himalayas, so certainly our climate would cause it little or no inconvenience. I noted with surprise that in the same paragraph that confounded Kellas Cat was mentioned - surprise that a magazine of this calibre should seemingly take it seriously. Let it be stated here and now, once and for all, that the KC is a large feral domestic cat PERHAPS with an admixture of *F.sylvestris* blood in it.

The mention of the Bee Eaters occasionally nesting in this country (page five) reminds me of the famous occasion in 1953 when, as far as is known for the first time, this species succesfully nested here in Sussex. Why a pair should have decided, aparrently apropos of nothing, to go about their domestic duties so far from home is a mystery, but do it they did. I frequently have to travel between my home in Surrey to Worthing, and as I pass a large sand pit near the village of Washington I always mentally nod to it and think of the event now so long ago.

At long last I have been vindicated - and I'm not being wise after the event either (although I can be pretty good at this) - but I have long maintained that the infamous "Beast of le Gevaudan" (page seven) was a Hyena, as the contemporary description, along with details of its dentition, clearly pointed to this. This animal caused such terror over such a wide area that the French government sent an official military mission down to the area on a 'search and destroy' operation. Incidentally, I was astounded to learn that, so far, the species hasn't been decided upon, as the three species of Hyena are all so different from each other I should have thought identification would have taken about three seconds flat. I wonder, I wonder, dare I suggest that it might have been the striped species, as this is the one least inclined to scavenge and more likely to kill its prey!

EDITOR'S NOTE: Full marks Clin! Karl Shuker confirmed in the October issue of *Fortean Times* that the skin was indeed that of a striped hyena. I have always been of the suspicion that whereas (as has been proved to be the case) the Le Gevaudan creature had its basis in a real live animal, that not even the most voracious hyena could have killed and eaten as many children as this one is said to have done. I have a sneaking suspicion, therefore, that there was another predator at large in that part of France at the same time ... probably a bipedal one with disturbing paedophilic and homicidal tsstes.

No, sorry - the story of the Duke of Wellington and the Sparrow Hawk *(Accipiter nisus)* - on page nine - has contrived to get a bit garbled. When the Crystal Palace was being glazed in 1851 it was discovered that many house sparrows were living in the large trees within the building, so Queen Victoria, fearing that they might turn out to be a nuisance, asked the Duke, (on whom she always relied in times of crisis), what ought to be done. *"Sparrow Hawks, Ma'am"*, was his damned silly and totally unpractical reply.

Regarding the 'Political Correctness Police' mentioned by "Mungo Park" on page eleven, this brings up an interesting instance of double standards. Political correctness is an utter and absolute embargo on the freedom of speech - yet 'freedom of speech' was one of the favourite cliches of the red raggers who instituted P.C in the first place. Surely they can't have it both ways - or perhaps they can, as they are always right. I always say, by the way, that I am an oasis in a desert of political correctness. Or if you want to put it another way, I'm so reactionary, I'm nearly radical!

Talking about MUNGO PARK, it isn't generailly known that this intrepid Scots physician/explorer travelled extensively through West Africa - *"that Turkish Bath provided by nature"*, as it has aptly been described - clad in frock coat and top hat, just as though he'd been strolling down Edinburgh's Claremont Street.

The article "The Dragons of Yorkshire" (pages fourteen to twenty) was interesting and well, indeed painstakingly researched - BUT has such material really a place in the all too few pages of 'Animals & Men'? Eighty percent of it dealt with mythology, or at least as I saw it, it did - and this is a very different subject from Cryptozoology!

EDITOR'S NOTE: Whils it is true that this magazine is most usually devoted to cryptozoology it has always been editorial policy to include elements of forteana, zoomythology (like Richard Freeman's article on dragons), fringe zoology, and indeed anything else that takes our fancy. However the whole editorial team would like to shout a resounding 'hoorah' in support of Clin's views about so-called Political Correctness. The editorial team includes some of the least politically correct people one could ever hope to meet!

"When the wolf was at large up north" (page twenty-two), clearly refers to the famous Allendale Wolf that caused so much damage to farmstock before it was accidentally killed on a railway line early this century. There are two fascinating points here:

a. The animal escaped from the zoological garden attatched to Shotley Junction railway station in County Durham; believe it or do the other thing, but the Station Master was a keen naturalist who kept an extensive collection of animals, and as passengers often had to wait for quite long periods at the junction, he allowed them to see his stock to alleviate their boredom! (Incidentally, and I feel extremely strongly about this, note well that I say RAILWAY station; it I had my way calling such a place a TRAIN station would be a punishable offence).

b. Here is a real mystery. When the said wolf was killed it was realised that it couldn't have been the one that got out a few months earlier, as that had been an adult animal, whereas this one was considerably younger. So work that one out if you can....

On page twenty-seven Gray's "Mammalian Hybrids" is mentioned - and oh, how disappointing that book is! It could be one of world importance, bearing in mind its subject, and one that I could well use and refer to and acknowledge in my work, but most of its entries are so tantalisingly brief even vague in some cases - that I've found it to be of little real value.

Now, what is one to make of "Big Cat Reports from Scotland" - seven solid columns of them taking up pages twenty-eight to thirty-one? Simply, very simply indeed, from the evidence available, that the Galloway region of Scotland is heaving with 'em. QED. So let's go back to page four where there are other such sightings, which combine to make me reiterate my perpetual cry of "Why, just why, do I never see the big cats with which the country is seemingly infested?"

Y'know, looking back, I can see that there are a couple of points earlier on that I'd overlooked, so here they are...

The mention of albinos on page six reminds me that in the excellent little Mansfield Museum in that not particularly congenial little town's Leeming Street, there's a section devoted to mounted albinos, including such unlikely species as Green Woodpecker and Waterhen, and there's another good section along the same lines at the superb Rothschild's Museum at Tring in Hertfordshire. Talking of Albinos, here's a good talking point to bring up at your next dinner party. Question - which English King was an albino? Answer - Edward the Confessor, the founder of Westminster Abbey!

On the same page there's some typical waffle from Leeds University about Gerbils (I PRESUME they mean the Mongolian Gerbil), pining and displaying grief when taken from their mates. Now, I began keeping that species back in 1968 (how the shadows are closing in...) so I suppose I was one of the first private individuals in the country to own them, and the first private individuals in the country to own them, and I can categorically state, as a firm believer in keeping notes and records, that I have never noted unhappiness in these animals under such conditions. Query: can you tell if a Gerbil is showing "*symptoms of Loss*"? -because I can't. Far more interesting is the fact that the Mongolian Gerbil was discovered by the famous Pere David, who also "discovered" (if that's the right term in this context) the Giant Panda and the Deer named after him. To go back to my mention of the Himalayas when talking about the Jungle Cat (you'll soon see the connection when I mention Mount Everest) - plus my firm conviction (which, I know, irritates some people) that there is no knowledge that is not valuable. Ask almost anyone you like the height of that said mountain, and the chances are they'll reply "*29,000 feet*" - but it ain't. In fact, although its summit is admittedly that height above sea-level, its a modest 7,000 feet as a mountain, which just happens to be sited on high ground.

Here's to our next merry meeting...

As is his custom Neil Nixon presents another selection from that wholly mythical compilation album:

NOW THAT'S WHAT I CALL CRYPTO

Stan Freburg meets the Abominable Snowman

Around about the time that most of today's middle aged monster fans were taking a torch under the bedclothes, and most of us just basically weren't, comedian, writer and purveyor of musical slaughter Stan Freburg was plying his trade on radio and on record. Freburg's comedy was a fast moving slapstick; mainstream in its reference points but still strong on vivid images and solid ideas. As part of his radio series Stan 'interviewed' the Abominable Snowman, discovering in the process that his subject was not a "*gentleman*" or a "*creature*" but "*a little bit of each actually*". He wore size 23 sneakers and stated that his work was to "*terrorise the mountain climbers that come up here. That is my trade and I'm proud of it*". Declaring himself to be an impressive ten and a half feet tall, the Snowman told Stan:

"If you think I'm tall, you should meet my brother, he jumped centre for Abominable State".

As comedy it works well, thanks mainly to Freburg's assured delivery and timing. As a sneak glimpse into crypto imagery it shows how far the world - and especially the USA - has come in forty years. In Freburg's world middle America laughed at a creature that was essentially a mutated middle American. These days Mulder and Scully struggle to understand layers of reality and barely hold on to their American ideals in the process. Sighted more rarely than the Thorganby Lion Freburg's gem did appear on the Capitol/EMI compilation LP : "*Best of the Stan Freburg Shows*".

NORTH OF THE BORDER
by Tom "wot no insult?" Anderson

It is to be hoped that the recent spate of animal mutilations nationwide has peaked, reports of little girl's ponies having nails hammered into them having receded since the spring. This decline has been variously reported as being due to the limited number of 'neanderthals' likely to engage in the passtime, the effects of the rural "Farm Watch" scheme and the perpetrators having such a limited attention span, most were one-offs. We now have an upsurge in rabbit strangling and the rapid growth of setting fire to and shooting cats with airguns, crossbows and ball bearings. This is a national problem, but Scotland seems to have a localised variation on the theme.

From the borders to the highlands, since the start of the year, there has been an epidemic of cats, (usually kittens) found crucified to trees, barn doors etc. There is no evidence of ritual or ceremonial, only the act itself, a life-form graffitti unique up here in its volume at least. As the "Shit for Brains" subculture seems to be targetting domestic animals, it could prove useful to hypothesise on future victims and take appropriate pre-emptive action. Ostrich farmers would seem to be safe, but cage birds could be taught to shriek "*Sod Off Pinhead*" and ring their bell at some volume. Plastic herons at garden ponds could be replaced by live maribous and pit vipers utilised as draught excluders behind closed doors. Any miscreant caught with a Stanley knife after dark would be tethered face down across the buttocks of a cow in oestrus. This should prevent any possible reci*divism*!

LETTERS TO THE EDITOR

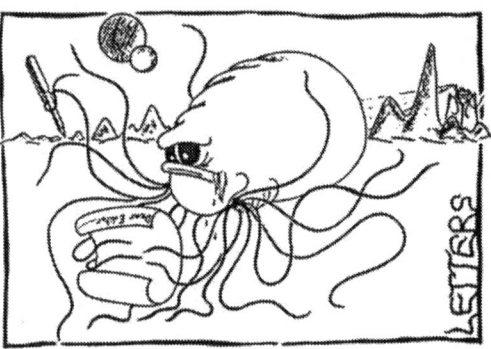

We welcome letters on any subject of interest to readers of this magazine, although we reserve the right to edit and omit where appropriate (or in the case of Tom Anderson, we reserve the right to add a string of completely unwarranted slurs on his character and dodgy innuendoes). Opinions expressed are those of the individual writer and not necessarily those of the Editor or his band of merrie men (and women). Every effort has been made not to infringe anyone's copyright, and libellous comments are always removed (unless they are funny enough). As the Editorial team haven't got any money anyway, even if we have libelled you its probably useless suing us!

INCONSISTENT DRAGON

Dear Jonathan,

Whilst reading the article "The Dragons of Yorkshire" by Richard Freeman in *Animals & Men* issue 14, I noticed an inconsistency.

In the section entitled 'Sea Dragons', the account of Filey coastguard Wilkinson Herbert states that the neck was *"rearing up three feet high"*. Later the neck is described as being a *"yard around"*.

It occurred to me that these would be strange measurements for the neck of a supposed sea-serpent.

I looked up the account in another publication *"There are Giants in the Sea"* by Michael Bright (Robson Books 1989). In his version of the coastguard's account, the creature's neck reared up eight feet high, not the three feet that Richard Freeman stated.

As neither of these authors gave any reference of where these accounts came from I was initially unable to check which of the neck measurements was correct.

I suspected that the eight foot neck was more likely for a creature with a neck circumpherence of three feet.

I recently obtained a copy of Bernard Heuvelmans's *"In the Wake of the Sea Serpents"* (Rupert Hart Davis 1968), which confirmed that the eight foot neck measurement was the correct one, and gave the information that the report was originally published in the *Daily Telegraph* on the 1st March 1934.

Yours sincerely,

Brian J. Godwin
Cumbria.

STOP HORSING ABOUT

Dear Jon,

Whilst thoroughly enjoying *Animals & Men* number 14, I feel that I must comment on the picture of the zebra/donkey cross that was reproduced on page five. Although from its appearance there is little doubt about its parentage, it is rather unfortunate that the Newsfile comment states that it looks like a Quagga.

Those readers who have studied the quagga, or who have read David Barnaby's book *"Quaggas and Other Zebras"* (Bassett Publications 1996), a book which I had the pleasure of editing, will realise that the animal, other than being an Equid has not a lot of similarity with the Quagga!

The Quagga had more definite body stripes starting at the spine, but usually failing to complete the

Letters

circuitous route around the body; the legs though were unstriped. One of Reinhold Rau's major concerns in the Quagga 'rebreeding' project is to get rid of the stripes on the leg, and to replace them with a brown background colour. Because of the stripes on the legs alone, the hybrid currently residing in Leominster would not be allowed anywhere near the Quagga Project zebras!

'Animals & Men' is not alone in this mistake though, the Daily Telegraph on the 25th September had an article on the Quagga Project and included a photograph of the same zebra/donkey hybrid, or one of it's contemporaries, suggesting that it was one of the Quagga Project Zebras.

With best wishes,

Chris Moiser,
Plymouth.

EDITOR'S NOTE: Ooops! Chris is, of course, quite right. As illustration compare the 19th Century engraving of the quagga used to illustrate Chris's article in *Animals & Men* several years ago with the contentious picture from the *Daily Telegraph* supplied by those jolly nice chaps at the Copyright Liberation Front.

Animals & Men # 15

Letters

FEET OF CLAY?

Dear Jon,

So Mark Chorvinsky thinks that 'Doc' Shiels faked the Morgawr and Nessie photos with modelling clay. If this is the case then it surely means that long necked lake monsters are genuine prehistoric survivors .... from the 'Plasticene Era'.

By the way, re. A&M 13. I'm not into Anne Rice and I'm not particularly into vampires except for real ones (a la Highgate and St. Leonards). Anyhow, dragons are much more interesting. But if you know any goth babes who want to bite my neck that's a different story!

Love

Richard Freeman,
Underneath the arches at Kings Cross.

ER .... ANOTHER POEM.

'Neath a moon of sickly pallor
Upstream of the fabled Exe
A coven vile of beings strange
Whose nightly deeds doth man perplex,

Eldritch forces drive them onwards
Females with revealing shifts on
Males, blank-eyed and incoherent
torn from the Arms of Clifton.

No incubas or chupacabras
no goat legged god on grisly throne
can impede their grim procession
Obesience at the Court of Holne.

From its battlements a night-bird screeches
the hour being twelve and ten,
their knock answered by roar so bestial
"Our Lord", they cry, "He's pissed again!"

Aflush with Tescos finest vintage
his mind with obscure concepts grapples
mutterings of "He's finally lost it"
seduced by ill fermented apples

Adopting then benign composure
with unctuous charm he proffers greetings
Hors d'oeuvres (one bag crisps, two pickles),
- a dreaded Editorial Meeting.

From scholarly to downright weird
a range of talents they presented
but almost without exception
all present there were half demented.

From Drake's old port an academic
at a loss to understand,
his presence at this Bacchanalia
instead of seiving Gambian sand.

A delegate from far off Gaul,
with piscine theories thought bizarre,
our forefathers, it seems were kippers
Yes, you've guessed it - F.de Sarre.

One of the few who've seen a mermaid,
though others search both far and frantic,
but never have brought home the bacon
like the enigmatic poet Scandic.

Garey Larsen I am not,
in fact I have produced some bummers,
how galling then to be upstaged,
by a youth of so few summers.

White-faced, the man from Guildford rises,
a traditional mammal, fish and fowl man.
Quoth he: "my life has been for naught,
last night I saw that bloody owlman".

Stunned by this news, the Ed. goes limp,
his fag end falls on cloak with stars on,
with Buckfast Sally puts it out,
there's no arson about with Parsons.

A late arrival fresh from Bristol,
where the MAFF consider him a snooper,
to spare his blushes I'll not name him,
but it rhymes with jackal pooper-scooper.

But where's the rest? I hear you wonder,
to be precise the famed Doc Shuker
then spake the Ed. "he 'phoned last night,
he lies abed with a Verruca".

Eyebrows raised, Ms. Williams queried,
"A new girlfriend? And no-one's seen her?"
The Ed. reposties "with THAT surname,
she must be a Russian ballerina".

While most remain stunned by all this twaddle,
a dark clad figure intercedes,
all clench their teeth, their buttocks, wallets,
awaiting the onslaught from Leeds.

He put his case and none demurred,
to finance his trip Antipodean,
to find the fabled Thylacine,
for was he not the Dick called Freeman?

Blearily the Ed. then focussed
and uttered as a born leader
"You've got more chance wiv the lottery mate,

Animals & Men # 15 — Reviews

I've got to upgrade my Amiga".

Apologies to Graham, Richard, Lisa, Tina,
Alyson and sundry coelacanths,
but the dreaded deadline is approaching,
and another one of JD's rants.

Endangered species stand alone,
quagga, dodo, beaked chelonian,
luckily no shortage yet,
of dire poets Caledonian.

To those not mentioned, the reason is one of the following:

You've already been vilified,
You don't rhyme,
Ypu are a hologram and automatically disqualified,
I've never really liked you,
All your biographical details were supplied by Jon and reflect badly on your literary abilities and personal hygeine.

You know who you are.

See you in court,

Tom Anderson,
Aberdeen.

EDITOR'S NOTE: At the risk of this being described by Aberdeen's Mr Entertainment as *"Another one of J.D.'s rants"* I have to apologise to him for revealing in print that he was the author of the poem in the last issue when he signed the poem 'Anon'.

However, as everyone he libelled last time thought that it was vastly amusing no harm was done and he has admitted liability for the above musings.

By the way, the libels perpetrated above are totally wrong. The Editorial Meeting that has produced this issue comprised purely of me and Graham, (and sundry other folk down the telephone), we were drinking wine not cider, and we were listening to Hawkwind on Samhain 1997. It seems appropriate.

For the third time this issue I am running a competition. I am offering a free lifetime's subscription to Animals & Men to the author of the most obscene (but printable) limerick that anyone can compose about ...... (you've guessed it) TOM ANDERSON!

BOOK REVIEWS

By Jonathan Downes unless otherwise stated

FAIRIES - Real Encounters with Little People by Janet Bord (Michael O'Mara Books, London, 1997) ISBN 1-85479-698-4
Indexed. £16.99

Up the airy mountain,
Down the Rushy Glen,
We daren't go a hunting,
For fear of Little men. (William Allingham)

Most Fortean Zoologists would probably regard fairies as being outside their zone of investigation. Nevertheless tiny hominoids have been believed in and seen by mmany 'normal' people, worldwide, over the centuries. I don't mean pretty picturebook sprites of the Cottingley kind, but the more potent 'Little People' of folk tradition. As Kipling's Puck was heard to say:

"Can you wonder that the People of the Hills don't care to be confused with that painty-winged, wand-waving, sugar-and-shake-your-head set of imposters?"

If there is a Bigfoot, then why not a Smallfooot? Size seems to matter so much to some chaps. It is as if the existence of the Mandrill should cancel out that of the Marmoset. Some 'Little People' could actually BE little people; the majority, however, must be viewed as supernatural beings.

Fairies (goblins, elves, brownies, pixies, leprechauns, spriggans and so on) - objective or otherwise - are still encountered. In recent years some ufologists have pointed out the often striking similarities between fairy lore and reports of UFO entities ('little green men' for instance). Abduction by aliens can be seen as a modern variation on the trip-to-Fairyland theme; an 'otherworld' experience.

Janet Bord's book covers many aspects of a complex and fascinating subject which she treats in a soundly Fortean way, building on the work of Evans Wentz, Lewis Spence, Katherine Briggs, and others. Janet knows that tales of fairies should not always be regarded as 'mere' fairytales.

I recommend this book as an excellent primer for those readers who are not just interested in mystery animals but also mysterious little men. Tony 'Doc' Shiels.

Borderlands - The Ultimate Exploration of the Unknown by Mike Dash (Heinemann £16.99 501pp)

I'm always in somewhat of a dilemma when I am sent something to review which has been written or recorded by a mate. What do you do if it turns out to be dreadful? This has happened on a number of occasions, especially with CDs, and I usually end up being non-committal and evasive about the product in an attempt to get it out of the way as quickly as possible. Mike Dash is a mate of mine so I was, in theory at least, faced with the usual dilemma. I needn't have worried, however, because this book is superb. *The Borderlands* referred to in the title are very similar to *The Outer Edge* described by John Keel, and in this exhaustive work, Dash deals with a wide range of fortean and paranormal phenomena with wit, style and aplomb. The sections on UFOs, Abductions and Cryptozoology (which cites yours truly in the references), are particularly interesting, and contain much material which is new to me. This is a concise and erudite overview of fortean research over the past 50 years or more, and as such cannot be reccomended highly enough.

Remember Belle Vue by C.H.Keeling (Clam Publications, 13 Pound Place, Shalford, Nr Guildford, Surrey) ISBN 1-874795-15-0

Clinton Keeling is a remarkable man, and I ain't just saying that because he is a regular contributor to this august journal where his healthy brand of scepticism is most welcome, but as I have said on a number of occasions, he has forgotten more about animals and their husbandry than most of us will ever know, and after half a century or so in the business he is still going strong.

He is also one of the foremost zoo historians in the UK if not the world, and this fascinating volume (the third that he has written about Manchester's ill fated zoological gardens) is full of fascinating scraps of information , and even more fascinating illustrations. This book cannot be recommended highly enough!

Walks in Mysterious Devon by Trevor Beer (Sigma Press £6.95) ISBN 1-85058-607-1

This excellent little book features twenty-eight different books around rural Devonshire, most of which have fascinating links to folklore and forteana. The fortean zoologist will be intrigued by accounts of spectral black dogs, rogue wolverines (which Trevor Beer claims here to have seen himself), and of course, the notorious Beast of Exmoor - the quasi-cryptid with which Trevor Beer is most widely linked. Excellent!

Louis S B Leakey - Beyond the Evidence. Martin Pickford. Janus, London 1997. ISBN 1 85756 396 4 £12 164pp Publication date: 1/12/97.

"It is difficult," says Pickford, *"to find a parallel in the scientific world where so much garbage has been spoken by so few for so long."*

It is the "few" who are chastised here, and not the

mainstream, so it might seem that the march of science has not been hindered too badly by this handful of rotters.

His book examines paleoanthropology between 1926 and 1972 and criticises excavational technique, geological assumptions, correlation of evidence and the persistence of unwarranted assumptions.

Topics covered include the Pliocene drought, Oldoway Man, the Kanam jaw, Ngira Man and Kanjera Man. There is no index.

Pickford apportions blame freely throughout the book, mainly (as the title of the book suggests), in the direction of Louis Leakey - and his son Richard, who *"kept up the family tradition concerning fossils themselves, principally by claiming that they were older than they were..."*

I'm not surprised. The phrase *"I want to believe"*, deriving as it does from the world of Ufology that is so despised by many "reputable" scientists, pervades much of science - where careers and research grants can depend so much upon having faith in current bandwaggons and schools of thought. GI

Kaptan June and the Turtles by June Haimoff. (Janus, London 1997.) ISBN 1 85756 229 1. £8.95 130pp

This is the true story of one person's fight to save the *Caretta caretta* turtle's Turkish beach - Dalyar - from tourist development.

Written in relaxed and chatty autobiographical style, it describes how June stayed there and made some friends; watched the turtles; how beach huts began springing up on and near the beach; her encounters with the authorities; the sounds of dynamite blasting signalled road construction; and the arrival of the cement mixers while discussions were still under way to make the beach a Special Protection Area.

David Bellamy, who visited the beach with a BBC documentary team, says in the foreword:

"Please read this book and then, like Kaptan June and all the other caring people recorded in it, join the fight to save the living world upon which we all depend." GI

The Unexplained - The Ultimate Gateway to the World of the Unknown. Ed: John and Anne Spencer. (Simon & Schuster, Sydney 1997). ISBN 0-684-81985-6 £20 192pp large format.

A brief run-down on just about all mysterious or unexplained phenomena one can imagine. None of the 300 or so items are in depth - it's breadth that's the point of this book. Hence 'gateway' in the title: it's not a reference tool, and doesn't have an index. Heavily illustrated, generally with two pictures on each page (most in colour). Topics range from underground lightning, auras and Kirlian electrophotography, sky cities, phantom ships, frog falls, hypnosis, precognition, black holes, crop circles, voodoo, Nessie, ghosts, and phone calls from the dead. A good present for the kids? GI

PERIODICAL REVIEWS

By Graham Inglis

The "new kid on the block" this time is Mystery Magazine and very good it is, too.

MYSTERY MAGAZINE

Apart from the physical appearance of the text (large font italics) this is a well-presented and uncluttered mag, starting with an informative contents page - with 'tasters' for each item.

There's a 'weird news' section and an interview with astronaut Gordon Cooper (who flew on two pre-Apollo missions) in which he gives his views on Roswell and whether any UFOs are alien craft. There's also an investigation of Spring-heeled Jack, panther sightings in the Sheffield area, and - here's something cute - a page of weird stories from very old newspapers (one's dated 1718). The cryptozoology page (to be a regular feature?) is by David Colman of S.U.P.R. on big cat sightings in Scotland.

I like this mag and I hope it doesn't fade away.

48a Bridge Street, Killamarsh, Derbyshire, S21 1BS. A4 Quarterly: £2 each. Side-stapled A4 24pp.

Animals & Men # 15 — Reviews

COVER-UP A4 16pp

S.U.P.R. (Scottish Unexplained Phenomena Research) mag. In issue 8 (Sep 97) David Colman battles his way through internecine pettiness amongst Scottish ufologists to include articles about UFO sightings (mainly in Fife) and to review a book, "The Bible Code" which claims that predictions of the future are embedded in Genesis. Cover-Up doesn't pull its punches, sometimes. An article about child abductions commences thus: *"Sightings is a news-stand magazine not greatly troubled by any search for objective evidence to support its claims..."*

David Colman, 39 Limefield Crescent, Bathgate, West Lothian, EH48 1RF, Scotland. £1.25

CRYPTONEWS near-A4 14pp

This newsletter of the British Columbia Scientific Cryptozoology Club (BCSCC) is, in their less-than-modest words, *"one of the most desirable cryptozoological publications available"* and is *"professionally produced and well-thought out"*. There is something distinctly un-English about such remarks: if true, then they should be self-evident.

The articles are well written but the lack of a contents page, and the concealment of subscription rates on p14 under a caption saying "Nessie" do strike me as rather odd.

Issue 29 adopts a decidedly watery theme as it includes items on a Turkish lake monster (in Lake Van), Caddy, Ogopogo, some Loch Ness caverns, and Lake Dakatau's Migo (explained as a convoy of crocodiles). On dry land there's the Yeti and the Thylacine (Tasmanian tiger).

BCSCC, Unit #89, 6141 Willingdon Ave, Burnaby, BC, V5H 2T9, Canada. Membership: $10 (US or Canadian)

THE DRAGON CHRONICLE

Anyone fascinated by or interested in dragons will, I believe, find this mag a real delight, packed with articles on dragon mythology, cultural influence and portrayal, artwork, links with ancient fossils, poetry, and adverts and services aimed at catering for dragon enthusiasts everywhere.

P O Box 3369, London, SW6 6JN. A4 44pp. Issue 11 (Sep 97) is out now.
Issues cost £2 ($5) from Jan 98; a 4-issue sub is £7 ($15) - but note the mag appears 3 times a year.

SIGHTINGS
(a.k.a. FROM BEYOND in the USA)

Issue 18 (Nov 97) includes an interview with Derrel Sims (by Jonathan Downes) on the subject of alien implants, which Sims claims to have surgically removed from various abductees. He (Jonathan, that is) also penned an article on the recent wave of sightings of unidentified phenomena in Devon.

There's the usual regular columns and reviews and news, too.

A 'news-stand' magazine: ISSN 1363-5166.
Published by Rapide, Roman Court, 48 New North Road, Exeter, Devon.

HERP LIFE A4 4pp

The newsletter of the South Western Herpetological Society, England. The September issue reviews Brian Eady's society lecture on tortoises he'd seen in Turkey - or sometimes not seen - thanks to construction work for the benefit of tourists... The newsletter also includes member's ads.

Info: Karen Tucker, 14 Shrubbery Close, Newport, Barnstaple, Devon, EX32 9DG

TORTOISE TRUST NEWSLETTER

Various news and features covering welfare, conservation, captive breeding and research.

ISSN 0963-9411. Info from:
Tortoise Trust, BM Tortoise, London, WC1N 3XX.

MAINLY ABOUT ANIMALS

Veteran zoologist (but not cryptozoologist - see Clinton's Cogitations in this issue) Clinton Keeling edits this A5 magazine. The latest issue features the first part of an article on mystery eagles by Darren Naish.

13 Pound Place, Shalford, Guildford, Surrey, GU4 8HH. Subscription £5 per year.

CRYPTOZOOLOGIA A4 20pp

The French language magazine of the Association for the Protection of Rare Animals, Brussels.

Square des Latins 49/4, B-1050 Bruxelles, Belgium

Animals & Men #15 — Sales

NEW BOOKS AVAILABLE

In the autumn of 1996 the CFZ announced the availability of the new Yearbook a mite prematurely, as circumstances (some of which should have been allowed for) conspired to delay its launch by about 4 extra months.

This time around, things are a lot better organised. Not perfect, but certainly better. The 1998 Yearbook is already in an advanced state of preparation and will be available before Christmas!

As will two books by Tony Shiels! Details of our "Winter Catalogue" follow....

(Please make all cheques payable to Jonathan Downes)

The 1998 YEARBOOK OF THE CENTRE FOR FORTEAN ZOOLOGY

Neil Arnold - more cryptozoological movies
Darrne Naish - brontosaurs
Michael Playfair - A-Z of monster-haunted lakes
Jonathan Downes - Mystery Kangaroos
Richard Muirhead - Strange snakes
Darren Naish - Ichthyosaurs
Chris Moiser - Nyaminyami *
Mike Grayson - The fortean fauna of Percy Fawcett
(the legendary Amazonian explorer)
Tom Anderson - Native American totem beasts
Richard Freeman - Giant crocodiles

...and much more!

*If you didn't know that this is the water god of the Kariba Dam, Zimbabwe, then you clearly are urgently in need of the Yearbook 1998!

Available soon from the CFZ for £10 - plus 75p p&p (UK) or £1 (overseas)

THE CANTRIP CODEX
by
Tony 'Doc' Shiels

Long out-of-print...

Now reissued, containing all of the original text and with additional material and pictures.

The Cantrip Codex was (at least, in 1988) deemed to be "the one and only, authentic and official sequel to The Shiels Effect" (1976).

Enter a bizarre semi-autobiographical world of witches, preparation of press releases, the little people, trickery and magic....

Available soon from the CFZ: £8 - plus 70p p&p (UK) or £1 (overseas)

THIRTEEN!
by Tony Shiels

A reissue of Tony's 1967 guide to 13 spooky and atmospheric tricks...

Available soon from the CFZ: £4 - plus 25p p&p UK and 50p overseas.

13

OUR OTHER PUBLICATIONS

Morgawr: The Monster of Falmouth Bay by A. Mawnan-Peller.

Now with a new introduction by Tony 'Doc' Shiels and an additional essay by Jonathan Downes, this seminal 1976 booklet is finally available again complete with a ridiculous cover. £ 1.50

The Smaller Mystery Carnivores of the Westcountry by Jonathan Downes.

Over a hundred pages of information on a range of small carnivores in this fascinating region of the British Isles.

Three species thought extinct, and tantalising hints of several species apparently new to science are detailed along with a revolutionary suggestion that a species of mammal known from mainland Europe also exists on these islands. Many illustrations. £ 7.50

The Owlman and Others by Jonathan Downes.

For the last twenty years girls and young women visiting Mawnan Old Church in southern Cornwall have reported sightings of a four to five foot tall humanoid creature covered in feathers.

This book discusses two decades of owlman evidence in meticulous detail and comes about as close as anyone ever will to the truth. Many illustrations. £ 10.00

The CFZ Yearbook 1996.

The first of our annual 'yearbooks' with nearly two hundred pages of research papers and longer articles. Karl Shuker writes about Sky Beasts, Jon Downes writes about mystery eagles, Richard Muirhead examines the flying snake of Namibia, and we even reprint the seminal Tony 'Doc' Shiels article *'The Nnidnidification of Ness'*. Francois de Sarre examines African Man Beasts, and Neil Arnold looks at the Loch Ness Monster. There is also plenty more and many illustrations. £ 12.00

The CFZ Yearbook 1997.

Another dose of cryptozoology, zoomythology and high strangeness. Francois de Sarre claims that humans are descended from bipedal fish, Rafael A Lara Palmeros discusses cattle mutilation in Mexico and hunts for the Chupacabras, Karl Shuker goes in search of anomalous aardvarks and the big grey man of Ben McDhui and Darren Naish (figuratively) takes a hatchet to the monster of Lake Dakataua. Tom Anderson examines the pros and cons of reintroducing extinxt mammal species to Scotland, and Michael Playfair provides an annonated list of cryptozoological movies. As always there is much more and many illustrations. £ 12.00

SPECIAL OFFER

We have strictly limited stocks of Dr Karl Shuker's excellent new book :

FROM FLYING TOADS TO SNAKES WITH WINGS

Which has only been published in the United States.

£10.99
+ 75p p&p

Animals & Men #15 — Sales

OUR OWN PUBLICATIONS

ANIMALS AND MEN
BACK ISSUES: £2

Back issues of "Animals & Men" are available at £2 each from the editorial address. Please see "methods of payment" below.

As well as the main features detailed below, all issues of "Animals & Men" have a "Newsfile" section and letters, reviews and other shorter pieces....

Issue

1. Relict Pine Martens, Giant Sloths, Sumatran and Javan Rhinos, Golden Frogs, Frog Falls.

2. Mystery bears in Oxford and The Atlas Mountains, Loch Ness reports, Green Lizards, Woodwose, The Tatzelwurm.

3. Giant Worm in Eastbourne, Lake Monsters of New Guinea, Giant Lizards in Papua, Mystery Cats, Black Dogs on Dartmoor, Scorpion Mystery

4. Manatees of St Helena, Migo: The Lake Monster of New Britain, The search for the Tasmania Thylacine

5. Mystery cats, Loch Ness, More on the "Migo Video", Boars and Pumas, The Hairy Hands of Dartmoor.

6. The Owlman Special; also the Humped Elephants of Nepal, Mystery Cats, Sabre-toothed cats, Mysterious hominids of Africa, The British Nandi Bear?, Bibliography of Cryptozoology books pt 1 (Shuker)

7. Mystery Whales, Strangeness in Scotland, On collecting a cryptid, Bodmin Leopard Skull, Bibliography of "Crypto" Books (Shuker) pt 2.

8. Green Cats and Dogs, Mystery Whales, Quagga Project, Bibliography of Cryptozoological books (3rd & concluding part), Malayan Man Beast.

9. Hong Kong Tiger, Horseman of Lincolnshire, Scottish BHM, Congo Peacock, Mystery whales.

10. Mystery Moth of Madagascar, Bengal Leopard Cats, The Derry, Wild Boars in Kent, a new Irish lake monster, mystery whales and the truth about the Essex Beach Corpses.

11. The "Walruses Special", also Feathered Dinosaurs, Ground Sloth Survival in North America, Mystery Whales, Initial Bipedalism

12. Lions: The Barbary Lion, etc. More Feathered Dinosaurs, Chinese Crabs in the Thames, Mystery Animals of Germany, News from New Zealand.

13. Pangolins; also Moby the Sperm Whale, Barking Beast of Bath, Yorkshire ABCs, Molly the Singing Oyster, Leatherback Turtles, Walruses

14. The Dragons of Yorkshire, Irish mystery animals, In Search Of "Gambo", Charlie Fort and the Vampire Sheep Slayer - and Jackals, and the first of Clinton#s Cogitations..

THE GOBLIN UNIVERSE
BACK ISSUES: £2 each

The sister mag to *Animals & Men* - it's the parish magazine of the outer edge!

Issues 4, 5 & 6 are available from the editorial address. Please see "methods of payment" below.

As well as the main features detailed below, all issues of "Goblin" have a "News from Nowhere" section and letters, record and book reviews...

4. St Neot: Weirdest village in the West?, Naked witches, hellhounds and Capel's tomb, the Vampire of St Leonards, Cattle Mutilation, and an account of psychic detective work.

5. Crop Circles and Animal Mutilations, Ghosts of Glamis Castle, Communication with UFOs, and The "Noosphere" and text semantics.

6. Jon and Tina are shown the Rendlesham UFO crash site. Also, Mystery Planets, Cannibalism in Scotland, and D.I.Y. countries and states.

METHODS OF PAYMENT

Postage and packing is extra; pleae add 25p (30p non-UK) per magazine and 75p (80p non-UK) per book.

Payment can be made in UK or US cash, by IMO (International Money Order), Eurocheque, or by a cheque drawn on a UK bank.

Please make all cheques payable to Jonathan Downes.

THE CENTRE FOR FORTEAN ZOOLOGY

So, what is the Centre for Fortean Zoology?

We are a non profit-making organisation founded in 1992 with the aim of being a clearing house for information, and coordinating research into mystery animals around the world. We also study out of place animals, rare and aberrant animal behaviour, and Zooform Phenomena; – little-understood "things" that appear to be animals, but which are in fact nothing of the sort, and not even alive (at least in the way we understand the term).

Why should I join the Centre for Fortean Zoology?

Not only are we the biggest organisation of our type in the world but - or so we like to think - we are the best. We are certainly the only truly global Cryptozoological research organisation, and we carry out our investigations using a strictly scientific set of guidelines. We are expanding all the time and looking to recruit new members to help us in our research into mysterious animals and strange creatures across the globe. Why should you join us? Because, if you are genuinely interested in trying to solve the last great mysteries of Mother Nature, there is nobody better than us with whom to do it.

What do I get if I join the Centre for Fortean Zoology?

For £12 a year, you get a four-issue subscription to our journal *Animals & Men*. Each issue contains 60 pages packed with news, articles, letters, research papers, field reports, and even a gossip column! The magazine is A5 in format with a full colour cover. You also have access to one of the world's largest collections of resource material dealing with cryptozoology and allied disciplines, and people from the CFZ membership regularly take part in fieldwork and expeditions around the world.

How is the Centre for Fortean Zoology organized?

The CFZ is managed by a three-man board of trustees, with a non-profit making trust registered with HM Government Stamp Office. The board of trustees is supported by a Permanent Directorate of full and part-time staff, and advised by a Consultancy Board of specialists - many of whom who are world-renowned experts in their particular field. We have regional representatives across the UK, the USA, and many other parts of the world, and are affiliated with other organisations whose aims and protocols mirror our own.

I am new to the subject, and although I am interested I have little practical knowledge. I don't want to feel out of my depth. What should I do?

Don't worry. We were *all* beginners once. You'll find that the people at the CFZ are friendly and approachable. We have a thriving forum on the website which is the hub of an ever-growing electronic community. You will soon find your feet. Many members of the CFZ Permanent Directorate started off as ordinary members, and now work full time chasing monsters around the world.

I have an idea for a project which isn't on your website. What do I do?

Write to us, e-mail us, or telephone us. The list of future projects on the website is not exhaustive. If you have a good idea for an investigation, please tell us. We may well be able to help.

How do I go on an expedition?

We are always looking for volunteers to join us. If you see a project that interests you, do not hesitate to get in touch with us. Under certain circumstances we can help provide funding for your trip. If you look on the future projects section of the website, you can see some of the projects that we have pencilled in for the next few years.

In 2003 and 2004 we sent three-man expeditions to Sumatra looking for Orang-Pendek - a semi-legendary bipedal ape. The same three went to Mongolia in 2005. All three members started off merely subscribers to the CFZ magazine.

Next time it could be you!

Project Kerinci, Sumatra - 2003
In search of the bipedal ape Orang Pendek

How is the Centre for Fortean Zoology funded?

We have no magic sources of income. All our funds come from donations, membership fees, works that we do for TV, radio or magazines, and sales of our publications and merchandise. We are always looking for corporate sponsorship, and other sources of revenue. If you have any ideas for fund-raising please let us know. However, unlike other cryptozoological organisations in the past, we do not live in an intellectual ivory tower. We are not afraid to get our hands dirty, and furthermore we are not one of those organisations where the membership have to raise money so that a privileged few can go on expensive foreign trips. Our research teams both in the UK and abroad, consist of a mixture of experienced and inexperienced personnel. We are truly a community, and work on the premise that the benefits of CFZ membership are open to all.

What do you do with the data you gather from your investigations and expeditions?

Reports of our investigations are published on our website as soon as they are available. Preliminary reports are posted within days of the project finishing.

Each year we publish a 200 page yearbook containing research papers and expedition reports too long to be printed in the journal. We freely circulate our information to anybody who asks for it.

Is the CFZ community purely an electronic one?

No. Each year since 2000 we have held our annual convention - the *Weird Weekend* - in Exeter. It is three days of lectures, workshops, and excursions. But most importantly it is a chance for members of the CFZ to meet each other, and to talk with the members of the permanent directorate in a relaxed and informal setting and preferably with a pint of beer in one hand. Starting this year-18-20 August 2006 - the *Weird Weekend* will be bigger and better and held in the idyllic rural location of Woolsery in North Devon.

We are hoping to start up some regional groups in both the UK and the US which will have regular meetings, work together on research projects, and maybe have a mini convention of their own.

Since relocating to North Devon in 2005 we have become ever more closely involved with other community organisations, and we hope that this trend will continue. We also work closely with Police Forces across the UK as consultants for animal mutilation cases, and during 2006 we intend to forge closer links with the coastguard and other community services. We want to work closely with those who regularly travel into the Bristol Channel, so that if the recent trend of exotic animal visitors to our coastal waters continues, we can be out there as soon as possible.

We are building a Visitor's Centre in rural North Devon. This will not be open to the general public, but will provide a museum, a library and an educational resource for our members (currently over 400) across the globe. We are also planning a youth organisation which will involve children and young people in our activities.

Apart from having been the only Fortean Zoological organisation in the world to have consistently published material on all aspects of the subject for over a decade, we have achieved the following concrete results:

- Disproved the myth relating to the headless so-called sea-serpent carcass of Durgan beach in Cornwall 1975
- Disproved the story of the 1988 puma skull of Lustleigh Cleave
- Carried out the only in-depth research ever into mythos of the Cornish Owlma
- Made the first records of a tropical species of lamprey
- Made the first records of a luminous cave gnat larva in Thailand.
- Discovered a possible new species of British mammal - The Beech Marten.
- In 1994-6 carried out the first archival fortean zoological survey of Hong Kong.
- In the year 2000, CFZ theories where confirmed when an entirely new species of lizard was found resident in Britain.
- Identified the monster of Martin Mere in Lancashire as a giant wels catfish
- Expanded the known range of Armitage's skink in the Gambia by 80%
- Obtained photographic evidence of the remains of Europe's largest known pike
- Carried out the first ever in-depth study of the *ninki-nanka*
- Carried out the first attempt to breed Puerto Rican cave snails in captivity
- Were the first European explorers to visit the `lost valley` in Sumatra

EXPEDITIONS & INVESTIGATIOINS TO DATE INCLUDE

- 1998 Puerto Rico, Florida, Mexico *(Chupacabras)*
- 1999 Nevada *(Bigfoot)*
- 2000 Thailand *(Giant snakes called nagas)*
- 2002 Martin Mere *(Giant catfish)*
- 2002 Cleveland *(Wallaby mutilation)*
- 2003 Bolam Lake *(BHM Reports)*
- 2003 Sumatra *(Orang Pendek)*
- 2003 Texas *(Bigfoot; giant snapping turtles)*
- 2004 Sumatra *(Orang Pendek; cigau, a sabre-toothed cat)*
- 2004 Illinois *(Black panthers; cicada swarm)*
- 2004 Texas *(Mystery blue dog)*
- 2004 Puerto Rico *(Chupacabras; carnivorous cave snails)*
- 2005 Belize *(Affiliate expedition for hairy dwarfs)*
- 2005 Mongolia *(Allghoi Khorkhoi aka Mongolian death worm)*
- 2006 Gambia *(Gambo - Gambian sea monster , Ninki Nanka and Armitage s skink*
- 2006 Llangorse Lake *(Giant pike, giant eels)*
- 2006 Windermere *(Giant eels)*
- 2007 Coniston Water *(Giant eels)*
- 2007 Guyana *(Giant anaconda, didi, water tiger)*

To apply for a <u>FREE</u> information pack about the organisation and details of how to join, plus information on current and future projects, expeditions and events.

Send a stamped and addressed envelope to:

**THE CENTRE FOR FORTEAN ZOOLOGY
MYRTLE COTTAGE, WOOLSERY,
BIDEFORD, NORTH DEVON
EX39 5QR.**

or alternatively visit our website at:
www.cfz.org.uk

Other books available from
CFZ PRESS

THE OWLMAN AND OTHERS - 30th Anniversary Edition
Jonathan Downes - ISBN 978-1-905723-02-7

£14.99

EASTER 1976 - Two young girls playing in the churchyard of Mawnan Old Church in southern Cornwall were frightened by what they described as a "nasty bird-man". A series of sightings that has continued to the present day. These grotesque and frightening episodes have fascinated researchers for three decades now, and one man has spent years collecting all the available evidence into a book. To mark the 30th anniversary of these sightings, Jonathan Downes has published a special edition of his book.

DRAGONS - More than a myth?
Richard Freeman - ISBN 0-9512872-9-X

£14.99

First scientific look at dragons since 1884. It looks at dragon legends worldwide, and examines modern sightings of dragon-like creatures, as well as some of the more esoteric theories surrounding dragonkind.

Dragons are discussed from a folkloric, historical and cryptozoological perspective, and Richard Freeman concludes that: "When your parents told you that dragons don't exist - they lied!"

MONSTER HUNTER
Jonathan Downes - ISBN 0-9512872-7-3

£14.99

Jonathan Downes' long-awaited autobiography, *Monster Hunter*...

Written with refreshing candour, it is the extraordinary story of an extraordinary life, in which the author crosses paths with wizards, rock stars, terrorists, and a bewildering array of mythical and not so mythical monsters, and still just about manages to emerge with his sanity intact.......

MONSTER OF THE MERE
Jonathan Downes - ISBN 0-9512872-2-2

£12.50

It all starts on Valentine's Day 2002 when a Lancashire newspaper announces that "Something" has been attacking swans at a nature reserve in Lancashire. Eyewitnesses have reported that a giant unknown creature has been dragging fully grown swans beneath the water at Martin Mere. An intrepid team from the Exeter based Centre for Fortean Zoology, led by the author, make two trips – each of a week – to the lake and its surrounding marshlands. During their investigations they uncover a thrilling and complex web of historical fact and fancy, quasi Fortean occurrences, strange animals and even human sacrifice.

**CFZ PRESS, MYRTLE COTTAGE,
WOOLFARDISWORTHY BIDEFORD,
NORTH DEVON, EX39 5QR
www.cfz.org.uk**

Other books available from
CFZ PRESS

ONLY FOOLS AND GOATSUCKERS
Jonathan Downes - ISBN 0-9512872-3-0

£12.50

In January and February 1998 Jonathan Downes and Graham Inglis of the Centre for Fortean Zoology spent three and a half weeks in Puerto Rico, Mexico and Florida, accompanied by a film crew from UK Channel 4 TV. Their aim was to make a documentary about the terrifying chupacabra - a vampiric creature that exists somewhere in the grey area between folklore and reality. This remarkable book tells the gripping, sometimes scary, and often hilariously funny story of how the boys from the CFZ did their best to subvert the medium of contemporary TV documentary making and actually do their job.

WHILE THE CAT'S AWAY
Chris Moiser - ISBN: 0-9512872-1-4

£7.99

Over the past thirty years or so there have been numerous sightings of large exotic cats, including black leopards, pumas and lynx, in the South West of England. Former Rhodesian soldier Sam McCall moved to North Devon and became a farmer and pub owner when Rhodesia became Zimbabwe in 1980. Over the years despite many of his pub regulars having seen the "Beast of Exmoor" Sam wasn't at all sure that it existed. Then a series of happenings made him change his mind. Chris Moiser—a zoologist—is well known for his research into the mystery cats of the westcountry. This is his first novel.

CFZ EXPEDITION REPORT 2006 - GAMBIA
ISBN 1905723032

£12.50

In July 2006, The J.T.Downes memorial Gambia Expedition - a six-person team - Chris Moiser, Richard Freeman, Chris Clarke, Oll Lewis, Lisa Dowley and Suzi Marsh went to the Gambia, West Africa. They went in search of a dragon-like creature, known to the natives as `Ninki Nanka`, which has terrorized the tiny African state for generations, and has reportedly killed people as recently as the 1990s. They also went to dig up part of a beach where an amateur naturalist claims to have buried the carcass of a mysterious fifteen foot sea monster named 'Gambo', and they sought to find the Armitage's Skink (*Chalcides armitagei*) - a tiny lizard first described in 1922 and only rediscovered in 1989. Here, for the first time, is their story.... With an forward by Dr. Karl Shuker and introduction by Jonathan Downes.

BIG CATS IN BRITAIN YEARBOOK 2006
Edited by Mark Fraser - ISBN 978-1905723-01-0

£10.00

Big cats are said to roam the British Isles and Ireland even now as you are sitting and reading this. People from all walks of life encounter these mysterious felines on a daily basis in every nook and cranny of these two countries. Most are jet-black, some are white, some are brown, in fact big cats of every description and colour are seen by some unsuspecting person while on his or her daily business. 'Big Cats in Britain' are the largest and most active group in the British Isles and Ireland This is their first book. It contains a run-down of every known big cat sighting in the UK during 2005, together with essays by various luminaries of the British big cat research community which place the phenomenon into scientific, cultural, and historical perspective.

CFZ PRESS, MYRTLE COTTAGE,
WOOLSERY, BIDEFORD,
NORTH DEVON, EX39 5QR
www.cfz.org.uk

Other books available from
CFZ PRESS

THE SMALLER MYSTERY CARNIVORES OF THE WESTCOUNTRY
Jonathan Downes - ISBN 978-1-905723-05-8

£7.99

Although much has been written in recent years about the mystery big cats which have been reported stalking Westcountry moorlands, little has been written on the subject of the smaller British mystery carnivores. This unique book redresses the balance and examines the current status in the Westcountry of three species thought to be extinct: the Wildcat, the Pine Marten and the Polecat, finding that the truth is far more exciting than the currently held scientific dogma. This book also uncovers evidence suggesting that even more exotic species of small mammal may lurk hitherto unsuspected in the countryside of Devon, Cornwall, Somerset and Dorset.

THE BLACKDOWN MYSTERY
Jonathan Downes - ISBN 978-1-905723-00-3

£7.99

Intrepid members of the CFZ are up to the challenge, and manage to entangle themselves thoroughly in the bizarre trappings of this case. This is the soft underbelly of ufology, rife with unsavoury characters, plenty of drugs and booze." That sums it up quite well, we think. A new edition of the classic 1999 book by legendary fortean author Jonathan Downes. In this remarkable book, Jon weaves a complex tale of conspiracy, anti-conspiracy, quasi-conspiracy and downright lies surrounding an air-crash and alleged UFO incident in Somerset during 1996. However the story is much stranger than that. This excellent and amusing book lifts the lid off much of contemporary forteana and explains far more than it initially promises.

GRANFER'S BIBLE STORIES
John Downes - ISBN 0-9512872-8-1

£7.99

Bible stories in the Devonshire vernacular, each story being told by an old Devon Grandfather - 'Granfer'. These stories are now collected together in a remarkable book presenting selected parts of the Bible as one more-or-less continuous tale in short 'bite sized' stories intended for dipping into or even for bed-time reading. `Granfer` treats the biblical characters as if they were simple country folk living in the next village. Many of the stories are treated with a degree of bucolic humour and kindly irreverence, which not only gives the reader an opportunity to re-evaluate familiar tales in a new light, but do so in both an entertaining and a spiritually uplifting manner.

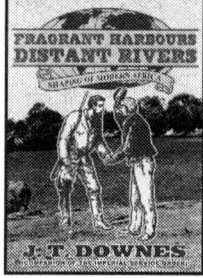

FRAGRANT HARBOURS DISTANT RIVERS
John Downes - ISBN 0-9512872-5-7

£12.50

Many excellent books have been written about Africa during the second half of the 19th Century, but this one is unique in that it presents the stories of a dozen different people, whose interlinked lives and achievements have as many nuances as any contemporary soap opera. It explains how the events in China and Hong Kong which surrounded the Opium Wars, intimately effected the events in Africa which take up the majority of this book. The author served in the Colonial Service in Nigeria and Hong Kong, during which he found himself following in the footsteps of one of the main characters in this book; Frederick Lugard – the architect of modern Nigeria.

CFZ PRESS, MYRTLE COTTAGE,
WOOLFARDISWORTHY BIDEFORD,
NORTH DEVON, EX39 5QR
w w w . c f z . o r g . u k

Other books available from
CFZ PRESS

ANIMALS & MEN - Issues 1 - 5 - In the Beginning
Edited by Jonathan Downes - ISBN 0-9512872-6-5

£12.50

At the beginning of the 21st Century monsters still roam the remote, and sometimes not so remote, corners of our planet. It is our job to search for them. The Centre for Fortean Zoology [CFZ] is the only professional, scientific and full-time organisation in the world dedicated to cryptozoology - the study of unknown animals. Since 1992 the CFZ has carried out an unparalleled programme of research and investigation all over the world. We have carried out expeditions to Sumatra (2003 and 2004), Mongolia (2005), Puerto Rico (1998 and 2004), Mexico (1998), Thailand (2000), Florida (1998), Nevada (1999 and 2003), Texas (2003 and 2004), and Illinois (2004). An introductory essay by Jonathan Downes, notes putting each issue into a historical perspective, and a history of the CFZ.

ANIMALS & MEN - Issues 6 - 10 - The Number of the Beast
Edited by Jonathan Downes - ISBN 978-1-905723-06-5

£12.50

At the beginning of the 21st Century monsters still roam the remote, and sometimes not so remote, corners of our planet. It is our job to search for them. The Centre for Fortean Zoology [CFZ] is the only professional, scientific and full-time organisation in the world dedicated to cryptozoology - the study of unknown animals. Since 1992 the CFZ has carried out an unparalleled programme of research and investigation all over the world. We have carried out expeditions to Sumatra (2003 and 2004), Mongolia (2005), Puerto Rico (1998 and 2004), Mexico (1998), Thailand (2000), Florida (1998), Nevada (1999 and 2003), Texas (2003 and 2004), and Illinois (2004). Preface by Mark North and an introductory essay by Jonathan Downes, notes putting each issue into a historical perspective, and a history of the CFZ.

BIG BIRD! Modern Sightings of Flying Monsters

£7.99

Ken Gerhard - ISBN 978-1-905723-08-9

From all over the dusty U.S./Mexican border come hair-raising stories of modern day encounters with winged monsters of immense size and terrifying appearance. Further field sightings of similar creatures are recorded from all around the globe. What lies behind these weird tales? Ken Gerhard is a native Texan, he lives in the homeland of the monster some call 'Big Bird'. Ken's scholarly work is the first of its kind. On the track of the monster, Ken uncovers cases of animal mutilations, attacks on humans and mounting evidence of a stunning zoological discovery ignored by mainstream science. Keep watching the skies!

STRENGTH THROUGH KOI
They saved Hitler's Koi and other stories

£7.99

Jonathan Downes - ISBN 978-1-905723-04-1

Strength through Koi is a book of short stories - some of them true, some of them less so - by noted cryptozoologist and raconteur Jonathan Downes. The stories are all about koi carp, and their interaction with bigfoot, UFOs and Nazis. Even the late George Harrison makes an appearance. Very funny in parts, this book is highly recommended for anyone with even a passing interest in aquaculture, but should be taken definitely *cum grano salis*.

**CFZ PRESS, MYRTLE COTTAGE,
WOOLSERY, BIDEFORD,
NORTH DEVON, EX39 5QR**

Other books available from
CFZ PRESS

BIG CATS IN BRITAIN YEARBOOK 2007
Edited by Mark Fraser - ISBN 978-1-905723-09-6

£12.50

People from all walks of life encounter mysterious felids on a daily basis, in every nook and cranny of the UK. Most are jet-black, some are white, some are brown; big cats of every description and colour are seen by some unsuspecting person while on his or her daily business. 'Big Cats in Britain' are the largest and most active research group in the British Isles and Ireland. This book contains a run-down of every known big cat sighting in the UK during 2006, together with essays by various luminaries of the British big cat research community.

CAT FLAPS! Northern Mystery Cats
Andy Roberts - ISBN 978-1-905723-11-9

£6.99

Of all Britain's mystery beasts, the alien big cats are the most renowned. In recent years the notoriety of these uncatchable, out-of-place predators have eclipsed even the Loch Ness Monster. They slink from the shadows to terrorise a community, and then, as often as not, vanish like ghosts. But now film, photographs, livestock kills, and paw prints show that we can no longer deny the existence of these once-legendary beasts. Here then is a case-study, a true lost classic of Fortean research by one of the country's most respected researchers.

CENTRE FOR FORTEAN ZOOLOGY 2007 YEARBOOK
Edited by Jonathan Downes and Richard Freeman
ISBN 978-1-905723-14-0

£12.50

The Centre For Fortean Zoology Yearbook is a collection of papers and essays too long and detailed for publication in the CFZ Journal *Animals & Men.* With contributions from both well-known researchers, and relative newcomers to the field, the Yearbook provides a forum where new theories can be expounded, and work on little-known cryptids discussed.

MONSTER! THE A-Z OF ZOOFORM PHENOMENA
Neil Arnold - ISBN 978-1-905723-10-2

£14.99

Zooform Phenomena are the most elusive, and least understood, mystery `animals`. Indeed, they are not animals at all, and are not even animate in the accepted terms of the word. Author and researcher Neil Arnold is to be commended for a groundbreaking piece of work, and has provided the world's first alphabetical listing of zooforms from around the world.

**CFZ PRESS, MYRTLE COTTAGE,
WOOLFARDISWORTHY BIDEFORD,
NORTH DEVON, EX39 5QR
w w w . c f z . o r g . u k**

Other books available from
CFZ PRESS

BIG CATS LOOSE IN BRITAIN
Marcus Matthews - ISBN 978-1-905723-12-6

£14.99

Big Cats: Loose in Britain, looks at the body of anecdotal evidence for such creatures: sightings, livestock kills, paw-prints and photographs, and seeks to determine underlying commonalities and threads of evidence. These two strands are repeatedly woven together into a highly readable, yet scientifically compelling, overview of the big cat phenomenon in Britain.

DARK DORSET
TALES OF MYSTERY, WONDER AND TERROR
Robert. J. Newland and Mark. J. North
ISBN 978-1-905723-15-6

£12.50

This extensively illustrated compendium has over 400 tales and references, making this book by far one of the best in its field. Dark Dorset has been thoroughly researched, and includes many new entries and up to date information never before published. The title of the book speaks for itself, and is indeed not for the faint hearted or those easily shocked.

MAN-MONKEY - IN SEARCH OF THE BRITISH BIGFOOT
Nick Redfern - ISBN 978-1-905723-16-4

£9.99

In her 1883 book, *Shropshire Folklore*, Charlotte S. Burne wrote: *'Just before he reached the canal bridge, a strange black creature with great white eyes sprang out of the plantation by the roadside and alighted on his horse's back'*. The creature duly became known as the `Man-Monkey`.

Between 1986 and early 2001, Nick Redfern delved deeply into the mystery of the strange creature of that dark stretch of canal. Now, published for the very first time, are Nick's original interview notes, his files and discoveries; as well as his theories pertaining to what lies at the heart of this diabolical legend.

EXTRAORDINARY ANIMALS REVISITED
Dr Karl Shuker - ISBN 978-1905723171

£14.99

This delightful book is the long-awaited, greatly-expanded new edition of one of Dr Karl Shuker's much-loved early volumes, *Extraordinary Animals Worldwide*. It is a fascinating celebration of what used to be called romantic natural history, examining a dazzling diversity of animal anomalies, creatures of cryptozoology, and all manner of other thought-provoking zoological revelations and continuing controversies down through the ages of wildlife discovery.

**CFZ PRESS, MYRTLE COTTAGE,
WOOLFARDISWORTHY BIDEFORD,
NORTH DEVON, EX39 5QR
w w w . c f z . o r g . u k**

Other books available from
CFZ PRESS

DARK DORSET CALENDAR CUSTOMS
Robert J Newland - ISBN 978-1-905723-18-8

£12.50

Much of the intrinsic charm of Dorset folklore is owed to the importance of folk customs. Today only a small amount of these curious and occasionally eccentric customs have survived, while those that still continue have, for many of us, lost their original significance. Why do we eat pancakes on Shrove Tuesday? Why do children dance around the maypole on May Day? Why do we carve pumpkin lanterns at Hallowe'en? All the answers are here! Robert has made an in-depth study of the Dorset country calendar identifying the major feast-days, holidays and celebrations when traditionally such folk customs are practiced.

CENTRE FOR FORTEAN ZOOLOGY 2004 YEARBOOK
Edited by Jonathan Downes and Richard Freeman
ISBN 978-1-905723-14-0

£12.50

The Centre For Fortean Zoology Yearbook is a collection of papers and essays too long and detailed for publication in the CFZ Journal *Animals & Men*. With contributions from both well-known researchers, and relative newcomers to the field, the Yearbook provides a forum where new theories can be expounded, and work on little-known cryptids discussed.

CENTRE FOR FORTEAN ZOOLOGY 2008 YEARBOOK
Edited by Jonathan Downes and Corinna Downes
ISBN 978 -1-905723-19-5

£12.50

The Centre For Fortean Zoology Yearbook is a collection of papers and essays too long and detailed for publication in the CFZ Journal *Animals & Men*. With contributions from both well-known researchers, and relative newcomers to the field, the Yearbook provides a forum where new theories can be expounded, and work on little-known cryptids discussed.

ETHNA'S JOURNAL
Corinna Newton Downes
ISBN 978 -1-905723-21-8

£9.99

Ethna's Journal tells the story of a few months in an alternate Dark Ages, seen through the eyes of Ethna, daughter of Lord Edric. She is an unsophisticated girl from the fortress town of Cragnuth, somewhere in the north of England, who reluctantly gets embroiled in a web of treachery, sorcery and bloody war...

CFZ PRESS, MYRTLE COTTAGE,
WOOLFARDISWORTHY BIDEFORD,
NORTH DEVON, EX39 5QR
w w w . c f z . o r g . u k

www.ingramcontent.com/pod-product-compliance
Lightning Source LLC
Chambersburg PA
CBHW062154080426
42734CB00010B/1686